Africa's Mineral Fortune

For too long Africa's mineral fortune has been lamented as a resource curse that has led to conflict rather than development for much of the continent. Yet times are changing and the opportunities to bring technical expertise on modern mining alongside appropriate governance mechanisms for social development are becoming more accessible in Africa.

This book synthesizes perspectives from multiple disciplines to address Africa's development goals in relation to its mineral resources. The authors cover ways of addressing a range of policy challenges, environmental concerns, and public health impacts and also consider the role of globalization within the extractive industries. Academic research is coupled with key field vignettes from practitioners exemplifying case studies throughout. The book summarizes the challenges of natural resource governance, suggesting ways in which mining can be more effectively managed in Africa. By providing an analytical framework it highlights the essential intersection between natural and social sciences, central to efficient and effective harnessing of the potential for minerals and mining to be a contributor to positive development in Africa.

It will be of interest to policy makers, industry professionals, and researchers in the extractive industries, as well as to the broader development community.

Saleem H. Ali is Blue and Gold Distinguished Professor of Energy and the Environment at the University of Delaware, USA. He also holds Adjunct Professor status for both the Global Change Institute and the Sustainable Minerals Institute at the University of Queensland, Australia, where he previously served as Chair in Sustainable Resource Development and Professor of Sustainability Science and Policy.

Kathryn Sturman is Senior Research Fellow at the Centre for Social Responsibility in Mining at the Sustainable Minerals Institute, University of Queensland, Australia. She is an international relations specialist in resource governance and policy considerations of mining for human security and development.

Nina Collins is the Mining Advocacy Coordinator at Oxfam Australia, focusing on the operations of Australian mining, oil, and gas companies operating in developing countries and their impacts on communities. She is also an Industry Fellow at the Centre for Social Responsibility in Mining, Sustainable Minerals Institute, University of Queensland, Australia.

Routledge Studies of the Extractive Industries and Sustainable Development

African Artisanal Mining from the Inside Out
Access, norms and power in Congo's gold sector
Sarah Geenan

Mountain Movers
Mining, sustainability and the agents of change
Daniel M. Franks

Responsible Mining
Key concepts for industry integrity
Sara Bice

Mining in Latin America
Critical approaches to the new extraction
Edited by Kalowatie Deonandan and Michael L. Dougherty

Industrialising Rural India
Land, policy, resistance
Edited by Kenneth Bo Nielsen and Patrik Oskarsson

Governance in the Extractive Industries
Power, cultural politics and regulation
Edited by Lori Leonard and Siba Grovogui

Social Terrains of Mine Closure in the Philippines
Minerva Chaloping-March

Mining and Sustainable Development
Current issues
Edited by Sumit K. Lodhia

Africa's Mineral Fortune
The science and politics of mining and sustainable development
Edited by Saleem Ali, Kathryn Sturman and Nina Collins

https://www.routledge.com/series/REISD

Africa's Mineral Fortune
The Science and Politics of Mining and Sustainable Development

Edited by Saleem H. Ali, Kathryn Sturman, and Nina Collins

First published 2019
by Routledge
2 Park Square, Milton Park, Abingdon, Oxon OX14 4RN

and by Routledge
52 Vanderbilt Avenue, New York, NY 10017

First issued in paperback 2020

Routledge is an imprint of the Taylor & Francis Group, an informa business

© 2019 selection and editorial matter, Saleem H. Ali, Kathryn Sturman and Nina Collins; individual chapters, the contributors

The right of Saleem H. Ali, Kathryn Sturman and Nina Collins to be identified as the authors of the editorial material, and of the authors for their individual chapters, has been asserted in accordance with sections 77 and 78 of the Copyright, Designs and Patents Act 1988.

All rights reserved. No part of this book may be reprinted or reproduced or utilised in any form or by any electronic, mechanical, or other means, now known or hereafter invented, including photocopying and recording, or in any information storage or retrieval system, without permission in writing from the publishers.

Trademark notice: Product or corporate names may be trademarks or registered trademarks, and are used only for identification and explanation without intent to infringe.

British Library Cataloguing-in-Publication Data
A catalogue record for this book is available from the British Library

Library of Congress Cataloging-in-Publication Data
Names: Ali, Saleem H. (Saleem Hassan), 1973- editor, author. | Sturman, Kathryn, editor, author. | Collins, Nina, 1980- editor, author.
Title: Africa's mineral fortune : the science and politics of mining and sustainable development / edited by Saleem Ali, Kathryn Sturman and Nina Collins.
Other titles: Routledge studies of the extractive industries and sustainable development.
Description: New York : Routledge, 2018. | Series: Routledge studies of the extractive industries and sustainable development | Includes bibliographical references and index.
Identifiers: LCCN 2018008072| ISBN 9781138606920 (hardback : alk. paper) | ISBN 9780429467424 (ebook)
Subjects: LCSH: Mineral industries—Environmental aspects—Africa. | Mineral industries—Government policy—Africa. | Mines and mineral resources—Africa.
Classification: LCC HD9506.A35 .A37 2018 | DDC 338.851096—dc23
LC record available at https://lccn.loc.gov/2018008072

ISBN 13: 978-0-367-58758-1 (pbk)
ISBN 13: 978-1-138-60692-0 (hbk)

Typeset in Bembo
by Swales & Willis Ltd, Exeter, Devon, UK

Contents

List of contributors viii

Introduction 1
SALEEM H. ALI AND KATHRYN STURMAN

PART I
The politics of African mining 7

1 Harmonizing African resource politics? Lessons from the African Minerals Development Centre 9
RODGER BARNES, KOJO BUSIA, AND MARIT KITAW

2 Evaluating conflict risks in Africa's resource governance 28
KATHRYN STURMAN AND FITSUM WELDEGIORGIS

3 Chinese mining in Africa and its global controversy 45
BARRY SAUTMAN AND YAN HAIRONG

4 *Field vignette.* Moving from prescriptive to performance-based regulation: the case of waste management 63
ANDY FOURIE, MWIYA SONGOLO, AND JAMES MCINTOSH

5 *Field vignette.* Ghana's policy on artisanal and small-scale mining 65
BENJAMIN ARYEE

PART II
Data and models: supporting strategic planning for Africa's minerals 69

6 Developing accurate and accessible geoscientific data for sustainable mining in Africa 71
JUDITH A. KINNAIRD AND RAYMOND J. DURRHEIM

vi Contents

7   Challenges in measuring the local and regional contributions
    of mining: lessons from case studies in Rwanda, Zambia,
    and Ghana   86
    JULIA HORSLEY, SHABBIR AHMAD, AND MATTHEW TONTS

8   Measuring transformative development from mining:
    a case study of Madagascar   119
    FITSUM WELDEGIORGIS AND CRISTIAN PARRA

9   *Field vignette.* The Extractives Dependence Index and
    its impact on Africa   146
    DEGOL HAILU AND CHINPIHOI KIPGEN

10 *Field vignette.* The West African Exploration Initiative (WAXI)
    as a model for collaborative research and development   148
    M. W. JESSELL AND THE WAXI TEAM

PART III
Environment, health, and innovation   151

11 Conservation priorities and extractive industries in Africa:
    opportunities for conflict prevention   153
    MAHLETTE BETRE, MARIELLE CANTER WEIKEL, ROMY CHEVALLIER,
    JANET EDMOND, ROSIMEIRY PORTELA, ZACHARY WELLS, AND
    JENNIFER BLAHA

12 Ebola and other emerging infectious diseases: managing
    risks to the mining industry   182
    OSMAN A. DAR, FRANCESCA VILIANI, HISHAM TARIQ, EMMELINE
    BUCKLEY, ABBAS OMAAR, ELOGHENE OTOBO, AND DAVID
    L. HEYMANN

13 Mineral investment decision-making in Africa: a real options
    approach in integrating price and environmental risks   195
    KWASI AMPOFO AND ALIDU BABATU ADAM

14 The potential of Zambian copper-cobalt metallophytes for
    phytoremediation of minerals wastes   208
    ANTONY VAN DER ENT, PETER ERSKINE, ROYD VINYA, JOLANTA
    MESJASZ- PRZYBYŁOWICZ, AND FRANÇOIS MALAISSE

15 *Field vignette.* South Africa's underground women miners   228
    ASANDA BENYA

16 *Field vignette.* Sapphire mining, water, and maternal health
    in Madagascar   230
    LYNDA LAWSON

PART IV
Reconciling scales of mining governance — 235

17 Strategies for working with artisanal and small-scale miners in sub-Saharan Africa — 237
NINA COLLINS AND LYNDA LAWSON

18 Artisanal and small-scale mining community health, safety, and sanitation: a water focus — 264
DANELLIE LYNAS, GERNELYN LOGROSA, AND BEN FAWCETT

19 Gauging the effectiveness of certification schemes and standards for responsible mining in Africa — 282
RENZO MORI JUNIOR

20 *Field vignette.* The Australia-Africa Minerals and Energy Group (AAMEG) — 297
TRISH O'REILLY

21 *Field vignette.* Sourcing "conflict-free" minerals from Central Africa — 300
STEVEN B. YOUNG

Conclusion: a multifaceted fortune — 302
KATHRYN STURMAN AND SALEEM H. ALI

Index — 305

Contributors

Alidu Babatu Adam is a Senior Mining Specialist for the World Bank in Ghana.

Shabbir Ahmad is a Postdoctoral Associate in Econometrics and Business at the University of Queensland, Australia.

Saleem H. Ali is Blue and Gold Distinguished Professor of Geography at the University of Delaware and Professorial Research Fellow at the University of Queensland, Australia.

Kwasi Ampofo is a Mineral Economist and Risk Analyst based in Brisbane, Australia.

Benjamin Aryee is the CEO of the Mineral Commission of Ghana.

Rodger Barnes is a Senior Research Associate at the University of Queensland, Australia.

Asanda Benya is a Lecturer in Sociology at the University of Cape Town, South Africa.

Mahlette Betre is an Independent Consultant in Conservation Planning in Washington DC, USA.

Jennifer Blaha is a Senior Manager for Responsible Mining and Energy at Conservation International.

Emmeline Buckley is Project Manager of the International Health Regulations Strengthening Project in London, UK.

Kojo Busia is Senior Mineral Sector Governance Advisor at the African Mineral Development Center.

Romy Chevallier is a Senior Researcher at the South African Institute of International Affairs.

Nina Collins is Mining Advocacy Coordinator at Oxfam Australia.

Osman A. Dar is director of the Centre on Global Health Security's One Health project at Chatham House in London, UK.

Contributors ix

Raymond J. Durrheim is Professor and Chair of the School of Geosciences at the University of Witwatersrand, South Africa.

Janet Edmond is Senior Director for Peace and Development Partnerships at Conservation International, Washington DC, USA.

Peter Erskine is Associate Professor at the Sustainable Minerals Institute, University of Queensland, Australia.

Ben Fawcett is a Senior Researcher at the International Water Centre in Brisbane, Australia.

Andy Fourie is Professor of Engineering Systems at the University of Western Australia.

Degol Hailu is Senior Advisor on Sustainable Development, United Nations Development Programme (UNDP) in Addis Ababa, Ethiopia.

Yan Hairong is Associate Professor of Anthropology at Hong Kong Polytechnic University.

David L. Heymann is Professor of Infectious Epidemiology at the London School of Hygiene and Tropical Medicine, UK.

Julia Horsley is a Research Fellow in Social Science at the University of Western Australia.

M. W. Jessell is Professor at the Centre for Exploration Targeting, University of Western Australia.

Renzo Mori Junior is a Senior Researcher in Certification Systems at Think Impact Consulting, Melbourne, Australia.

Judith A. Kinnaird is Professor of Economic Geology at the University of Witwatersrand, South Africa.

Chinpihoi Kipgen is a Research Officer at the United Nations Development Programme in Addis Ababa, Ethiopia.

Marit Kitaw is Technical Advisor for the United Nations Development Programme in Mozambique.

Lynda Lawson is Senior Researcher for the Gemstones and Sustainable Development Knowledge Hub and Knowledge Transfer Manager, Sustainable Minerals Institute, University of Queensland, Australia.

Gernelyn Logrosa is a Doctoral Candidate at the Minerals Industry Safety and Health Centre, Sustainable Minerals Institute, University of Queensland, Australia.

Danellie Lynas is a Research Fellow at the Minerals Industry Safety and Health Centre, Sustainable Minerals Institute, University of Queensland, Australia.

François Malaisse is an Emeritus Professor and a Senior Researcher in Botany at the University of Liege, Belgium.

James McIntosh is an Executive Officer at Curtin University, Australia.

Jolanta Mesjasz-Przybyłowicz is a Professor in the Department of Botany and Zoology Stellenbosch University, South Africa.

Abbas Omaar is a Researcher at the Center for Global Health Security at Chatham House, London, UK.

Trish O'Reilly is a Consultant in Mineral Governance and former CEO of the Australia-Africa Mineral and Energy Group in Perth, Australia.

Eloghene Otobo is a Researcher at the London School of Hygiene and Tropical Medicine, UK.

Cristian Parra is Founding Director of Analysis for Development Group in Brisbane, Australia.

Rosimeiry Portela is a Senior Director in Ecological Economics at Conservation International and Professor of Practice at Arizona State University, USA.

Barry Sautman is Professor of Political Science at the Hong Kong University of Science and Technology.

Mwiya Songolo is Head of the Mining Engineering Department at Copperbelt University, Zambia.

Kathryn Sturman is Senior Research Fellow at the Sustainable Minerals Institute at the University of Queensland, Australia.

Hisham Tariq is a Researcher at London School of Hygiene and Tropical Medicine, UK.

Matthew Tonts is Pro Vice Chancellor and Executive Dean in the Faculty of Arts, Business, Law and Education at the University of Western Australia.

Antony van der Ent is ARC Postdoctoral Research Fellow in phytochemistry at the University of Queensland, Australia.

Francesca Viliani is Public Health Director at International SOS, Copenhagen, Denmark.

Royd Vinya is a Lecturer in Environment and Geography at Copperbelt University, Zambia.

Marielle Canter Weikel is Senior Director in Energy and Mining at Conservation International, Washington DC, USA.

Fitsum Weldegiorgis is a Senior Researcher at the International Institute for Environment and Development in London, UK.

Zachary Wells is Technical Director of the Conservation Stewards Program at Conservation International, Washington DC, USA.

Steven B. Young is Professor of Industrial Ecology at the University of Waterloo, Canada.

Introduction

Saleem H. Ali and Kathryn Sturman

For many centuries Africa's rich mineral endowment has shaped the fortunes of its peoples and the environment. Beginning in the early twenty-first century, the commodities price boom and concerns about human impacts on the environment have seen increasing value placed on the continent's renewable and non-renewable natural resources. Operators explored and extracted mineral and energy resources at an unprecedented pace during the boom, with regulators scrambling to keep up. The limitations of the so-called commodities super cycle soon became apparent to Africa's policymakers and investors alike, however, prompting a need for revised approaches to resource governance in the past five years.[1] At the same time, competing land and water uses, deforestation, and loss of biodiversity remain key concerns for climate change mitigation and environmental conservation. This raises an enduring question: How does Africa develop its mineral fortune sustainably, both in environmental and in socio-economic terms? The concentration of low-income, resource-dependent countries in Africa places it at the center of global debates about sustainable development and the extractive industries.

In the spirit of the Routledge Extractive Industries and Sustainable Development series, this book aims to synthesize perspectives from multiple disciplines to address Africa's sustainable development challenges. Mineral development is a highly technical arena where geologists, mining engineers, and environmental scientists have traditionally held ascendancy. However, the social impact of minerals and the long history of their linkage to colonialism have posited serious development questions that extractive industries and governments need to address. Mineral development is not merely a matter for engineers and scientists to negotiate, but also the start of a complex supply chain of materials that have immediacy for human and environmental impacts in Africa and worldwide.

Several chapters in this book were informed by work under the auspices of the Sustainable Minerals Institute of the University of Queensland and the University of Western Australia, with financial support from the Australian Aid Initiative on International Mining for Development that ran for three years (2012–2015). The Sustainable Minerals Institute's holistic approach to environmental and social sustainability is reflected in the book's unique conceptualization of this topic. We have also invited authors from other leading research centers in science and policy to contribute, however, to give full topical coverage to key issues around African mineral development. Wherever possible we have teamed African researchers with non-African researchers in chapter authorship to ensure lessons and insights that transcend geographic biases. We also invited environmental organizations such as Conservation International as well as international policy think tanks such as Chatham House to ensure timely relevance of the material with key stakeholders outside academia. We convened a workshop with the authors in December 2016 to ensure connectivity between chapters and to provide authors and editors with more-robust communication that would better integrate the themes of the volume.

The book is divided into four parts, the delineation of which were determined through interactions at the authors' workshop. To sharpen the focus of key themes we also decided to include field vignettes within each part, with short exemplifying cases written by practitioners. Part I starts by considering the political template on which mineral extraction occurs in Africa with chapters on resource governance and the impact of globalization on development paths within the continent. Part II considers key areas of data deficits that can impact mineral development paths within Africa and how these can be addressed through a range of natural science and social science methods. Part III focuses on the key environmental challenges that are linked to mining in Africa, and some novel ways of addressing these challenges. In addition, the Ebola crisis in West Africa in 2015 prompted us to include a chapter on public health and mining. Finally, in Part IV we come full circle with the challenge of resource governance and present key ways in which scales of mining development can be more effectively managed in the African context of multiple jurisdictions and uneven regulatory capacity between and within states. We are conscious of the tension between aspirations for African unity as an antidote to a fracturing colonial legacy on the one hand, and the desire of African states to distinguish themselves culturally and eschew a simplistic homogenization on the other. In the context of mineral development, however, there has been a concerted effort by the African Union to recognize cross-cutting lessons across Africa, hence the continental cadence of this volume is appropriate. This volume presents an eclectic set of methods to show how environmental science and resource governance require a diverse skill set and consequent capacity building within Africa.

Our goal is to provide an accessible and high-quality anthology that can assist African policymakers and link Africa's challenges to global sustainability conversations. Minerals are an essential part of our lives because of their use in technologies we use daily. The supply chains of so many of these essential elements can be traced back to Africa; this gives the countries of the continent some leverage but also makes them vulnerable to what has been termed "the resource curse." The lessons gleaned from this volume on how to harness Africa's mineral wealth for development will undoubtedly have implications for material use and flows worldwide.

This book has two key conceptual frames that have assisted us in selecting and developing chapters. First, we consider the life cycle of mineral extraction as conducted by mineral developers and regulated by governments. Figure 0.1 lays out the chain of value that goes through the four key phases of an extractive operation. The two contravening arrows at the bottom indicate that the value chain has to constantly contend with costs and benefits that require various forms of scientific data, risk analysis, and policy considerations to balance.

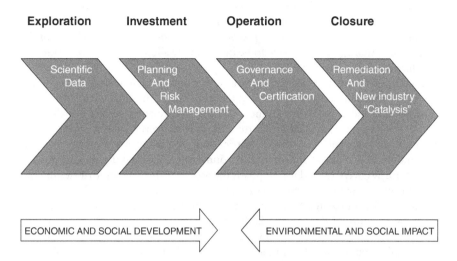

Figure 0.1 Life cycle of mineral extraction

We are preparing this book for publication at a time when there is considerable anxiety regarding the developmental benefits of mining. The resource boom from 2000 to 2012 that was spurred by the rapid infrastructure developments in China has abated, and resource economies are waiting for the next demand cycle to spur economic activity. China's investment in Africa has led to a growing literature of dissent regarding the scale and scope of the enterprise and its overall costs and benefits for the continent.[2] Furthermore, investment in exploration is currently being hampered by lack of capital and a dire legacy of errant closure of past projects with serious environmental legacies that governments and industry are now trying to remediate.[3] At the same time, the extractive enterprise can be sustainable only in its contributions toward development if the capital harnessed from the extractive enterprise can catalyze other sectors of development. Minerals as a primary sector might be deemed non-renewable on human time scales, but the minerals sector certainly has the potential to stimulate other sectors of the economy that might otherwise remain dormant. The economic linkages accorded by the extractive sector need to be considered at every phase of the life cycle. Governments need to consider more carefully the investment of the broader wealth to catalyze other industries beyond the life of the mineral extraction phase.

With this backdrop, we have also considered a second conceptual frame for the book, one that advances the discussion of mineral contributions to Africa's development across the realms of the natural sciences and public policy. This is a novel approach that aims to bridge an apparent divide in studies of mineral development. Much of the literature on extractive industries is either highly technical earth sciences and engineering research on the one hand, or political economy and anthropological analyses of minerals and development (or lack thereof) on the other.[4] There is also a stark disconnect between the environmental conservation discourse on Africa and the vast economic development literature.

Where integrative anthologies have been attempted, they have focused on very broad questions of sustainability rather than on focusing on analyses by a range of disciplinary practitioners.[5] We have provided an analytical frame that allows for the science and public policy interface to be more closely aligned and analyzed through the chapter selections for this book.

Figure 0.2 shows the basic data premises on which the science and policy interface must be predicated at the center of the diagram. In this figure we have linked key aspects of the mineral development challenge represented by the book's chapters. It is important to note that we are ultimately concerned about moving away from a linear view of extractive industries and are particularly inclined to highlight industrial ecological approaches to connecting science and policy in mineral extraction. We have circled for emphasis the aspect of the diagram that ties in with the nascent concept of a circular economy.

Existing literature on Africa's extractive industries does not adequately address the cyclical nature of this most volatile of markets.[6] Academic and applied policy analysis alike tells the story of opportunities and threats facing African countries from heightened demand for minerals, but pays little attention to what could happen when demand drops. Two discourses stand out: (1) the optimistic "Lions on the Move" thesis from the McKinsey Global Institute and (2) the pessimistic "scramble for Africa."[7] Neither addresses the inevitable situation facing many resource-dependent countries when price fluctuations force operators to put megaprojects on ice at short notice, and transnational corporations divest assets as quickly as they acquired them. How might geology and technology intersect with these economic challenges to allow for more-sustainable opportunities to emerge? Timing, flexibility, and long-term planning are key themes explored in this book, considered at each stage of extractives development.

The cyclical nature of the extractive industries has been shown to have a significant impact on political dynamics in resource-rich countries. For example,

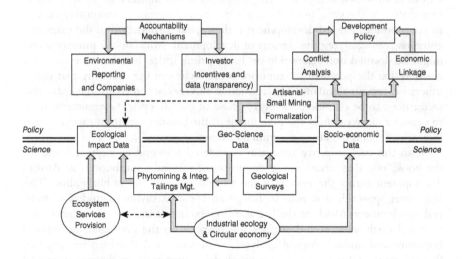

Figure 0.2 Themes linking science and public policy

Anthony Bebbington notes that political elites in resource-rich countries are more vulnerable to commodity price fluctuations than to any other exogenous factor related to extractive industries.[8] Political scientist Miriam Lowi states that in Algeria, "When oil rents become important in an environment in which cleavages are deep and the 'national question' is contested, political stability is sacrificed in periods of resource contraction and distribution crises."[9] The cyclical nature of mineral and energy commodity prices undermines the stability of resource-dependent political settlements. This is especially the case for neo-patrimonial systems that use resource rents to co-opt political support. A sharp decline in mineral, oil, or gas prices strips this type of power away.

For example, Zambia's exposure to the external shock of declining copper prices and rising oil prices in the 1970s has been a well-documented case of the curse of single commodity dependence. The prolonged economic recession eventually led to significant public protests and social mobilization led by the trade union movement. The most frustrating aspect of copper price dependence for the Zambian government has been the disruption of long-term development plans.[10] We have further augmented such socioeconomic case research on Zambia with in-depth analysis of science-based approaches to addressing the country's mineral quandaries through novel approaches to mine waste management, remediation of sites, health and safety concerns and managing social risk.

A boom in commodity prices increases non-tax revenue to resource-rich countries, which can have a stabilizing effect on ruling coalitions. Whether this stability is achieved by channeling windfall revenues into inclusive social spending, or into more-exclusive patron–client payments, depends on the nature of the political settlement. A stable political settlement is a basic condition for success of countries that have attained development from the mineral and energy resources. Developmental states might be able to achieve this level of stability without political inclusivity. Political and economic marginalization of actors living within resource-rich regions has been seen to fuel conflict in many countries, however. This risk is most acute at the early stages of exploration, licensing, and construction of new projects when they can be derailed by localized conflict, and before revenues have been shared with these subnational actors. Realistic and incremental efforts to achieve broad-based development from resource extraction are needed to balance inclusivity with stability.

The literature on the politics of mining also tends to focus on the interaction between political elites and large, state-owned, and multinational enterprises. There is a gap in considering artisanal and small-scale mining (ASM) from this perspective. This manuscript attempts to address this challenge through chapters that consider the impacts of a broad range of mining development efforts in Africa. Even where official government policy might be to encourage small-scale mining, the political dynamics within a country or region can undermine trust in licensing and regulation procedures. Accentuating these factors are the array of natural variables of ecosystem fragility, disease vector prevalence, and climatic variation across the several environmental zones in Africa. We have situated this anthology within this very broad context of natural resource governance research wherein Africa is the archetype for dealing with the most complex array of ecological and social challenges. We are conscious of the need for clear solutions rather than the recurring lament that often characterizes discourse on African development.

As the "African dream" is reimagined with the advent of the United Nations' seventeen Sustainable Development Goals and their ambitious 2030 achievement timeline, we hope that this book will be of use to the broader development community.[11] Mineral extraction is an evocative topic wherein the narratives become polarized between renewability and non-renewability.[12] As green technology researchers frequently remind us, however, even clean energy sources such as wind and solar require minerals for infrastructure development.[13] A sound understanding of the natural and social science that underpins mineral extraction is imperative. The costs and benefits of various techniques of extraction, the opportunities and limits of recycling, and the role of technological change in this milieu needs integrative analysis. Our aim in this volume is to highlight the essential intersection of science and politics in efficiently and effectively harnessing the potential for minerals as a contributor to development in Africa.

Notes

1. Commodity super cycles are periods of about forty years when commodity prices steadily climb for a decade or two, then fall slowly back to where they were.
2. Howard W. French, *China's Second Continent: How a Million Migrants Are Building a New Empire in Africa* (New York: Vintage, 2014); David H. Shinn and Joshua Eisenman, *China and Africa: A Century of Engagement* (Philadelphia: University of Pennsylvania Press, 2012).
3. Ravi Jain, *Environmental Impact of Mining and Mineral Processing: Management, Monitoring, and Auditing Strategies* (London: Butterworth-Heinemann, 2015).
4. Pádraig Carmody, *The New Scramble for Africa* (Cambridge, UK; Malden, MA: Polity, 2011); Bonnie Campbell (ed.), *Modes of Governance and Revenue Flows in African Mining* (Basingstoke: Palgrave Macmillan, 2013).
5. J. Richards (ed.), *Mining, Society, and a Sustainable World* (Heidelberg; New York: Springer, 2009).
6. See, for example, D. Bryceson et al. (eds.), *Mining and Social Transformation in Africa* (London: Routledge, 2014); J. A. Grant et al. (eds.), *New Approaches to the Governance of Natural Resources: Insights from Africa* (Basingstoke: Palgrave MacMillan, 2014). Bonnie Campbell, (ed.), *Modes of Governance and Revenue Flows of African Mining*.
7. McKinsey Global Institute, "Lions on the Move: The Progress and Potential of African Economies" (McKinsey & Company, 2010), www.mckinsey.com/global-themes/middle-east-and-africa/lions-on-the-move. Pádraig Carmody, "Cruciform Sovereignty, Matrix Governance and the Scramble for Africa's Oil: Insights from Chad and Sudan," *Political Geography* 28:6 (2009): 353–361.
8. A. Bebbington, "Governing Natural Resources for Inclusive Development" in S. Hickey, K. Sen, and B. Bukenya (eds.), *The Politics of Inclusive Development: Interrogating the Evidence* (New York; Oxford: Oxford University Press, 2015).
9. M. R. Lowi, "Oil rents and Political Breakdown in Patrimonial States: Algeria in Comparative Perspective," *Journal of North African Studies* 9:3 (2004): 85.
10. M. Hinfelaar and J. Achberger, "The Politics of Natural Resource Extraction in Zambia" (Lusaka, Zambia: Southern African Institute for Policy and Research, 2016).
11. For a discussion of the relationship of extractives to the Sustainable Development Goals refer to: World Economic Forum, *Mapping Mining to the Sustainable Development Goals: A Preliminary Atlas* (Geneva: World Economic Forum, 2016), http://unsdsn.org/wp-content/uploads/2016/01/160115-Atlas_full.pdf.
12. For a broad review of minerals and sustainability discourse from a historical and contemporary perspective refer to Saleem H. Ali, *Treasures of the Earth: Need, Greed and a Sustainable Future* (New Haven; London: Yale University Press, 2009).
13. David Abraham, *Elements of Power* (New Haven; London: Yale University Press, 2016).

Part I
The politics of African mining

1 Harmonizing African resource politics?

Lessons from the African Minerals Development Centre

Rodger Barnes, Kojo Busia, and Marit Kitaw

> **In brief**
>
> - The Africa Mining Vision offers a framework for resource exploitation that achieves sustainable development in countries with diverse mineral prospectivity, development histories, and institutional and human capital.
> - The vision's origins and principles for broad-based socioeconomic development across the continent are being implemented by the African Minerals Development Centre.
> - The Africa Mining Vision is significant both as a framework for guiding countries that are seeking to maximize the contribution of mining to sustainable development and in educating practitioners and scholars who are working on inclusive social and economic development opportunities from mining in Africa.

Domesticating the Africa Mining Vision (AMV) at the country level through Country Mining Vision (CMV) entails deliberate interventions more than an "invisible hand." It requires strong national ownership of the process, which, although government-led, must be inclusive enough to promote broad consensus on how to ensure that mining policies outlast political and electoral cycles. A multi-stakeholder approach that considers the perspectives of civil society and community-based organizations, local administrations, and the private sector is fundamental, as shown in the examples of Lesotho and Mozambique. A trusted broker with technical and financial capacity is needed at the country level to effectively facilitate the beginning of the process, through a mechanism that can consequently be organically adopted by the country.

Implementing the AMV at the country level ultimately requires a profound understanding of the institutional and agency dynamics in the country; the incentives to adapt to change and a continuous mechanism to mobilize the multiple stakeholders to collaborate across sectorial and jurisdictional divides. It relies on high-level government leadership to sensitize actors to change and articulate clear and effective pathways for transformation. The buy-in of key actors and agencies is needed so that ministries responsible for mineral development along with national development authorities become owners and strong implementing agents of the AMV and domesticated CMVs.

Introduction

In February 2009, the African Union Assembly of Heads of State and Government (AU) adopted the Africa Mining Vision (AMV). For the first time, African nations devised a strategy for harnessing the extractive resources sector that would spur economic development and structural transformation across the continent. Based on foundational ideals of equity and fairness, the AMV delivered a comprehensive blueprint for governments, private enterprises, and non-government agencies alike to pursue the optimal exploitation of resources.

In adopting the AMV, the AU called on the international community and Africa's development partners to support the efforts of member states "towards enhancing the contributions of mineral resources to the achievement of the MDGs (Millennium Development Goals), the eradication of poverty and the promotion of sustainable economic growth and development."[1] Through the AMV, the AU set a bold and ambitious agenda to endow the extractive resources sector with a prominent role in socioeconomic development.

This chapter examines the implementation of the AMV as a roadmap for change, both in the governance of extractive resources and in the linkage of the mining industry with other parts of the economy. The chapter first reviews the AMV's historical and political origins along with its main principles and objectives. Next, the chapter describes the AMV's implementation by the institution entrusted with the task, the African Minerals Development Centre (AMDC).

The inception of the AMV coincided with China's spectacular growth and sustained demand for mineral resources. Lower commodity prices and the ensuing fiscal crises are currently challenging many resource-dependent African countries. Questions have arisen over whether the AMV's development objectives can be maintained in a depressed commodity cycle. Busia and Akong contend that the AMV is forward-looking and robust enough "to bring about a lasting paradigm shift for Africa's extractive sector, anchored on its broad-based development."[2] This chapter examines the approaches, successes, and challenges of implementing the AMV as the primary framework for practitioners and scholars in ensuring that extractive resources contribute to sustainable economic and social development in Africa.

Box 1.1   Text of the Africa Mining Vision

"Transparent, equitable and optimal exploitation of mineral resources to underpin broad-based sustainable growth and socioeconomic development."
  This shared vision will comprise

- A knowledge-driven African mining sector that catalyzes and contributes to the broad-based growth and development of, and is fully integrated into, a single African market through

- o   Downstream linkages into mineral beneficiation and manufacturing;
- o   Upstream linkages into mining capital goods, consumables, and services industries;
- o   Sidestream linkages into infrastructure (power, logistics, communications, water) and skills and technology development (HRD and R&D);
- o   Mutually beneficial partnerships between the state, the private sector, civil society, local communities, and other stakeholders; and
- o   A comprehensive knowledge of its mineral endowment;

- A sustainable and well-governed mining sector that effectively garners and deploys resource rents and is safe, healthy, gender and ethnically inclusive, environmentally friendly, socially responsible, and appreciated by surrounding communities;
- A mining sector that has become a key component of a diversified, vibrant and globally competitive industrializing African economy;
- A mining sector that has helped establish a competitive African infrastructure platform, through the maximization of its propulsive local and regional economic linkages;
- A mining sector that optimizes and husbands Africa's finite mineral resource endowments and that is diversified, incorporating both high-value metals and lower-value industrial minerals at both commercial and small-scale levels;
- A mining sector that harnesses the potential of artisanal and small-scale mining to stimulate local/national entrepreneurship, improve livelihoods, and advance integrated rural social and economic development; and
- A mining sector that is a major player in vibrant and competitive national, continental, and international capital and commodity markets.

Harnessing Africa's resource endowment

The narrative of Africa rising has dominated the first decade of the twenty-first century, with the continent's GDP rising on average nearly 5 percent a year.[3] This growth, however, has not led to improvements in broad-based measures such as the Human Development Index (HDI) or in reduced rates of poverty. Instead, Africa's share in world trade has remained very low, with its exports concentrated in natural resources and minerals. At the same time, African economies have not changed in ways that would sustain social and economic development. World economic trends such as volatile commodity prices have underscored the perils of strong economic growth without concurrent industrial development and structural transformation.[4]

Yet even though demand for mineral commodities is falling in this post-boom commodity cycle, the diverse range of commodities and extent of development needs across Africa present extraordinary opportunities for wealth creation. As the world's second largest continent, Africa is home to an estimated 30 percent of global mineral reserves, with several countries hosting world-class mineral deposits.[5] Thirty-four of Africa's fifty-four countries have economies that are mineral dependent, with minerals composing at least one-quarter of their exports.[6] Africa ranks among the world's largest suppliers of certain minerals, with three-quarters of the global platinum supply, half of the world's diamonds and chromium, and up to one-fifth of gold and uranium.[7] Furthermore, it has globally important reserves of bauxite, iron ore, cobalt, tantalum, copper, and tin. As of 2013 Africa received about 14 percent of mining investments worldwide, with $110 billion invested in 26 projects around the continent.[8] In 2012, 17 percent of the world's planned worldwide exploration budgets, or $3.4 billion, was earmarked for Africa.[9]

The continent is also becoming an important petroleum exporter. Countries in sub-Saharan Africa have 5 percent of global production and 5 percent of the world's oil reserves. Nigeria and Angola are among the top 20 oil producers in the world, and between 2001 and 2010 the two countries' oil reserves increased by 20 percent and 100 percent, respectively. Other significant oil reserves have been discovered in Ghana, Uganda, the Democratic Republic of the Congo, and Kenya. The U.S. Geological Survey estimates that Kenya, Mozambique, and Tanzania hold greater reserves of natural gas than the combined reserves of the United Arab Emirates and Venezuela. Mozambique's natural gas reserves are double those of Libya.[10]

The McKinsey Global Institute's outlook for growth in Africa thus remains promising, with the continent's production of oil, gas, and most minerals predicted to continue to grow steadily by between 2 and 4 percent per year. At this rate of growth, the value of resources production would rise from $430 billion in 2016 to $540 billion by 2020.[11]

While the rich resource endowment of Africa is unquestioned, uncertainty remains over the extent to which African countries can leverage exploitation of their mineral and petroleum resources to transform their economies and societies.

The link between mineral endowment and economic growth has been the subject of debate for years. Many scholars argue that developing countries dependent mainly on resources for export earnings have experienced relatively slow rates of economic growth.[12] For several decades, economists have warned of the phenomenon known as the resource curse, in which imprudent expenditure of resource rents destabilizes the economy and renders much agricultural and manufacturing activity internationally uncompetitive.[13] This occurred, for example, in Mexico, Nigeria, and Venezuela during the 1979–1980 oil boom. Some researchers also believe that mining has the propensity to increase local conflict and even spread violence regionally.[14]

One emerging paradox in Africa is that some of the countries that are richest in resources rank among the lowest on the HDI. For instance, oil-dependent countries such as Angola and Chad and mineral-dependent states such as the Democratic Republic of the Congo and Guinea have low HDI. Niger is the world's fourth-ranking producer of uranium and yet also trails the index.[15] Yet there is no conclusive correlation between resource dependence and HDI: relatively high HDI levels are found in oil-exporting countries such as Algeria, Gabon, and Libya and mineral-rich South Africa and Botswana. Furthermore, whereas Africa in general has struggled to leverage resource development for the benefit of its people, some mineral-rich countries such as Botswana successfully use mining to induce economic growth and reduce poverty.[16]

The evidence suggests that the abundance of a country's resource endowment or reliance on mineral or petroleum exports is not a clear determinant of either prosperity or impoverishment.[17] While achieving development from minerals is clearly not automatic, neither is it the case that poor socioeconomic development is a necessary outcome of a country's resource endowment. The above analysis of differing development outcomes across various African countries suggests that the management of a country's resources is critically important.

The need for African-driven solutions

Prior to the AMV, no effective template existed for optimizing resource exploitation to achieve sustainable development. Resource economics drew on lessons from the economic transformation of northern European countries after North Sea oil was exploited in the late 1970s. Scholars warned of potential pitfalls such as Netherland's Dutch Disease, and cited examples of best practices such as Norway's use of oil revenue to lift it to the highest levels of development.

But these lessons did not translate neatly to the African context, with its vastly different history and political conditions. The impact of colonialism and the exploitative precedents it established affected the way countries developed their resources as they gained independence.

The global steel requires large volumes of coal and iron ore, and higher demand is an indicator of economic growth. Three global steel cycles since World War II also influenced mineral resource development. These cycles, outlined in the technical documentation supporting the AMV, are shown below:[18]

- Phase 1 (1950–1984): high intensity. Postwar reconstruction efforts and increasing buying power within the developed world produces strong mineral demand and high prices. Negligible impact in the developing world.
- Phase 2 (1984–2000): low intensity. Developed world infrastructure installed. Oversupply and low prices for most minerals.
- Phase 3 (2000–ca. 2011): high intensity (higher than phase 1). Developing world takes off and trade rules are revised, reflecting a partial loss of developed world hegemony over global trade systems. High demand and high prices.

Many African states were still colonies during phase 1, and on gaining independence there was a strong push for national sovereignty. Extractive resources came under state control in the 1960s and 1970s, leading to the nationalization of large private companies in Ghana, Guinea, Zambia, and other countries. The timing of this development was unfortunate, however, because it occurred just before the onset of phase 2, a period of weak demand and low prices. State control of the industry served to further weaken it through political interference in business decisions, disrespect for managerial and technical expertise, low reinvestment leading to capital consumption, and inability to access finance.[19]

By the late 1980s, generally at the instigation of the World Bank, many African countries instituted reforms to curb abuses and actively sought foreign direct investment. They also privatized state-owned enterprises and revised national mineral policies to emphasize security of tenure (security of the rights to the deposits the miners wish to mine) and to strengthen mineral rights. The new legal, regulatory, and administrative frameworks reflected a shift from government as an owner-operator to government as regulator-administrator, with the private sector assuming the lead role in mineral development projects.[20] Governments offered comprehensive incentive packages to attract mining investors, particularly in the form of reduced taxes and royalties. In the late 1990s the industry witnessed less regulation, lower state share of resource rents, and limited linkages between the resources sector and the rest of the domestic economies.

These reforms accompanied a rise in mineral prices in phase 3, along with increased foreign direct investment and an influx of mining capital, technology, and skills. By the turn of the century, however, critics questioned whether the resources boom, and the ensuing rise in export earnings in many mineral economies in Africa, had a desirable impact on domestic resource development. They criticized the reforms as being narrow-minded and geared toward attracting foreign investment and promoting exports rather than fostering domestic development.[21]

This progression in Africa coincided with an increasing global focus on sustainable development. In 2002 the World Summit on Sustainable Development in Johannesburg shined a spotlight on mining, particularly on how the industry could contribute to global sustainable development. The mining industry came under tremendous pressure to improve its performance in several areas, such as environmental, social, and economic impacts; stakeholder participation, including local and indigenous communities and women; and technical capacity in host countries.

Following the World Summit on Sustainable Development the mining industry sponsored a major initiative to frame its response to the global focus on sustainable development. The Mining, Minerals and Sustainable Development project sought to recast the minerals sector as a major contributor to sustainable development. It acknowledged the dilemma of achieving sustainability from exploitation of nonrenewable resources and argued that natural (mineral and petroleum) resource capital could be channeled into financial, human, and institutional capital to propel economic and social development. In its 2002 report, the Mining, Minerals and Sustainable Development project proposed a strategy for implementing sustainable development principles in the mining and minerals industry.[22] The International Council on Mining and Metals followed suit the next year and adopted the Sustainable Development Framework to help members improve their sustainable development performance.[23]

The extensive scrutiny of the global minerals industry revealed that a complex array of coordinated and interrelated conditions was necessary for resource development to produce desirable social and economic outcomes. Even experiences in developed parts of the world—such as central Australia, where indigenous communities hold significant statutory rights to control access to traditional lands and enter beneficial agreements—showed the need for proactive interventions to ensure that financial and other agreement benefits translated into human and institutional development.[24] Without effective linkages through local employment, training, and business development, there was little prospect of skills transfer, wealth creation, or community development.

Industrialization and resource development in Africa thus did not lead to diversified industrial economies. High levels of continental debt, poverty, capacity constraints, and lack of infrastructure only exacerbated the situation.[25]

For Africa, creating better linkages between the extractive sector and other sectors of the local economy was essential to ensuring that long-term benefits accrued to host countries. Most of Africa's minerals were exported as ores, concentrates, and metals without significant value addition, so there was a great potential for mineral beneficiation. Beyond value addition and increased local processing, the regimes in resource-rich African states also needed to change to integrate the minerals sector within a diversified economy.[26]

In formulating a long-term strategy to eradicate poverty and establish sustainable growth and development across the continent, it became clear that structural transformation of economies was a central component.

Formulating the Africa Mining Vision

The AMV emerged against the backdrop of externally imposed reforms, including those promoted by the Bretton Woods institutions, that were seen as lacking legitimacy and accountability. These reforms, based on unequal relationships between African countries and powerful multinational and financial institutions, privileged the interests of privately owned companies and elites over those of the majority of the population.[27] In formulating the AMV, policymakers recognized that little progress had been made to restructure the mining sector, which had operated as an enclave since colonial times. What was needed was a framework for integrating the sector more coherently into the continent's economy and society.[28]

Busia and Akong describe the priorities of the AMV as "the product of long-standing aspirations of African governments to chart an alternative development path that is owned by countries."[29] Its evolution was stimulated by several initiatives for resource development that were implemented at the sub-regional, continental, and global levels.

Linking the extractive and other industries through economic clusters and infrastructure corridors formed the basis of initiatives such as the AU's New Partnerships for Africa's Development and the Big Table dialogue in February 2007 between African ministers of finance and economic planning and their counterparts in the Organisation for Economic Cooperation and Development.[30] Titled "Managing Africa's Natural Resources for Growth and Poverty Reduction," the Big Table was jointly organized by the United Nations Economic Commission for Africa (UNECA) and the African Development Bank.[31] The Big Table noted that the reforms promoted by the Bretton Woods institutions in the 1990s had not led to development. Instead, it found that the resource industry continued to operate in economic enclaves and that poverty had not been reduced.[32] The Big Table recommended that a study group be established to pursue the issue of mineral sector reforms.

In 2007, UNECA set up the International Study Group on Africa's Mineral Regimes (ISG), comprising experts on mineral development issues, including leading policymakers, researchers, academics, and practitioners, along with representatives of the AU, and regional economic communities; UN agencies; the Commonwealth Secretariat United Kingdom; and civil society organizations.[33] The ISG noted the potential for African mineral resources to anchor an industrialization strategy that would secure sustainable growth and development. Toward this end, the ISG sought to move the focus of mineral policy beyond extracting resources and sharing revenue to encompass a more broadly based restructuring of African economies.

The work of the ISG informed the formulation of the AMV, which was endorsed by the first session of the AU Conference of Ministers Responsible for Mineral Resources Development held in Addis Ababa in October 2008. This conference was notable for the Addis Ababa Declaration on the Development and Management of Africa's Mineral Resources, which reaffirmed the group's "commitment to prudent, transparent and efficient development and management of Africa's mineral resources to meet the Millennium Development Goals (MDGs), eradicate poverty and achieve rapid and broad-based sustainable socio-economic development."[34]

In adopting the AMV four months later, the AU Assembly requested that the AU conference of ministers develop a concrete action plan for implementing the AMV through the AU Commission and in partnership with UNECA, the African Development Bank, regional economic communities, and other stakeholders.[35]

The AMV Action Plan was finalized during the second conference of AU mining ministers held in Addis Ababa in December 2011, with input from policymakers, the private sector, non-governmental organizations, artisanal and small-scale miners, academic institutions, and international development organizations.[36] As shown in Table 1.1, the AMV Action Plan features nine program clusters structured around the seven pillars of the AMV.[37]

Table 1.1 Key pillars of the AMV and AMV Action Plan program clusters

AMV pillars	AMV Action Plan program clusters
Fostering a transparent and accountable mineral sector in which resource rents are optimized and utilized to promote broad economic and social development	Mining revenues and mineral rents management
Optimizing knowledge and benefits of finite mineral resources at all levels of mining and for all minerals	Geological and mining information system
Building human and institutional capacities toward a knowledge economy that supports innovation, research, and development	Human and institutional capacities; Research and development
Harnessing the potential of small-scale mining to improve livelihoods and integration into the rural and national economy	Artisanal and small-scale mining
Promoting good governance of the mineral sector in which communities and citizens participate in mineral assets and in which there is equity in the distribution of benefits	Mineral sector governance
Fostering sustainable development principles based on environmentally and socially responsible mining, which is safe and includes communities and all other stakeholders	Environmental and social issues

Developing a diversified and globally competitive African mineral industry that contributes to broad economic and social growth through the creation of economic linkages	Linkages and diversification; Mobilization of mining and infrastructure investment

Source: AMV Action Plan, www.africaminingvision.org/amv_resources/AMV/AMV_Action_Plan_dec-2011.pdf

Implementing the Africa Mining Vision through the African Minerals Development Centre

The scale of change envisioned by the AMV was ambitious; to carry out its implementation, the second conference of AU mining ministers established the African Minerals Development Centre (AMDC) supported by development partners including Australia and Canada.[38] The AU formally launched the AMDC in December 2013 at the Mineral Resources Ministers' Conference in Maputo, Mozambique, with the AU Commission, UNECA, the African Development Bank, and the United Nations Development Programme (UNDP) as implementing partners.[39] The AMDC was commissioned to "work with member States and their national and regional organizations to enable mineral resources [to] play a greater transformative role in the development of the continent through increased economic, social linkages and improved governance."[40] The AMDC's business plan laid out a number of roles for the new center:[41]

- tracking and coordinating the implementation of the AMV and activities of the Action Plan;
- identifying gaps and areas of need in the member states and addressing them by tapping expertise and information resources from a broad range of local and international sources;
- undertaking and coordinating policy research to develop policy strategies and options for realizing the AMV;
- conducting an advocacy and information dissemination campaign including websites and discussion fora to engage various stakeholders and help create a movement for achieving the AMV;
- monitoring, evaluation, and review of initiatives undertaken to advance the AMV; and
- generally providing a think-tank capacity for embedding the AMV into Africa's long-term development.[42]

Effecting change required the AMDC to work with countries that have distinct histories and development trajectories where mineral resources either play or potentially play a significant role in the economy. Africa is a diverse continent, and many countries face considerable political, social, and economic challenges, including poverty, disease, and civil strife. Implementation of the AMV also had to address additional challenges in a range of areas:

- Integrating mining operations into the domestic economy. Most mining operations were developed as enclaves with weak integration into the rest of the economy. In Africa many factors hampered the development of economic linkages in the mineral sector, including inadequate infrastructure; low levels of industrialization, which translates into weak markets for mineral products; procurement policies of mining companies that do not prioritize local content; and, in general, a poor knowledge base and inadequate technological capabilities.
- Establishing effective regulatory frameworks founded on sound legal systems that provide accountability and equity. Such frameworks entail public sector institutions that are transparent in their dealings with the private sector and civil society. Transparency includes fighting corruption using legal recourses as well as adherence to clearly articulated policies.
- Fairly structuring the revenue flows and distributing them locally and nationally, as well as reconciling conflicting interests and managing expectations. The process of transforming revenue flows into permanent wealth that outlasts finite resources requires good planning and making well-informed judgments on savings versus investment.
- Dealing with inadequate capacity and weak coordination of actions within the mineral institutional chain.[43] Without strong institutions it is difficult to generate resource-driven broad-based development.
- Addressing permanent adverse impacts of mining on the land and the environment. The effects on the environmental and social impacts of mining are strongly linked. Adverse impacts are worsened by poor governance, a weak regulatory environment, and insufficient enforcement capacity. One of the key challenges is the capacity to enforce impact management instruments, such as environmental and social impact assessments.
- Using participatory and consultative approaches in societies that normally follow a more hierarchical pattern. This challenge is worsened by asymmetry of power, knowledge, capacity, and resources between mining companies and local communities. Involving women and addressing gender issues could be particularly difficult because women are at a disadvantage due to the lack of legal frameworks, policies, and programs that consider their needs and protect their rights; limited access to resources; lack of a political voice; and disproportionate power relations between the genders in households and communities.
- Normalizing the artisanal and small-scale mining sector, which constitutes an extremely important constituency in many mineral jurisdictions. Often artisanal and small-scale miners are trapped in a poverty loop that they have difficulty escaping. They use rudimentary equipment and techniques because their access to knowledge and finance is limited. Many miners also lack the business skills and information needed to enhance the economics of their operations. Environmental impacts are serious and labor conditions harsh. Access to extension services and other administrative support seldom exist. Incidents of child labor, crime, and gender-based violence are widespread in places.
- Facilitating cooperative action at the regional level and moving away from competition for foreign investment. This allows neighboring countries to pursue common development objectives.[44]

Domestication of the Africa Mining Vision through Country Mining Vision

The tenets of the AMV encompass a broad array of economic sectors, well beyond the extractive resources sector. The vision seeks to engage government institutions and stakeholders in changing the way agencies function to ensure well-designed, predictable processes and consistency in dealings among government, businesses, and communities. Collaboration, coordination, transparency, and communication underpin the strategic planning necessary for long-term sustainable development. Multiple stakeholders, often with competing interests, need to have an understanding of the stakes at play.

To implement the AMV at the country level, the AMDC engaged with leading experts to develop the *CMV Guidebook*, which articulates clear-cut guidelines and options for aligning a country's mining policies to the principles of the AMV.[45] Developing a CMV involves multi-stakeholder consultations to formulate a shared vision on how resource exploitation can promote broad-based development and structural transformation of the country's economy.

The *CMV Guidebook* provides a step-by-step method for strategic assessment, policy dialogue, and sector analysis, as well as mechanisms for conducting stakeholder consultations, including steps for policy design and formulation of a CMV implementation, monitoring, and evaluation tool. The CMV is intended to enrich, not replace, countries' existing economic strategies and national development plans.[46] The goal is to achieve institutional cohesiveness, whereby sectors pursue coherent policies that ensure a transformative role for extractives in the economy.

The CMV seeks to establish a change process. The trajectory of change and reform often hinges on the maturation of mining or petroleum industry in the country. In a mature regime, there might be a predisposition to conducting business in the usual way; a resistant mindset could emerge among influential policymakers. In other economies where resource development is an emerging sector, starting fresh can have fewer barriers to introducing AMV-compliant petroleum and mineral regimes.

Lesotho case

Lesotho represents a model case of formulation of a CMV where the AMDC was able to conceptualize, design, and institute the process toward a CMV from the outset and maintain an ongoing commitment toward building the relevant institutions. The AMDC was able to work collaboratively with Lesotho's agencies and introduce best practices for relevant policies, processes, and procedures.

Lesotho is a small land-locked country and is not historically a mineral-based economy. The government commissioned new mining operations, but existing regulations were not adequate for good governance of the sector. Development of a CMV began with a government review of the 1962 Mining and Mineral Policy, with the support of the AMDC and UNDP. The review identified the need to improve mineral resource management, maximize revenue collection and utilization, and increase impact of mineral extraction both at the national and local level.[47] An inclusive working group was established to drive the CMV in Lesotho. The Multi-Stakeholder Working Group comprised forty people from several ministries including the Mining Ministry, Economic Development and Finance, as well as non-governmental organizations, local communities, the private sector, and small-scale miners. Workshops were held in 2013 that identified issues and proposed themes to guide preparation of an AMV-compliant Mining and Mineral Policy.

Using a draft Mining and Mineral Policy for discussion, the working group facilitated extensive consultations in 2014 with more than 600 people from 10 administrative districts. The new mineral policy was adopted in June 2015 and is expected to provide strategic direction for the development of Lesotho's mineral resources and ensure that the sector contributes to socioeconomic development and transformation. Many of the new elements of the policy will require further development of the existing legal and regulatory framework. The AMDC is assisting with preparing proposed mining legislation. The process proved very effective partly because of the absence of already developed policies in the mineral sector. The AMDC was able to work collaboratively with Lesotho's agencies and introduce best practice

Experience shows that no one-size-fits-all approach works in the CMV process, but some key implementation lessons have emerged.[48] The importance of strategic framing for dialogue is paramount. This includes understanding the characteristics and dynamics of the political and economic institutions and how they operate in the country. It is important to appreciate how the linkages between ministries, commissions, and other relevant agencies function. In developing the CMV, it is critical to disentangle the various agencies and to allocate respective roles for implementing the AMV. Initial engagement with a country's top leadership is essential so that leaders can assign an agency, preferably one linked to the head of state, to drive the process.

Another lesson is to appreciate how the country's administration and regulatory processes intersect with mining, including allocation of mineral rights and management of impacts and revenue. While adequate legislation might exist, domestic political interests affect the coordination and capacity of institutions to function effectively, and this effect can become more pronounced where forceful regional or ethnic interests are at stake. Understanding the principles of the AMV is sometimes not enough, since these forces can take sway over proper project planning and allocation of resources. For example, the AMV emphasizes the need for infrastructure development, but domestic politics might influence the location or nature of that development. These same forces can act to supersede principles for regional cooperation in favor of satisfying a political or ethnic constituency.[49]

The nature of the strategic planning exercize is critical in this regard. If the voices advocating framing policy around the AMV are few, the prospects for transformation will be dim. Strong domestic champions are needed to sway the policy elite. In many ways, the policy champions are as important as advocates with technical knowledge of the sector.

Strong national ownership of the CMV process also must be achieved. Although government-led, it must be inclusive enough to promote broad consensus on how to ensure that mining policies outlast political and electoral cycles. A multi-stakeholder approach that takes into account the perspectives of civil society and community-based organizations, local administrations, and the private sector is fundamental to promoting a realistic view of what benefits the extractives sector can deliver and how they can deliver them. The development of a CMV works best when it is steered by a multi-stakeholder coordinating body or task force dedicated to overseeing the entire process.

Mozambique case

Mozambique exemplifies a country that has embarked on policy and regulatory reform in the mining sector and has used the AMV to contribute to the process. Prior to adopting the AMV in 2009, Mozambique had ratified the Southern African Development Community Protocol on Mining in 2000 to create a sector that would contribute to economic development, poverty alleviation, and an improved standard and quality of life in the region.[50] Then in 2013, the World Economic Forum, in collaboration with UNECA and UNDP Mozambique, held the Responsible Mineral Development Initiative roundtable in Maputo. The dialogue identified five areas for action: (1) increasing the skilled workforce through training programs; (2) making the Extractive Industries Transparency Initiative (EITI) a permanent forum for multi-stakeholder dialogue on the extractive sector; (3) prioritizing work on local content issues; (4) funding and implementing the Ministry of Mineral Resources communication strategy and outreach; and (5) improving integrated land-use and infrastructure planning.[51]

The outcomes of the roundtable informed the formulation of the country's Mineral Resources Policy and Strategy (PERM), which was approved in 2013. The policy was drafted after discussions with a variety of ministries and institutions, including provincial governments, civil society organizations, and private sector partners and associations.

The Mining Law was approved by parliament in 2014 with the objective to regulate the use of mineral resources in accordance with the PERM.

As part of its program Extractive Industries for Sustainable Development in Mozambique, UNDP Mozambique conducted stakeholder policy dialogues in 2016, one of which addressed the impact of declining mineral commodity prices on sustainable development.[52]

As a follow-up to the dialogue, the Ministry of Mineral Resources and Energy in Mozambique, with the support of UNDP Mozambique, developed the Implementation Plan of the Mineral Resources Policy and Strategy (PI PERM). The PI PERM was the fruit of consultations of many partners, including government at central and provincial levels and all relevant ministries, the private sector, civil society organizations, women's organizations, and academia.

The PI PERM, approved in 2017, was seen as an efficient tool for planning, strategy setting, program delivery, identification of partners and partnership building, and division of labor.[53] The PI PERM can be considered as a CMV for Mozambique since it follows, to a large extent, the CMV process as described in the *CMV Guidebook*.[54] Indeed, the Mozambican process started with the organization of (1) a high-level roundtable and dialogue on extractives; followed by (2) a review of existing legal, institutional and regulatory frameworks that led to the formulation of the Policy and Strategy of Mineral Resources, significantly compatible with the AMV principles. The Implementation Plan (3) for integrating mining into national development visions and plans (the PI PERM) was then elaborated, informed by the outcomes of (4) dialogues, followed by (5) the elaboration of an effective communication strategy, and (6) the provision of technical support to stakeholders at local and national levels to facilitate consultations with local communities and national stakeholders.[55] This led to the (7) enhancement of capacity for long-term visioning, strategy setting and integrated development and planning.

Several factors contributed to Mozambique's success at domesticating the AMV: (1) high-level ownership of the process; (2) the presence of an active UNDP at the country level that served as a trusted broker and provided valuable technical and financial support to foster interinstitutional and intra-agency collaborations; multi-stakeholder dialogues; and intersectoral synergies to promote the AMV and to oversee the development of the PI PERM (the CMV); and (3) the emergence and proliferation of AMV champions at several levels, including government institutions and civil society organizations. The challenges ahead are to secure coordination commitment from all stakeholders in implementing the workplans agreed on; to earmark resources for interventions included in the PI PERM, and actual implementation of the PI PERM to achieve the sustainable development desired.

Shifting paradigms under the Africa Mining Vision

Busia and Akong stress that institutional processes under the AMV do not call for a business-as-usual approach. The AMV policy discourse goes beyond optimizing the revenue potential of mining by making improvements to legal and regulatory frameworks, tweaking mineral fiscal regimes, or creating a stable business environment. It opens space for a fundamental, structural change away from maximizing revenues toward harnessing mineral resources for broad-based development.[56]

Market volatility is recognized by scholars as the key problem plaguing an extractive-led development agenda. Commodity prices rise and fall, often in unpredictable patterns that create vulnerability, risks, and crises.[57] Busia and Akong argue that the current commodity downturn presents an opportunity for African countries to implement the AMV as a forward-looking, multidimensional instrument for change. Their analysis of the current crises facing the extractive sector in Africa concludes that disruptive events open windows of opportunity and are important drivers of paradigm shifts. "The outcomes depend on varied factors including how the AMV ideational foundation is capable of organizing interests, institutions, actors and power for change."[58]

The trajectory of change under the AMV is inherently complex, dynamic, and difficult to predict. The AMDC's evaluations of scenarios for change find that implementation of AMV in different countries will be non-linear and incremental. Although the ideational foundation of the AMV is sound, resistance to change by entrenched interests and actors is likely.[59]

For instance, development partners remain influential actors in policy development and implementation across Africa. Whereas donor countries might advocate for global principles in line with the AMV, at the bilateral level they are wedded to the enclave approach and push agendas disconnected to the goals of the AMV by pursuing specific interests on behalf of their national companies, such as tax concessions or shelter from indirect taxes (including customs or import duties). It is critical to have effective, consistent coordination and adherence to the principles of the AMV among all multilateral and bilateral donors and partners supporting the resources sector at the country level.

Transforming relationships from rents to sustainable revenue mobilization through productive and higher-value-adding activities also has the potential to threaten settled arrangements between business and politics. Leveraging the transformative potential of minerals will require "nudging domestic elites—ruling, business and bureaucrats towards a shared vision and interests in taking advantage of mineral value chains."[60]

Implementing the AMV at the country level ultimately relies on high-level government leadership to sensitize actors to change and articulate clear and effective pathways for transformation. The buy-in of key actors and agencies is needed so that ministries responsible for mineral development along with national development authorities become owners and strong implementing agents of the AMV and domesticated CMVs.

Conclusion

The AMV has emerged as the primary strategy for positioning Africa's extractive resources as a force for transforming the continent's social and economic development. The AMV is a unique undertaking: it represents the only continental initiative to garner a cross-country, cross-sectorial, transformative plan of action for the mining sector.

In adopting the AMV, the AU set an ambitious and coordinated agenda that requires innovative ways to encourage the structural transformation needed to revise the externally driven revenue-first model of the mining sector. The scale and scope of the undertaking will take time, and the AMDC observed, its adoption across the continent is likely to be incremental and non-linear.[61] Key to the AMV's success is its presentation as a framework to guide governments, cross-sectorial agencies, and the private sector toward sustainable development outcomes from resource extraction. The multifaceted aspects of the AMV, which incorporate socioeconomic linkages, infrastructure development, diversification, and environmental and social impacts, give it broad legitimacy among stakeholders.

The foundational proposition in formulating the AMV is that extractive resources in themselves are neither a blessing nor a curse. The most important factor in determining whether they will be a blessing or a curse is the country's level of governance capacity and the existence of robust institutions. The AMV recognizes, however, that governance is only one element in the array of challenges that must be addressed in formulating a policy for a development-oriented sector.[62] The AMV also highlights the need for effective stakeholder engagement and participation, including industry and the private sector, international and regional financial institutions, government and social agencies and institutions, local communities, and non-governmental organizations.

Mining minerals can evoke strong sentiments in African politics because of its historic linkage to colonialism and poor contribution to the continent's development. Given such a history, a high-level platform was necessary to effect a significant departure from externally driven mineral development to an African-owned agenda. The AMV accomplished this challenge through the efforts of its founding leaders and the broad imprimatur that it has from the AU, UNECA, the African Development Bank, and UNDP. The implementation of the vision has been delegated to various institutional structures such as the AMDC and, at the national level, through the CMV process. Midlevel CMV processes must also have good communication with one another to ensure that resources are efficiently harnessed and to increase the potential for cross-border mineral infrastructure and development synergies in Africa.

The AMV has shifted the traditional discourse from a narrow narrative on maximizing revenues to the exploration of mechanisms that link mining with other sectors to maximize social and economic benefit. The AMV continues to gain legitimacy as the essential continent-wide framework to guide African countries in using their mineral endowment for sustainable economic and social development and to enable Africa to achieve a better place in the global economy.

Notes

1 African Union Commission (AU), African Development Bank, and United Nations Economic Commission for Africa (UNECA), "Building a Sustainable Future for Africa's Extractive Industry: From Vision to Action: Action Plan for Implementing the AMV" (Addis Ababa: AU, African Development Bank, and UNECA, December 2011), 7.
2 Kojo Busia and Charles Akong, "The African Mining Vision: Perspectives on Mineral Resource Development in Africa," *Research Paper Series No. 3/17* (Ottawa: University of Ottawa, Centre for Governance, 2017).
3 McKinsey Global Institute, "Lions on the Move: The Progress and Potential of African Economies" (McKinsey & Company, 2010), 9, www.mckinsey.com/global-themes/middle-east-and-africa/lions-on-the-move (accessed May 15, 2016).
4 Marit Kitaw, "Africa's Minerals for Development: The Role of Transformational Leadership" (Addis Abba: ECA Publications, 2015), http://repository.uneca.org/bitstream/handle/10855/22967/b11553340.pdf?sequence=1.
5 Marit Kitaw, "Africa's Minerals for Development."
6 Seventeen countries have no mining industry at all. These include some of the small island nations such as Seychelles and Mauritius.
7 Carlos Lopes, "Leveraging Africa's Extractive Sector for Inclusive Economic Transformation," *GREAT Insights Magazine* (2014): 3–7, http://ecdpm.org/great-insights/extractive-sector-african-perspectives/africas-extractive-sector-economic-transformation/ (accessed May 15, 2016).
8 Viktoriya Larsson and Magnus Ericsson, "E&MJ's Annual Survey of Global Metal-Mining Investment," *Engineering and Mining Journal* (January 6, 2014), www.e-mj.com/features/3674-e-mj-s-annual-survey-of-global-metal-mining-investment.html#.V0xsZpF96hc Annual Survey of Metal-Mining Investment (accessed May 15, 2016).
9 United States Geological Survey, *Minerals Year Book 2012: Africa* (Washington, DC: U.S. Department of the Interior, 2012), 1–2.
10 Cited in Marit Kitaw, "Africa's Minerals for Development."
11 McKinsey Global Institute, "Lions on the Move," 8.
12 Jeffrey. D. Sachs and Andrew M. Warner, "Natural Resource Abundance and Economic Growth," *Development Discussion Paper No. 517a.* (Cambridge, MA: Harvard Institute

for International Development, 1995); T. L. Karl, *The Paradox of Plenty: Oil Booms and Petro-States* (Berkeley; Los Angeles: University of California Press, 1997).
13 Jeffrey D. Sachs, and Andrew M. Warner, "The Curse of Natural Resources." *European Economic Review* 45 (2001): 828.
14 For example, see Nicolas Berman, Mathieu Couttenier, Dominic Rohner, and Mathias Thoenig, "This Mine Is Mine! How Minerals Fuel Conflicts in Africa," *Research Paper 141* (Oxford: Oxford Centre for the Analysis of Resource Rich Economies).
15 Carlos Lopes, "Leveraging Africa's Extractive Sector."
16 Cited in Marit Kitaw, "Africa's Minerals for Development."
17 Antonio Pedro, "Mainstreaming Mineral Wealth in Growth and Poverty Reduction Strategies," *Policy Paper No. 1* (Ethiopia: Economic Commission for Africa), 17.
18 African Union (AU), "Africa Mining Vision" (Addis Ababa: African Union and United Nations Economic Commission for Africa, February 2009), 9.
19 AU. "Africa Mining Vision," 11.
20 UNECA, "Africa Review Report on Mining: Summary" (Addis Ababa: October 27–30, 2009, Committee on Food Security and Sustainable Development Regional Implementation Meeting for CSD-18 Sixth Session), 6.
21 UNECA, "Minerals and Africa's Development: The International Study Group Report on Africa's Mineral Regimes" (Addis Ababa: UNECA, 2011), 17, www.uneca.org/sites/default/files/publications/mineral_africa_development_report_eng.pdf (accessed December 13, 2015).
22 International Institute for Environment and Development, "Breaking New Ground: The Report on The Mining, Minerals and Sustainable Development Project" (London: Earthscan Publications Ltd, May 2002), www.iied.org/mmsd-final-report (accessed May 26, 2016).
23 The key aspect of the International Council on Mining and Metallurgy SD framework are the "10 Principles," which include corporate governance; health and safety; human rights; responsible product design; environment and biodiversity; social, economic, and institutional development; appropriate materials choice; public engagement; and independently verified reporting arrangements. See www.icmm.com/our-work/sustainable-development-framework/10-principles.
24 Rodger Barnes, "Building an Implementation Framework for Agreements with Aboriginal Landowners: A Case Study of The Granites Mine" (MPhil thesis, University of Queensland, 2014).
25 For example, see Antonio Pedro, "Mainstreaming Mineral Wealth," 28.
26 Marit Kitaw, "Africa's Minerals for Development: The Role of Transformational Leadership" (Addis Abba: Paper presented at the symposium on Extractive Industries for African Development: A Paradigm Shift, organized by the African Studies Program, Penn State University, University Park, PA, March 27, 2015, ECA Publications, 2015), http://repository.uneca.org/bitstream/handle/10855/22967/b11553340.pdf?sequence=1.
27 Kojo Busia and Charles Akong, "The African Mining Vision: Perspectives on Mineral Resource Development in Africa."
28 AU, "Africa Mining Vision."
29 Kojo Busia and Charles Akong, "The African Mining Vision: Perspectives on Mineral Resource Development in Africa."
30 New Partnerships for Africa's Development is an AU program that seeks to eradicate poverty, place African countries, both individually and collectively, on a path of sustainable growth and development, build the capacity of Africa to participate actively in the world economy and body politic, and accelerate the empowerment of women.
31 "Managing Africa's Natural Resources for Growth and Poverty Reduction: The 2007 Big Table," http://repository.uneca.org/handle/10855/561.
32 Daniel Franks, *Mountain Movers: Mining, Sustainability and the Agents of Change* (London: Routledge, 2015), 70.

33 See Appendix A of the ISG report, www.uneca.org/publications/minerals-and-africas-development.
34 Kojo Busia and Charles Akong, "The African Mining Vision: Perspectives on Mineral Resource Development in Africa," *Research Paper Series No. 3/17* (Ottawa: University of Ottawa, Centre for Governance, 2017).
35 In adopting the AMV four months later: A "Tentative Framework for Action" attached to the AMV set out a matrix of actions related to achieving the AMV goals at the national, sub-regional and continental levels over three time frames, namely short-term (first five years), medium-term (five to twenty years), and long-term (twenty to fifty years). Roles and responsibilities were suggested for key actors including national bodies, regional economic communities, and continental bodies, "and other stakeholders": AU, African Development Bank, and UNECA, "AMDC Business Plan," 15.
36 Ibid., 16.
37 The conference added a cluster on policy, regulations, regional cooperation, and harmonization as part of the implementation strategy of the plan. These became are cross-cutting themes throughout AMV Action Plan.
38 The AMDC was established at the ECA as an initial five-year project with grants from Australia and Canada, further funding being considered by other partners. UNECA provided the initial staff of three (Wilfred Lombe, Kojo Busia, and Marit Kitaw).
39 UNDP joined the AU Commission, UNECA, and the African Development Bank in 2013 through the launch of a regional program on extractives in Africa titled, "Harnessing extractive industries for human development in sub-Saharan Africa," which indicates that UNDP will design activities to support and complement the seven AMDC result areas, www.undp.org/content/dam/undp/documents/partners/civil_society/miscellaneous/2014_UNDP_Extractives-and-Human-Development-in-Africa_14Apr2014.pdf.
40 AU, African Development Bank, and UNECA, "AMDC Business Plan," 17.
41 The AMDC outputs are organized around seven "Work Streams" that relate to the AMV program clusters: (1) Policy and licencing; (2) Geological and mining information systems; (3) Governance and participation; (4) Artisanal and small-scale mining; (5) Linkages, investment and diversification; (6) Human and institutional capacities; and (7) Communication and advocacy. Each Work Stream has an overarching goal and a number of specific expected outcomes.
42 AU, African Development Bank, and UNECA, "AMDC Business Plan," 18.
43 Institutions that are directly and indirectly involved in the administration of the sector include ministries of mines, geological surveys, environment authorities, land administration, central banks, revenue and customs authorities, and ministries of finance.
44 Marit Kitaw, "Africa's Minerals for Development," 8–9.
45 African Minerals Development Center (AMDC), *A Country Mining Vision Guidebook: Domesticating the AMV* (Addis Ababa: AMDC, 2014), www.uneca.org/sites/default/files/PublicationFiles/country_mining_vision_guidebook.pdf (accessed June 22, 2015). The experts met in Debre Zeit, near Addis Ababa in September 2014 for five days to produce the *Country Vision Guidebook*.
46 For example, *Poverty Reduction Strategy Papers*, which are documents required by the International Monetary Fund and World Bank before a country can be considered for debt relief within the Heavily Indebted Poor Countries initiative.
47 Two reports were produced in the course of the review. One, a government led detailed analysis of national economic and social landscape, mapping the geological potential and review of mining policies. The second was a regional and international report sponsored by UNDP and AMDC, which reviewed regional and global mineral development trends and identified key external drivers of the extractives sector and their impact on Lesotho's economy.
48 Antonio Pedro, "The Country Mining Vision: Towards a New Deal" in *Austrian Development Policy 2015: Commodities and Development* (Vienna: Austrian Foundation for Development Research, 2015), 33.

49 For example, see Kojo Busia and Charles Akong, "The African Mining Vision: Perspectives on Mineral Resource Development in Africa."
50 Further details can be found at: www.ecolex.org/details/treaty/protocol-on-mining-tre-001347/; www.sadc.int/documents-publications/show/Protocol_on_Mining.pdf.
51 World Economic Forum, "Responsible Mineral Development Initiative (RMDI)" (Geneva: WEF Publications, 2013), www3.weforum.org/docs/WEF_MM_RMDI_Report_2013.pdf.
52 On September 25, 2015, countries adopted a set of goals to end poverty, protect the planet, and ensure prosperity for all as part of a new sustainable development agenda. Each goal has specific targets to be achieved over the next fifteen years. For more information: www.un.org/sustainabledevelopment/sustainable-development-goals/.
53 UNDP calls for a people-centered and planet sensitive implementation of the mining policy, www.mz.undp.org/content/mozambique/en/home/presscenter/pressreleases/2017/03/13/mozambique-calls-for-a-people-centered-and-planet-sensitive-implementation-of-the-mining-policy-.html.
54 AMDC, *Country Mining Vision Guidebook*, 22.
55 UNDP Mozambique held training for journalists at central on Extractive Industries and investment opportunities during this time.
56 Kojo Busia and Charles Akong, "The African Mining Vision: Perspectives on Mineral Resource Development in Africa."
57 C. Bain, *Guide to Commodities: Producers, Players, and Prices, Market Consumers and Trends* (London: The Economist and Profile Books Ltd, 2013).
58 Kojo Busia and Charles Akong, "The African Mining Vision: Perspectives on Mineral Resource Development in Africa."
59 Ibid.
60 Ibid.
61 Ibid.
62 UNECA, "Minerals and Africa's Development: The International Study Group Report," 19.

2 Evaluating conflict risks in Africa's resource governance

Kathryn Sturman and Fitsum Weldegiorgis

In brief

- The risks of conflict associated with extractive industries in Africa need to be factored in to policymaking in a more rigorous and informed manner. More-accurate, disaggregated data is needed to evaluate the link between extractive activities and incidents of conflict at sub-national, national, and transnational levels.
- To understand the varied types of resource conflict across Africa, it is necessary to grasp how political settlements shape the extractive industries (and vice versa). Politics plays a role in how resource conflict is perceived, addressed, and ultimately either resolved or made worse by regulation and policy.
- Case studies of three countries in east Africa demonstrate the broad spectrum of conflict risks related to mineral and energy resources exploration and production that exists in the continent.
- Multi-stakeholder mechanisms for more-inclusive governance of natural resources can be used throughout the natural resource governance value chain to address the conflict risks at every level.

Introduction

The African Mining Vision (AMV) has become part of the "Africa rising" narrative of the twenty-first century, in which demand for mineral resources is an economic driver of growth and sustainable development, breaking the cycle of poverty and conflict. The AMV Action Plan includes the objective to "eliminate human rights abuses and the possibility of natural resources fueling conflict." To achieve this, it calls for action to "develop methodologies and tools for conflict risk analysis and mainstream them into planning frameworks."[1] Furthermore, the plan suggests that tenets of good resource governance such as principles of transparency, stakeholder engagement, inclusivity, and benefit sharing (including revenues) also serve as peace-building strategies.

This chapter reviews the conflict risks associated with extractive industries in Africa at different levels of analysis: the local, regional, national, and transnational. It argues that the "good governance" mantra does not fully account for the complexity of the relationship between extractive industries and conflict, and that preventing conflict between companies, communities, and governments is, above all, a delicate political balancing act. As we will see, models for engaging a broad range of actors in natural resource governance reach beyond government regulation of the environmental and social impacts of the sector to include multi-stakeholder efforts.

The balancing of interests needed to prevent conflict takes place in different venues, from the mine site and community meeting place to national and global forums. We will look at several examples in east Africa to see the variety of conflict risks associated with extractive industries in this region. Opportunities to reduce conflict in several countries depend on the relationship between stakeholders in government, the foreign and domestic private sectors, and civil society.

Between 2003 and 2013 east Africa became a frontier for nascent extractive industries, with new discoveries of oil in the Great Rift Valley, significant new offshore blocks of natural gas, and increased demand for minerals across the region. These opportunities arose in countries with less experience in the governance of mining and energy resources than the African mining and oil giants of southern, west, or north Africa. As a result, the risk analysis in some of those countries distinguishes between preexisting conflict dynamics and new tensions around mineral development. The new investments are being developed primarily within fragile post-conflict societies and sometimes within zones of ongoing conflict. We will examine conflict risks and opportunities for conflict transformation in Mozambique, Kenya, and Rwanda.[2]

The conflict analysis in this chapter includes oil and gas exploration as well as mining, although the two sectors are regulated separately in most African jurisdictions and have different impacts, positive and negative, on local communities. No regional blueprint like the AMV exists for the development of oil and gas in Africa, but many policy prescriptions for mining are relevant to the petroleum sector.

Revisiting resource conflict

Since the early 1990s a sizeable literature has challenged the previously dominant view that natural resource endowments were an asset to developing countries and a prerequisite for industrialization.[3] Termed the "resource curse" literature, these studies contend that the abundance of minerals or oil in a country increases the likelihood that it will experience negative economic, political, and social outcomes. A subset of this literature pertains to resource conflict, and a review of that literature shows three main approaches:[4]

1   A geopolitical perspective, focusing on foreign interference in political settlements and territorial disputes over resource-rich regions or offshore resources, as well as outcomes such as coups d'état, foreign backing of repressive regimes, or regime change.

2   A political economy perspective, consisting primarily of quantitative political science and econometric studies that look for statistically proven patterns to mineral resource and conflict variables.
3   An ethnographic perspective, providing case studies at the country, community, or site level and often yielding findings that are more contextually rich and nuanced.

Each type of research is relevant to a broad contextual understanding of conflict risks associated with extractive industries. The studies typically use the term *resource conflict* to describe either conflict over access to the revenues or rents generated by minerals, or conflict that is funded by these resources.[5] When the conflict is protracted, both types of conflict can be present; an example would be Angola's war economy of rebel-held diamonds and government-owned oil.[6] These studies generally refer to conflict on the scale of a war, which can be referred to as "armed conflict" to distinguish it from other forms of conflict. Armed conflict is defined in the Uppsala conflict data program as "a contested incompatibility that concerns government and/or territory where the use of armed force between two parties results in at least 25 battle-related deaths in one calendar year."[7]

Another branch of the literature deals with social conflict over the impacts of mining activities. Such conflict includes demonstrations, strikes, and blockades; damage to private or public infrastructure, equipment, and buildings; and injuries or deaths.[8] These studies tend to focus on the role and responsibilities of large-scale, formal extractive industries in dealing with conflict that is within their sphere of influence. Companies are seldom seen as directly implicated in armed conflict on the scale of a civil war, such as the crisis in Bougainville, Papua New Guinea. Compared to studies of these types of conflict in Latin America and Asia Pacific, there is arguably less attention being paid to conflict associated with large-scale mining in Africa than to the "resource conflict" associated with government corruption and exploitation of small-scale mining by armed groups.

A comprehensive analysis of conflict risks associated with extractive industries needs to span both of these branches of literature about the extractive industries. It needs to consider all stakeholders in the various forms of conflict, including the various levels of government and all forms of resource extraction from private sector mega-projects to informal, artisanal mining.

For example, Berman et al. find that mining activity in an area initially increases the probability of protests and low-level violence at the local level and then leads to violence across the territory as the financial capacity of armed groups increases. The authors' analysis of data on conflict events and mining of fifteen minerals in all African countries from 1997 to 2010 concludes that the mining boom has contributed significantly to an increased risk of conflict in Africa.[9]

The use of data in this study is significant, in that it focuses on local conflict in close proximity to mining activities, across a wide sample. The analysis can thus drill down below the national level by putting two datasets together. The first is the armed conflict location and event dataset, based on reports from news agencies and humanitarian organizations and from research publications.[10] The second is the raw materials data on the location of mining companies since 1980.[11] The limitations of both datasets are acknowledged in the study—namely, that news and other reporting of conflict varies by country and location and that artisanal and small-scale mines are not included in the mines data. Nevertheless, the importance of this kind of empirical study is that it allows for more evidence-based policymaking.

Understanding the full range of conflict risks associated with mining and oil and gas requires more accurate data about both the extent of resource exploration and extraction in Africa and the incidence of conflict.[12] In addition to good data, contextual understanding of conflict dynamics is an essential part of risk analysis for the extractive industries.[13] Most existing guidance for companies and governments makes this point about risk assessment before offering practical advice for conflict transformation.[14] We will now consider a conceptual framework for a more comprehensive approach to understanding conflict risks associated with the extractive industries.

Conceptual framework: why politics matters

The concept of political settlements helps explain how stakeholders within government, industry, and civil society interact in ways that increase or reduce conflict associated with the extractive industries.[15] As defined by the United Kingdom's Department for International Development, a political settlement is the "expression of a common understanding, usually forged between elites, about how power should be organised and exercised."[16] It is the outcome of bargaining and negotiation between elites and often the outcome of peace processes in war-to-peace transitions.[17] A political settlement is not a one-off event, but rather an ongoing process.

Khan has proposed a typology of political settlements to use in analyzing how different configurations of power determine levels of stability and growth in developing countries.[18] The typology describes four configurations:

1 a weak dominant party that is vulnerable to strong lower-level factions within its ranks;
2 an authoritarian coalition in which a strong dominant party is vulnerable to strong excluded elites;
3 competitive parties alternating within a clientalist system, in which a weak democracy with numerous factions is inclusive but unstable; and
4 a potentially developmental coalition, which is stable with strong implementation capabilities.

While stressing that each context is different, the political settlements approach suggests that more-stable, more-inclusive agreements have better developmental outcomes than insecure ruling coalitions that need to distribute economic opportunities and benefits to ensure stability in the short term. The stability of the settlement is determined both by the horizontal inclusion of powerful elites and by vertical relations between strong elites and relatively weak (or acquiescent) constituencies.

Conflict is an aspect of change to political settlements. Figure 2.1 summarizes the relationship between political settlements, extractive industries, and conflict. Moving clockwise from the lower left of the figure, we can make several observations about that relationship:

- Conflict can be a pathway to a new political settlement involving several elite interests. The more inclusive the political settlement, and the weaker the actors excluded from that settlement (either horizontally at the elite level or vertically between elites and lower ranks), the more stable and enduring the settlement is likely to be.[19] Conversely, the more exclusive the political settlement and the stronger the excluded actors, the more likely it is that conflict will continue or reignite.
- The nature of an existing political settlement determines how new oil or mining sectors are developed, and this development in turn can change political settlements as new elite actors, interests, and ideas emerge. Power dynamics might shift between the ruling elite and a new business elite that benefits from extractives growth, or between those in control of the central government and sub-national actors in the resource-rich regions.[20]
- The development of new oil, gas, or mining projects can cause or exacerbate conflict. Alternatively, it can provide incentives and opportunities to end or reduce conflict.[21]

The spatial dimension of mining is important to conflict analysis.[22] Mining produces minerals from geographically fixed, often remote locations; these locations are unlikely to be the same place where benefits from mining are realized. In post-conflict countries, new mining developments can revive old animosities, particularly when the past conflict is a civil war with strong territorial dimensions. Mining can thus be a source of friction between national and sub-national actors.[23]

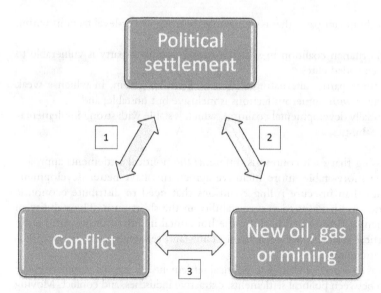

Figure 2.1 Relationship between the political settlement, extractive industries, and conflict

In east Africa, the risk analysis focuses on four levels of conflict, which can be imagined as ripples spreading out from extractive projects: (1) localized conflict in the immediate vicinity of a single mining or hydrocarbons concession, or in a nearby town that is affected by the cumulative negative impacts (such as rapid in-migration) of several extractive projects in the region; (2) sub-national conflict within a resource-rich province, county, or state within a country, such as a secessionist conflict; (3) wider-ranging conflict by armed groups that are sustained by or in pursuit of mineral revenues or rents and that pose a national security threat; and (4) illicit cross-border trade in minerals from zones of conflict, which fuels further conflict by armed groups operating at a transnational level, such as in the Great Lakes region.

While in theory the distinction between national and sub-national dynamics of power, distribution, and conflict threats might be clear in the resource-conflict literature, a further distinction within the sub-national analysis is needed. Between the macro level of the nation-state in Africa and the micro level of the individual mine and its immediate surrounds, the cumulative impacts of multiple mining companies operating in an area—that is, their combined effect on society, the economy, and the environment—are not well known.[24] Considering the spatial dimensions of conflict associated with extractive industries, as well as the types of political settlement influencing stakeholders at each of the levels of analysis, provides for a more comprehensive analysis of conflict risks associated with extractive industries.

The next section analyzes conflict risks in three countries in east Africa and looks at how political settlements in each of these countries affect these risks in relation to extractive industries.

Cases of conflict risks in east African extractive industries

Mozambique

Although Mozambique is a country relatively new to extractive industries, the war that divided north from south after independence decades ago still influences governance of this sector. Coal mining and gas developments in the north are tightly controlled by the central government located in the south. Independence from Portugal in 1975 was followed by civil war until 1992 between the Mozambican Liberation Front (FRELIMO) and the Mozambican National Resistance (RENAMO). A one-party dominant political settlement emerged, with FRELIMO becoming weaker over time and vulnerable to renewed threats from RENAMO in recent years.

Mining of the rich coal deposits in Tete Province has brought both opportunities and challenges to the northwestern region of Mozambique. Three major mining companies, two of which operate side by side in the Moatize district, resettled communities that were located near the mines, sparking protests.[25] In January 2009 the local conflict came to a head when protesters blocked the railway line that provided the only coal transportation route from Tete Province. Community leaders and civil society organizations attributed the cause of the unrest to grievances over the resettlement process, particularly the poor quality of housing that was provided for the resettled families in Cateme.[26] As a result of these protests, community resettlement has been the focus of attention for all stakeholders concerned with the social impacts of mining, as well as for the mining companies themselves. Civil society organizations have identified resettlement as the primary source of tension around mining in Tete Province. Related concerns include the lack of access to water and fertile land in the resettlement sites, as well as the loss of livelihoods caused by having to move to remote locations.[27]

Without detracting from the seriousness of existing resettlement challenges and other impacts of mining on a few local communities, it is safe to say that these disputes are likely to be only the tip of the iceberg for Tete Province. The large number of mining concessions granted across the region will make it very difficult to find suitable land for resettled communities. Being dependent on a single industry will make the city of Tete vulnerable to the cyclical nature of commodity prices, causing it to alternate between boom town and ghost town, depending on global commodity market.[28]

The city of Tete has already felt the cumulative negative impacts of mining exploration, construction, and production. Tensions have developed over the rising cost of living and poor quality of life (for example, increasing traffic volume and accidents, strain on water use, lack of housing, spread of infectious diseases from overcrowding, and increase in prostitution). The conflict risks at this level include sporadic protests over the rising cost of living and deteriorating social services, violence triggered by migration and overcrowding, crime associated with unemployment and weakening of social norms, and political mobilization of disaffected groups, particularly during elections in 2014.

More broadly, the resumption of armed attacks allegedly by the RENAMO on key transport corridors in 2013 raised the specter of renewed north–south conflict in Mozambique more than two decades after the civil war ended.[29] The effects of politically motivated conflict on mining activities became apparent when two armed attacks took place on a truck and a bus. Although the incidents were unrelated to mining, they had the effect of stopping production. Rio Tinto suspended coal transportation from the Benga mine in Tete, fearing similar attacks.

The resurgence of organized resistance to the FRELIMO government, while not directly linked to mining, underscores the risk that the negative impacts of mining will foment discontent in a fragile, post-conflict society, one that lacks the political culture and institutions that would accommodate dissent from one-party dominance and elite interests.

Although Tete Province was always at the periphery of the war between RENAMO and FRELIMO, it is now at the center of Mozambique's economic recovery and development plans. Media reports have drawn a link between the new revenues generated by mining for the Mozambique government and the opportunistic attacks on the road and rail infrastructure by groups that have not benefited from the mining boom.[30] Situated far away from the seat of government in Maputo, Tete is vulnerable to the politicization of unrest motivated by both greed and grievance. Timely and accurate analysis of the cumulative social impacts of coal mining throughout the northwest region of Mozambique is vital to averting wider conflict.

Handled badly, the changing dynamics between national and sub-national interests might cause conflict that is more serious and widespread than has been seen so far. Context-specific policy measures that are well implemented could prevent conflict and ensure that these mining investments become a source of national, regional, and local development. But without multi-stakeholder efforts at economic inclusion and political accommodation of marginalized actors, conflict is a serious risk to the resource-led growth and development strategy of Mozambique.

Kenya

Kenya faces unique constraints of history and geography that cannot be overcome merely by applying "good governance" prescriptions. Postelection violence in 2008 led to a new constitution in 2010, which called for a two-tier system of government designed to improve the distribution of state resources and counter perceptions of ethnic power imbalances. The political settlement in Kenya is a competitive democracy, with elements of clientelism that political economists refer to as "crony capitalism."[31]

Minerals, oil, and gas are located mostly in arid areas left largely underdeveloped by successive Kenyan governments. With the discovery of new resources, these areas remain susceptible to potential violence and armed conflict. Northern Kenya in particular has suffered from scarcity of water and pasture, which can lead to conflict between pastoralist groups in the region, especially in Turkana.[32]

Kenya's proximity to warring neighbors also poses risks. Kenya is next door to countries with a history of armed conflict, and small arms are regularly smuggled across borders. The proliferation of arms feeds into ethnic violence and crime such as the theft of cattle, and the abundance of weapons has rendered the conflicts more protracted and complex. The discovery of resources could thus make matters worse, especially if local communities are left without a share of the attendant rents.

Oil has been discovered in the disputed region known as the Ilemi Triangle, which borders Ethiopia, Kenya, and South Sudan. The area has never been clearly demarcated, and press reports have suggested that South Sudan has been keen on instituting an international claim to the border area.[33] Despite this tinderbox of problems, oil-based development could be a stabilizing factor if it increases employment opportunities and changes the region's historically marginalized status.[34] Although residents of Turkana have articulated their concerns and grievances with oil producers, their engagement remains fragmented and structurally weak, particularly because most communities are situated in impoverished, remote, and often unstable locations.

The security of northeastern Kenya has been precarious since the early 1990s, when the war in Somalia sent waves of refugees into Kenya. Somali nationalism fanned irredentist claims for the unification of "Greater Somalia," destabilizing the border region of Kenya.[35] The continued insecurity of this region was cited by the Ugandan government in 2016 as one of the reasons for pulling out of an agreement with the Kenyan government to jointly develop an oil pipeline via a northern route to the port of Lamu.[36]

Offshore oil and gas exploration in northern Kenya has also been affected by a dispute with Somalia. The federal government of Somalia has taken Kenya to the International Court of Justice to determine the maritime boundary in the Indian Ocean. The Somali government claims six oil and gas exploration blocks that Kenya has already licensed to multinational companies.[37] Regional security analysts argue, however, that Kenya's role in keeping Somalia's current government in power is likely to dampen this conflict.[38]

Kenya's political accommodation of ethnic conflict following the 2007 election is being tested by oil and gas discoveries that have been linked to insecurity in coastal regions such as Lamu County and in arid northern regions such as Turkana, and it could be further threatened by terrorist attacks that have been carried out by Somalia's al-Shabaab. Kenya's boundary dispute with Somalia in the Indian Ocean encompasses the oil and gas exploration blocks referenced above, which are located in the disputed area of Lamu Basin; a quick resolution of this controversy will be necessary to avoid resource-fueled disputes that are even harder to mediate. The Ilemi Triangle, moreover, lies in the Tertiary Rift Basin, which stretches over three oil and gas exploration blocks. The discovery of minerals in Kwale County, among other regions, is likely to result in conflict and weak national cohesion if land use associated with mining or drilling negatively affects food security or water supply. All these concerns could act as catalysts to conflict.

Rwanda

Although it is not as resource rich as some of its neighbors, Rwanda has significant deposits of tungsten, coltan, and tin; it also enjoys relative stability and the capacity to implement long-term policy goals. The post-genocide political settlement in Rwanda has avoided the adversarial style of competitive party politics in favor of consensus building and suppression of ethnic mobilization.[39] The result is a political arrangement that pursues development as an important tool to maintain power and ensure stability.[40] Rwanda's political settlement has been described as developmental patrimonialism, referring to the ability of the ruling elite to centralize and channel economic rents into development.[41] This most closely resembles the developmental coalition in Khan's typology of political settlements.

Rwanda has extremely scarce land resources and is the second-most densely populated country in Africa, with 471 persons per square kilometer as of 2016.[42] In addition to population pressure, the inequitable distribution of land usually controlled by elite groups has historically and increasingly been a cause of tension.[43] About 80 percent of the population survives off agriculture, using 78 percent of the total arable land area in Rwanda. The pressure on land use is a conflict risk to be factored in to plans to develop Rwanda's mining industry, because approving land for mineral exploration and development is likely to displace rural communities.

Artisanal and small-scale mining in Rwanda has yet to emerge as a well-organized and fully regulated economic sector. Formalizing the sector by establishing cooperatives or small private companies could provide a significant contribution to inclusive development, but only if the minerals mined on a small scale can be guaranteed as originating from within the country's borders.

Over the past fifteen years, Rwanda has been under intense international scrutiny over the cross-border minerals trade from the eastern Democratic Republic of the Congo (DRC). After Rwandan and Ugandan troops withdrew from the Kivus and Ituri regions at the end of the Second Congo War in 2003, there were reports of minerals and gold from these regions being smuggled into Rwanda, where they were then exported to other countries, with the proceeds used to support armed groups' attacks on civilians in the DRC.[44] It has since become imperative for Rwanda to disassociate its mining industry from the narrative of conflict dynamics that has emerged in the region.[45] Perks argues that security concerns "dominate the policy dialogue concerning mineral development in Rwanda" and divert attention away from the economic potential of the sector.[46]

In November 2006 the twelve members of the International Conference on the Great Lakes Region, including Rwanda, signed a protocol against illegal exploitation of natural resources.[47] Four years later the Rwandan government, in collaboration with the International Tin Research Institute (ITRI), adopted the ITRI Tin Supply Certification initiative to meet the new United States' Dodd-Frank Act requirements.[48] ITRI reported that, as of 2013, all registered mine sites in Rwanda were covered by the initiative and received certification tags. For the artisanal and small-mining sector, a voluntary Certified Trading Chains in Mineral Production program was launched in Rwanda in 2008 to encourage transparency and responsible mining practices.[49]

In complying with the tin certification and other initiatives, Rwanda's government and mining industry stand to gain the confidence of investors, buyers, and other stakeholders.[50] The interests of these political and business elites are thus aligned with both the national interest and the livelihood needs of miners.

Yet because the stakes are so high for Rwanda to prove that it is not laundering minerals from the DRC, there might be incentives to suppress evidence to the contrary. Like anticorruption campaigns in the public sector, the certification process needs to be able to withstand overt and covert political pressure as well as having the technical and institutional capacity to detect wrongdoing.

As long as the price for certified minerals from Rwanda is higher than the price for uncertified minerals from the DRC, the incentive to disguise one for the other exists. With the cost of the tagging process passed on to producers within Rwanda, there might also be a perverse incentive for some to recoup these costs on the black market. Rather than shy away from the politics of managing the conflict risks, policymakers should address the underlying dynamics of the problem in addition to supporting technical solutions.

Governance mechanisms for addressing conflict risks at different levels

The regulation of mining and petroleum exploration and production is an important task for the legislative and executive arms of government in each of the countries discussed above. A different level of governance is required, however, to address most of the complex conflict risks associated with the extractive industries. Multi-stakeholder engagement distinguishes "governance" from "government" in these cases. Public participation can be fostered in many ways to increase transparency and perceptions of fairness and to reduce resource conflict.

The Natural Resource Governance Institute has developed a governance value chain to describe the stages through which the full value of an extractive resource can be managed.[51] As discussed below, the value chain has five stages, starting with the decision to mine and ending with the investment of the proceeds. In each of the five stages, conflict risks can be mitigated through various mechanisms for greater transparency and public participation in resource governance.

Deciding to extract

Recognition of indigenous or communal land and water rights, and in a few countries mineral rights, can be regarded as the first step in the value chain for ensuring community participation in natural resource governance. This legal recognition gives communities the right to decide whether to permit extraction on their land and to determine use of their natural resources such as water. The International Labour Organization's Indigenous and Tribal Peoples Convention is the key instrument enshrining these rights, although controversy over the definition of "indigenous" in the African context has impeded adoption of this instrument.

In 2009 the Economic Community of West African States issued a directive on mining, which held that the principle of free, prior, and informed consent applies to all communities in Africa (since they're all considered indigenous). In west Africa, Oxfam and the Wassa Association of Communities Affected by Mining (WACAM) are lobbying national governments to ratify the mining directive and incorporate it into national law to empower local communities.

In addition to legal recognition of land rights, integrated land use planning is a political and administrative process designed to guide the orderly occupation and sustainable use of land considering economic, sociocultural, environmental, and institutional factors. It is a vital tool for managing competing interests and mitigating conflict over scarce natural resources. Integrated land use planning should include community development concerns at the earliest stage of deciding whether to extract minerals within a defined area. Local government has an especially important role to play in integrating extractive projects and their economic linkages with competing land uses and environmental concerns in the region under its jurisdiction.

The importance of engaging with communities as early as possible in extractive projects is now widely recognized by industry. Strong relationships based on trust and mutual benefit usually evolve over a long period of time. Engagement during the exploration stage also provides an opportunity for companies to positively influence the community's perception of a project. Furthermore, the longer that companies or governments wait to consult with communities, the more likely it is that expectations about impacts (both positive and negative) will be unrealistic. This is particularly the case for large and complex projects.

Getting a good deal

Public participation in designing the regulatory framework for extractive industries occurs primarily through government consultation, formal dialogue, and parliamentary public hearings. In post-conflict countries, public participation might have taken place in reaching a peace agreement or political settlement that includes natural resource governance arrangements.

Community development agreements between extractive companies and affected communities are becoming increasingly common. The negotiation process for these agreements can build relationships that not only help parties benefit from project development, but also promote equality, fairness, and conflict avoidance and resolution.[52] The agreements take many forms, depending on the jurisdictions in which they are negotiated. Indigenous land use agreements have been used to preserve native title rights in Australia, and other voluntary agreements are used in Canada, Ghana, Papua New Guinea, and South Africa. Community development agreements are being mandated as part of national resource governance frameworks in Guinea, Kenya, Mozambique, Nigeria, Sierra Leone, South Sudan, and South Africa.[53]

Community-based natural resource management allows the community to decide how to manage and capture the benefits of resources located on communal lands. It has been applied mostly to renewable natural resources such as land, water, forests, fishing grounds, and wildlife. Examples include community forests in Cameroon, village forest reserves in Tanzania, and community-run wildlife tourism ventures in Kenya. Community-based management can be used in dealing with the environmental impacts of mining and oil and gas extraction, and in resolving conflicts between communities and extractive industries over fishing grounds (especially offshore oil and gas), hunting grounds, arable land, and water.

Grievance mechanisms are now widely adopted by extractive companies and are recommended by organizations such as the International Finance Corporation and the UN. The UN Guiding Principles on Business and Human Rights, for example, promotes the use of grievance mechanisms by companies at the operational level (at the site of extraction for mining and petroleum companies), and offers eight criteria for judging their effectiveness. Some scholars caution, however, that community grievance mechanisms are unlikely to be effective in politically repressive contexts where people are extremely fearful of speaking out against national development projects.[54] There also remains a need for grievance mechanisms that are less reactive and more preventive in nature. The issues of anonymity and whistleblower protection need to be addressed more effectively as well.

Ensuring transparency

The Extractive Industries Transparency Initiative (EITI) provides for multi-stakeholder governance of extractive industries, in order to promote greater transparency among government, industry, and civil society on how natural resources can be harnessed for sustainable development.[55] Half of the countries implementing the EITI are in Africa, and multi-stakeholder groups developed by the EITI operate in key mining countries such as the DRC, Ghana, Tanzania, and Zambia. As a natural progression from its revenue transparency initiative, the EITI has inspired a global campaign for contract transparency in the extractive industries.

Certification schemes for the traceability of minerals in conflict-affected and post-conflict areas have become important instruments for ensuring greater transparency and due diligence of mineral supply chains. The jewelry market for diamonds and colored gemstones, including emeralds from Zambia, sapphires from Madagascar, and rubies from Mozambique, is particularly well suited to these consumer-focused campaigns. It is more difficult to trace gold to a single source than gemstones, but its widespread availability offers many opportunities to engage with conflict-affected communities through certification projects.

Managing volatile resources

Sovereign wealth funds are designed to save a portion of natural resource revenues for future generations, and stabilization funds are macroeconomic tools for stabilizing public expenditure in resource-dependent countries during the boom-bust commodity cycles. These funds provide important policy options for most resource-rich countries, although it has been difficult to implement these policies successfully in countries such as Chad with highly insecure governments.[56] Stabilization funds might also be used to secure sub-national development and to reduce conflict arising from fluctuations in resource revenues allocated to local government through revenue-sharing arrangements. Community trusts, foundations, and funds set up by mining and oil and gas companies provide for community investment, compensation to mitigate negative impacts, and the distribution of government payments such as royalties to communities. Community participation in these mechanisms can be achieved through representative governance structures and co-financing arrangements and in the project design. Community participation is also important in planning for closure at the end of the mine life or for unexpected events such as commodity price drops, natural disasters, or social unrest.

Investing for sustainable development

Public participation in government budget processes helps ensure that resource revenues are invested for sustainable development. Such engagement should happen at the national, state or provincial, and local government level, although national budgets, given their complexity, should be summarized and explained to facilitate community participation. Participatory planning methods, which are widely used internationally, help identify development needs, set priorities, and reduce conflict over benefit sharing within the community. By involving a broad range of community members, such as women, youth, older people, and people from minority groups, participatory methods can ensure that development plans are relevant to many sectors of the community.

Finally, local content provisions in mining and petroleum laws, contract requirements, and company policies function as key mechanisms both for mitigating conflict risks posed by community opposition to extractive projects and for peace building through economic development of conflict-affected regions or countries. For example, procurement from local suppliers provides income for local communities, and also strengthens local business capacity to in turn promote peace building.

Conclusion

The examples of conflict risks discussed in this chapter illustrate the variety and complexity of risks within a country, region, and even over the course of a project, from exploration to production. The types of conflict risks include the following:

- Armed attacks on exploration activities in regions where sub-national groups perceive the national government and foreign companies to be stealing what is rightfully theirs.
- Protests over resettlement issues by local communities displaced by the construction of mines.
- Threats to property and violence when large-scale mining companies seek to secure their concessions from artisanal, small-scale miners and when informal miners are displaced by the licensing and development of formal projects.
- Conflict over the threat of negative environmental impacts by extractive industries, such as land and water stress and loss of pasture for livestock.
- Minerals mined illegally in zones of conflict, then smuggled across borders so that they can enter the legal chain of production in a neighboring country, thereby harming the export industry of that country and prolonging the war economy where the minerals originate.
- Disputes between national and sub-national levels of government over the share of resource revenues allocated to each, as well as local protests over wasteful or corrupt spending of resource revenues.

A focus on political settlements provides a new lens through which to evaluate issues raised in the resource-curse and resource-conflict literature. The political settlements perspective also clarifies the need for a more open, inclusive national dialogue about how to prevent conflict associated with large- and small-scale mining operations, whether formal and informal, and with oil and gas exploration and production. This dialogue is an inherently political process that should be acknowledged as such by internal and external actors alike.

African governments, domestic and foreign companies, and other stakeholders need to adopt more multi-stakeholder governance mechanisms to address the conflict risks associated with resource extraction. Further research in this area could contribute to the AMV Action Plan's call for the African Union Commission "to integrate the peace and security dimensions of natural resources into its existing conflict prevention and early warning" mandate.[57] This policy development could be a valuable step toward ensuring that Africa's extractive industries are an engine for economic growth and for building peace.

Notes

1 African Union (AU), "Africa Mining Vision" (Addis Ababa: African Union and United Nations Economic Commission for Africa, February 2009), 34.
2 Two of the case studies for this chapter (Kenya and Rwanda) are based on a research project commissioned by the UK Department for International Development's east Africa Research Hub, titled "Evidence Synthesis of The Impact of Extractive Industries on Political Settlements and Conflict." The project was led by the authors, with input from research partners at the Kenya Institute of Public Policy, Research and Analysis and the Institute for Security Studies. The case study of Mozambique is based on the lead author's observations while conducting research on mining-induced resettlement in Tete Province, which included research collaboration between SMI and Oxfam Australia, funded by the Australian Government's International Mining for Development Centre.

3 W. W. Rostow, *The Stages of Economic Growth: A Non-Communist Manifesto*. (Cambridge, UK: Cambridge University Press, 1961), 4–16; A. Krueger, "Trade Policy as an Input to Development," *Working Paper No. 466* (Cambridge, MA: National Bureau of Economic Research, 1980).
4 P. Le Billon, *Wars of Plunder: Conflicts, Profits and the Politics of Resources* (New York: Columbia University Press, 2012).
5 P. Le Billon, "The Political Ecology of War: Natural Resources and Armed Conflicts," *Political Geography* 20:5 (2001): 561–584.
6 J. Cilliers and C. Dietrich (eds.), *Angola's War Economy: The Role of Oil and Diamonds* (Pretoria: Institute for Security Studies, 2000).
7 Uppsala Conflict Data Programme, ucdp.uu.se.
8 R. Davis and D. Franks, "Costs of Company-Community Conflict in the Extractive Sector," *Report No. 66* (Cambridge, MA: Harvard Kennedy School, Corporate Social Responsibility Initiative, 2014).
9 "armed groups increases": N. Berman, M. Couttenier, D. Rohner, and M. Thoenig, "This Mine is Mine! How Minerals Fuel Conflicts in Africa," *CESifo Working Paper No. 5409* (2015), 2. "conflict in Africa": ibid., 1.
10 ACLED, 2013, www.acleddata.com/.
11 RMD-InterraRMG, "Raw Material Data," 2013, www.spglobal.com/marketintelligence/en/campaigns/metals-mining.
12 N. Berman, M. Couttenier, D. Rohner, and M. Thoenig, "This Mine is Mine!"
13 R. Bailey, J. Ford, O. Brown, and S. Bradley, *Investing in Stability: Can Extractive-Sector Development Help Build Peace?* (London: Chatham House, 2015).
14 See, for example, the "Voluntary Principles on Security and Human Rights," 2000; A. Grzybowski, *Extractive Industries and Conflict: Toolkit and Guidance for Preventing and Managing Land and Natural Resources Conflict* (New York: UN Interagency Framework Team for Preventive Action, 2012).
15 M. Khan, "Political Settlements and the Governance of Growth-Enhancing Institutions" (unpublished monograph, 2010); A. Bebbington, *Natural Resource Extraction and the Possibilities of Inclusive Development: Politics Across Space and Time* (Manchester, UK: The University of Manchester, 2013); S. Hickey, B. Bukenya, A. Izama, and W. Kizito, "The Political Settlement and Oil in Uganda," *ESID Working Paper No. 48* (Manchester, UK: Effective States and Inclusive Development Research Centre, 2015).
16 Department for International Development, "Building Peaceful States and Societies," *Practice Paper* (London: Department for International Development, 2010). See also C. Barnes, "Renegotiating the Political Settlement in War-To-Peace Transitions" (London: Conciliation Resources, 2009), 20; A. Rocha Menocal, "Citizens Voice and Accountability: Understanding What Works and Doesn't Work in Donor Approaches: Lessons and Recommendations Emerging from a Donor Evaluation," *Briefing Paper* (2009).
17 "between elites": J. di John and J. Putzel, "Political Settlements: Issues Paper" (Governance and Social Development Resource Centre (GSDRC), International Development Department, University of Birmingham, 2009). "war-to-peace transitions": A. Rocha Menocal, "Citizens Voice and Accountability."
18 M. Khan, "Political Settlements and the Governance of Growth-Enhancing Institutions," 65.
19 Ibid.
20 S. Hickey, et al., "The Political Settlement and Oil in Uganda."
21 A. Bebbington and L. Hinojosa, "Contention and Ambiguity: Mining and the Possibilities of Development," *Development and Change* 39:6 (2008): 887–914.
22 See, for example, A. Bebbington, *Natural Resource Extraction*, 18.
23 P. Collier and A. Hoeffler, "Greed and Grievance in Civil War," *Oxford Economic Paper 56* (Washington, DC: The World Bank, 2001); M. Ross, "What Do We Know About Natural Resources and Civil War?" *Journal of Peace Research* 41:3 (2004): 337–356; P. Le Billon, "The Political Ecology of War: Natural Resources and Armed Conflicts"; P. Le Billon, *Wars of Plunder: Conflicts, Profits and the Politics of Resources*.

24 Daniel M. Franks, Jo-Anne Everingham, and David Brereton, "Governance Strategies to Manage and Monitor Cumulative Impacts at the Regional Level" (Brisbane: Centre for Social Responsibility in Mining, University of Queensland, 2012).
25 C. Kabemba and C. Nhancale, *Coal Versus Communities: Exposing Poor Practices by Vale and Rio Tinto* (Johannesburg: Southern Africa Resource Watch, 2012); Human Rights Watch (HRW), *"What is a House Without Food?" Mozambique's Coal Mining Boom and Resettlements* (USA: Human Rights Watch, 2013).
26 Kabemba, C. and Nhancale, *Coal Versus Communities: Exposing Poor Practices by Vale and Rio Tinto*; HRW, *"What is a House Without Food?"*
27 D. Kemp, S. Lillywhite, and K. Sturman, *Mining, Resettlement and Lost Livelihoods: Listening to the Voices of Mualadzi, Mozambique* (Melbourne: Oxfam Australia, 2015).
28 J.W. Gartrell, H. Krahn, and T. Trytten "Boomtowns: The Social Consequences of Rapid Growth" in *Resource Communities: A Decade of Disruption*, D. D. Detomasi and J.W. Gartrell, (eds.) (Boulder: Westview Press, 1984), 85–100.
29 Author's interviews in Tete, March 2013.
30 John C. K. Daly, "Mozambique Guerrillas Threaten Country's Energy Infrastructure," *Oilprice.com*, 1 July, 2013, http://oilprice.com/Geopolitics/Africa/Mozambique-Guerrillas-Threaten-Countrys-Energy-Infrastructure.html.
31 D. Booth, B. Cooksey, F. Golooba-Mutebi and K. Kanyinga, "An Update on The Political Economy of Kenya, Rwanda, Tanzania and Uganda," *East African Prospects* (2014).
32 K. A. Mkutu, "Small Arms and Light Weapons Among Pastoral Groups in the Kenya–Uganda Border Area," *African Affairs* 106:422 (2007): 47–70; K. A. Mkutu, "Ungoverned Space and the Oil Find in Turkana, Kenya," *The Roundtable* 103:5, (2014): 497–515; K. Menkhaus, *Conflict Assessment 2014, Northern Kenya and Somaliland* (Nairobi: Danish Demining Group, 2015); Cordaid, "Oil Exploration in Kenya: Success Requires Consultation: Assessment of Community Perceptions of Oil Exploration In Turkana County, Kenya" (Nairobi: Cordaid, 2015).
33 "clearly demarcated": D. M. Anderson and A. J. Browne, "The Politics of Oil in Eastern Africa" *Journal of Eastern African Studies* 5:2 (2011): 369–410. "the border area": D. Ochami and P. Orengo, "Silent War Over the Elemi Triangle," *The Standard*, August 17, 2009.
34 E. Johannes, L. Zulu, and E. Kalipeni, "Oil Discovery in Turkana County, Kenya: A Source of Conflict or Development?" *African Geographical Review* (2014): 1–23.
35 K. Sturman, "Somaliland: An Escape from Endemic Violence," in *The Ashgate Research Companion to Secession*, A. Pavkovic and P. Radan (eds.) (Farnham, UK: Ashgate, 2011).
36 A. Anyimadu, "Tanzania Pipeline Deal Reflects Uganda's Practical and Strategic Concerns" (London: Chatham House, 2016).
37 T. Reitano and M. Shaw, "Briefing: Peace, Politics and Petroleum in Somalia" *African Affairs* 112:449 (2015): 666–675.
38 T. Reitano and M. Shaw, "Briefing: Peace," 675.
39 F. Golooba-Mutebi and D. Booth, "Bilateral Cooperation and Local Power Dynamics: The Case of Rwanda" (London: ODI, 2013); D. Booth and F. Golooba-Mutebi, "Developmental Patrimonialism? The Case of Rwanda," *African Affairs* 111:444 (2012): 379–403.
40 Jonathan Fisher and David M. Anderson, Authoritarianism and the Securitization of Development in Africa, *International Affairs* 91:1 (1 January, 2015): 131–151, https://doi.org/10.1111/1468-2346.12190.
41 D. Booth and F. Golooba-Mutebi, "Developmental Patrimonialism?"
42 National Institute of Statistics of Rwanda, "Rwanda Data Portal," http://rwanda.opendataforafrica.org/xmdxob/population-density.
43 J. Bigagaza, C. Abong, and C. Mukarubuga, "Land Scarcity, Distribution and Conflict in Rwanda," in *Scarcity and Surfeit: The Ecology of Africa's Conflicts*, J. Lind and K. Sturman (eds.) (South Africa: Institute for Security Studies, 2002), 50–82.
44 N. Garrett and H. Mitchell, "Trading Conflict for Development: Utilising the Trade in Minerals From Eastern DR Congo for Development" (Resource Consulting Services,

2009); Global Witness, "New Investigation from Global Witness Reveals High-Level Military Involvement in Eastern Congo's Gold Trade," (Global Witness, 2013); H. Sunman and N. Bates, "Trading for Peace: Achieving Security and Poverty Reduction Through Trade in Natural Resources in The Great Lakes Area" (Department for International Development Research, 2007).
45 R. Perks, "Digging into the Past: Critical Reflections on Rwanda's Pursuit for a Domestic Mineral Economy," *Journal of Eastern African Studies* 7:4 (2013): 732.
46 Ibid., 733.
47 G. Franken, J. Vasters, U. Dorner, P. Schütte, D. Küster, and U. Näher, "Certified Trading Chains in Mineral Production," in S. Hartard & W. Liebert (eds.), *Competition and Conflicts on Resource Use* (Berlin: Springer, 2015), 177–186.
48 T. R. Yager, "The Mineral Industry of Rwanda," *Minerals Yearbook 2014* (USA: U.S. Department of the Interior, U.S. Geological Survey, 2012).
49 M. Taka, "A Critical Analysis of Human Rights Due Diligence Processes in Mineral Supply Chains: Conflict Minerals in the DRC," *Occasional Paper 208* (Johannesburg: South African Institute of International Affairs, 2014), 10.
50 Other relevant initiatives include the OECD Due Diligence Guidance for Responsible Supply Chains of Minerals from Conflict-affected and High-risk Areas; the World Gold Council's Conflict-free Gold Standard; G8 project on Certified Trading Chains in Mineral Production in Rwanda; Motorola Solutions Inc. and AVX Corporation's closed-pipe supply line for conflict-free tantalum Solution for Hope Project; and the Conflict Free Smelter Programme (M. Taka, "A Critical Analysis of Human Rights Due Diligence Processes").
51 Natural Resource Governance Institute, "The Value Chain," 2010, www.resourcegovernance.org/analysis-tools/publications/value-chain.
52 S. Ali, D. Brereton, G. Cornish, B. Harvey, D. Kemp, J. Everingham, and J. Parmenter, *Why Agreements Matter* (Queensland: Rio Tinto, 2016).
53 C. Nwapi, "Legal and Institutional Frameworks for Community Development Agreements in The Mining Sector in Africa," *The Extractive Industries and Society* 4:7 (2017): 202–215.
54 Kemp, D. & Owen, J. "The Reality of Remedy in Mining and Community Relations: An Anonymous Case Study from Southeast Asia," in *Business and Human Rights in Southeast Asia: Risk and the Regulatory Turn*, M. Mohan, and C. Morel (eds.) (London: Routledge, 2015). Although the case study focuses on a country in Southeast Asia, the authors' point could apply to African countries in which political repression is justified by the state as necessary in the context of active or recent armed conflict.
55 "The EITI Standard," 2016, https://eiti.org/.
56 Oil governance plans in Chad went awry in 2005 when President Idriss Deby breached an agreement with the World Bank. The agreement was to keep revenues from the Chad-Cameroon oil pipeline in a trust fund administered by international trustees. An amendment to Chad's oil revenue law abolished the fund, while Deby used a signature bonus from oil companies partly to buy arms. See P. Carmody, "Cruciform Sovereignty, Matrix Governance and the Scramble for Africa's Oil: Insights from Chad and Sudan," *Political Geography* 28:6 (2009): 357.
57 AU, "African Mining Vision," 34.

3 Chinese mining in Africa and its global controversy

Barry Sautman and Yan Hairong

In brief

- Chinese mining in Africa has been widely misconceived in terms of scale and characteristics. Based on research and fieldwork in Zambia's Copper-Cobalt Belt from 2007 to 2015, we find that Chinese mining firms in Africa exploit resources and labor in much the same way that non-Chinese companies do. Chinese firms have distinctive features, however.
- We dispute the centerpiece of the discourse on Chinese mining in Africa, a 2011 report on Chinese copper mining in Zambia by Human Rights Watch, and show the report is empirically inaccurate and conceptually flawed.
- Chinese mining in Africa is not what Human Rights Watch claims it to be, but that does not mean that all is well. Fundamental problems exist and pertain to mining generally and mining in Africa specifically.
- Our 2014 survey of mine workers in Zambia found that the overwhelming majority see national ownership of mining assets as beneficial for overall development in Zambia and for the livelihood of mineworkers.

Introduction

Media reports of Chinese mining in Africa since the beginning of the twenty-first century have accused China of many evils, such as plundering the continent's resources, denying jobs to Africans, harming the environment, and botching infrastructure projects.[1] Controversy over Chinese firms has focused especially, but not exclusively, on copper mining in Zambia, the oldest and most developed Chinese investment in African minerals production, which began in the late 1990s.[2]

Having conducted research and fieldwork in Zambia's Copper-Cobalt Belt from 2007 to 2015, we find that Chinese mining firms in Africa exploit resources and labor in much the same way that non-Chinese companies do. Chinese firms have distinctive features, however. Some of these are legacies of socialism, while others reflect current policies, such as a reluctance to reduce employment even during price downturns for fear of impairing relations with African countries.[3] Chinese firms have established mineral processing facilities in Africa and built infrastructure to support mining operations and transportation of minerals for export.[4] Some of these ventures might not always abide by Western norms of profit maximization.

In this chapter we dispute the centerpiece of the discourse on Chinese mining in Africa, a 2011 report on Chinese copper mining in Zambia by Human Rights Watch (HRW), a prominent non-governmental organization based in the United States.[5] That report, which described labor practices by Chinese state-owned firms, received global media coverage and is viewed as a reliable appraisal of how Chinese firms operate in Africa. We show that the report is empirically inaccurate and conceptually flawed. The discourse of China in Africa, we argue, is based on Western perceptions of a global strategic competition with the Chinese state.

Chinese investment in the world and in Africa

The notion that China is buying up the world is persistent.[6] In fact, Chinese investment abroad, as a proportion of its domestic economy and in absolute terms, is far less than that of Western countries. In 2015 the stock, or cumulative assets, of U.S. outbound foreign direct investment (OFDI) equaled 33 percent of U.S. GDP; in France and the United Kingdom, the proportion was 54 percent, and in Canada, 70 percent. China's OFDI, however, was only 9 percent of its GDP. The stock of Chinese OFDI in that year was $1 trillion, or 4 percent, of the almost $25 trillion in OFDI worldwide. This amount was less than that of France ($1.25 trillion; 5 percent) and the United Kingdom ($1.56 trillion; 6 percent) and much less than that of the United States (about $6 trillion; 24 percent).[7] Yet, despite the common assertion of an overwhelming Chinese presence in African economies, the share of Chinese OFDI in sub-Saharan Africa's aggregate gross fixed capital stock was only 0.78 percent in 2012.[8]

The same investment profile applies to Chinese assets in Africa. Most foreign direct investment in the continent is, as it has long been, from Western states. In 2014 China's investment in Africa was $32 billion, whereas France's was $52 billion, the United Kingdom's was $66 billion, and the United States' was $64 billion.[9] Thus, of the $712 billion stock of foreign direct investment in Africa in 2014, China's share was 4. percent and less than 5 percent of China's stock of investment around the globe.[10]

Both Chinese and Western investors are drawn to resource-rich countries in Africa, but Chinese OFDI is *not* mainly in natural resources, no more so than Western investment is. For example, 66 percent of U.S. OFDI in Africa goes into mining, while only 28 percent of Chinese OFDI in Africa goes to that sector.[11] As discussed further below, Chinese firms invest chiefly in the manufacturing and services sectors, preferring politically stable countries with skilled labor markets and divided almost equally between countries with good governance and poor governance, the latter being those that Western firms tend to avoid.[12] Chinese OFDI in

Africa is also geographically concentrated, with 62 percent of its stock in 8 states: South Africa ($4.4 billion), Zambia ($2.2 billion), Nigeria ($2.1 billion), Angola ($1.6 billion), Algeria ($1.5 billion), Sudan ($1.5 billion), Zimbabwe ($1.5 billion), and Democratic Republic of the Congo (DRC; $1.1 billion).[13]

Chinese investment in Africa in 2013 was spread across various sectors: extractive industry (30.6 percent), construction (16.4 percent), finance (19.5 percent), manufacturing (15.3 percent), and other (technology, media, telecoms, agriculture, etc.; 18.2 percent).[14] Privately owned enterprises accounted for 10 percent of Chinese investment stock in Africa in 2014 but performed 53 percent of all Chinese projects there. The discrepancy derives from sectoral differences: Chinese state-owned enterprises invest mainly in construction (42 percent) and mining (22 percent), whereas private firms concentrate in trade and logistics (24 percent) and manufacturing (31 percent).[15]

Chinese investment in medium- to large-scale African mining

The largest firms in African mining are BHP Billiton (United Kingdom/Australia), Rio Tinto (United Kingdom/Australia), Anglo American (South Africa/United Kingdom), Xstrata (United Kingdom/Swiss), and Barrick (Canada).[16] Three-fourths of the world's mining companies (about 1,300 firms) are based in Canada.[17] In 2015 Canadian companies had $31.3 billion invested in mining assets in Africa.[18] In 2014–2015, 236 Australian companies had $30 billion invested in African mining.[19] At the end of 2013, Chinese mining investments in Africa represented about $7 billion of the $26 billion invested there.[20] Chinese mining investment in Africa was thus less than one-fourth the amount invested by Canadian and Australian mining firms.[21]

Chinese OFDI stock in African mining in 2013 was less than 1 percent of China's worldwide OFDI and 12 percent of Chinese OFDI mining investment worldwide.[22] Fairly low investment in African mining arguably fits China's profile as a mining country: China consumes a lot of metal, but it produces a lot of metal; it thus does not actually need to acquire a lot of metal from Africa for domestic consumption. China consumed 25 percent of world metals production in 2011, and domestically produced, in 2009, 15 percent of the world's iron ore and half its coking coal.[23] China's domestic production of gold, zinc, lead, molybdenum, iron ore, coal, tin, tungsten, rare earths, graphite, vanadium, antimony, and phosphate were, in 2014, first in the world; its production of copper, silver, cobalt, manganese, and bauxite/alumina was second in the world.[24] Yet in 2011 Chinese companies accounted for just 3 percent of the global mining value, compared to firms from the European Union (14.3 percent), Australia (8.9 percent), Brazil (8.3 percent), Canada (7.2 percent), and the United States (5.6 percent).[25] Chinese firms control less than 1 percent of total world mine production outside China.[26]

The scale of Chinese mining in Africa has grown since the start of the twenty-first century in terms of major assets in Africa, number of projects (registered and informal), and geographical spread (see Figures 3.1 and 3.2) and includes both

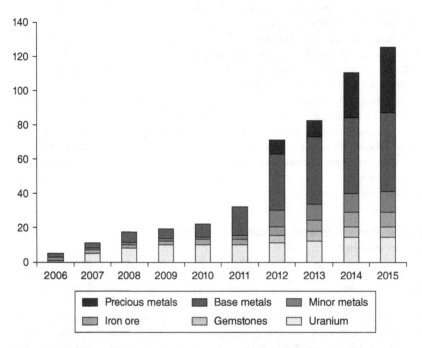

Figure 3.1 Numbers of Chinese company major mining projects in Africa by year

Source: Vladimir Basov, "The Chinese Scramble to Mine Africa," December 15, 2015, www.mining.com/feature-chinas-scramble-for-africa/

registered and informal or unregistered operations. The increase has occurred in part because China's own mining resources have faced depletion and because Chinese companies have greater tolerance for risk and political uncertainty than their Western counterparts.[27]

Chinese investment in African mining is in diverse minerals: bauxite, chrome, copper, iron ore, manganese, nickel, platinum, and uranium. As Figure 3.2 shows, Chinese mining projects are in at least eighteen African countries; about one-third of Chinese mining investment is in one country—Zambia—concentrated in four operations. Since Chinese firms are latecomers to Africa, their mining assets tend to be more marginal and less profitable than those of Western companies. That is in keeping with a general trend of Chinese companies abroad: they are more likely to lose money than other foreign investors—some 65 percent of Chinese investments do, as against a world average for OFDI of 50 percent.[28]

An example is the China Non-Ferrous Metal Mining Co. (CNMC). In 1998 it set up the Non-Ferrous Company Africa (NFCA) and bought the first Chinese-owned mine on the continent, Zambia's mothballed Chambishi mine. Rehabilitation of the mine, low copper prices, and inexperience in Africa caused NFCA to lose money until 2005. From 2006 to 2011 matters improved, and

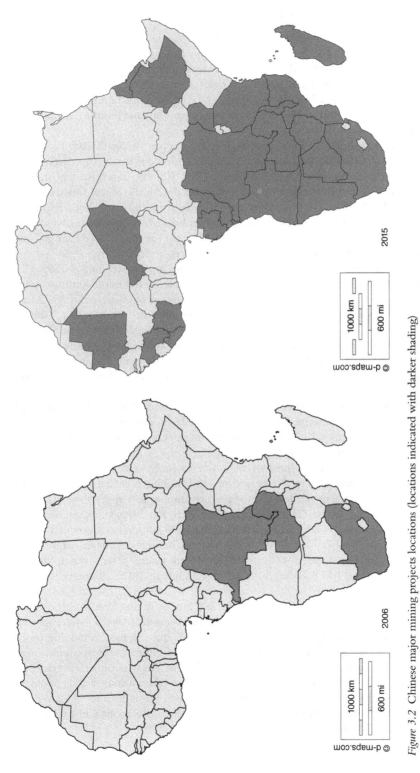

Figure 3.2 Chinese major mining projects locations (locations indicated with darker shading)

Source: Vladimir Basov, "The Chinese Scramble to Mine Africa," December 15, 2015, www.mining.com/feature-chinas-scramble-for-africa/

NFCA made a total of $200 million in profits.[29] From 2012 to 2014 NFCA, in addition to CNMC's other properties in Zambia (Luanshya mine, Chambishi copper smelter, and Sino Metals Leaching), made a total of $312 million in profits, but in 2015 it sustained a $280 million loss.[30] Overall, then, CNMC in Zambia has at best made a small profit, but given its expenditure of hundreds of millions of dollars for mine and smelter rehabilitation, expansion, and building, it has more likely sustained a loss.

Due to low prices of non-precious metals—copper yielded $10,000 a ton in 2011 but fell to $4,300 in 2016—many Chinese mining firms in Africa have been in straitened circumstances since at least 2015.[31] Chinese investment in mining in Africa is nevertheless going forward, as it did during the global financial crisis of 2008–2009.[32] Chinese companies then acquired mining assets from retreating Western firms, such as Luanshya mine in Zambia, which had been abandoned almost overnight by its Swiss owner. CNMC bought the mine, hired back all laid-off employees, and added a thousand more workers.[33]

In the current commodities downturn, China's Zijin Mining Group Co. acquired from Canada's Ivanhoe mines in 2015 a half interest in the Kamoa copper project in the DRC. It also bought a 60 percent ownership interest in Australia's NKWE Platinum, owner of the Bushveld Complex in South Africa. China's Shandong Iron and Steel purchased from the United Kingdom's African Minerals an additional 75 percent stake in Sierra Leone's Tonkolili iron ore mine and now owns 100 percent. In 2016 CNMC announced it would build two major smelters in the DRC.[34] Chinese projects in the DRC were then almost the only ones hiring in the mining sector.[35]

Contrary to popular belief in the West, extensive local hiring does take place at Chinese-owned enterprises in Africa, including mines, as shown by a recent survey of one thousand Chinese enterprises in Africa, where local employment averaged 86 percent of workforces.[36] Manufacturing is the most labor-intensive sector—in 2013–2014, 61 percent of jobs created by 156 Chinese greenfield investments in Africa were in that sector—and mining is capital intensive, with a $250,000 investment needed to produce one mining job in Zambia.[37] CNMC had 6,841 employees in 2014, of which 6,335 were Zambians and Congolese and 506 were Chinese—a 93 percent localization rate.[38] Another state-owned enterprise, Jinchuan, had 779 employees in 2013 at its Chibuluma mine in Zambia: 777 Zambians and 2 Chinese, a 99.9 percent localization rate.[39] The privately owned Collum coal mine had 800 Zambians and 70 Chinese in 2012, a 91 percent localization rate.[40]

Large state-owned firms remain central to Chinese mining in Africa. Recently, however, private concerns of various sizes have become active. Chinese mining firms were once strangers in the eyes of Africans but are now becoming more common as investors. They have joined chambers of mines and increasingly operate like other foreign firms in terms of labor relations and environmental practices. Whether this observance of Western norms benefits Africans is debatable, but even as they conform to international practices, Chinese mining firms in Africa remain controversial, as seen through the long-running discourse about Chinese mining in Zambia, discussed below.

The discourse of Chinese copper mining in Zambia

In Europe's financial crisis, Greece's practices have been called the worst. Northern European politicians and media portrayed Greeks as tax evaders, antibusiness, lazy, and corrupt. Although Greece has its problems, Richard Parker, Harvard professor of public policy, did a study that found such "moralizing clichés" inaccurate, because (1) Greek taxes are more than a third of GDP, but that is near Europe's average; (2) Greece has more small entrepreneurs per capita than any other European country; (3) Greek government workers are a fifth of Greece's total labor force, but that is Europe's average; and (4) corruption in Greece is not generalized, but highly concentrated in the health sector.[41]

Not just Greece, but other southern European countries are also blamed for the European financial crisis. Northern European elites, self-representing their peoples as virtuous and casting southern Europeans as problematic, effectively divided Europe into two "classes" and "center-right" northern European governments (Germany, France, United Kingdom) denigrated southern European "center-left" ideas.[42] Instances of racism against southern Europeans in northern European media have occurred; for example, a column in the major German news magazine *Der Spiegel* blamed the fact that a cruise ship ran aground in the Mediterranean on its having an Italian captain.[43] The northern European discourse of southern Europeans exemplifies the inaccuracies that underpin sweeping generalizations about supposed practices of whole countries and peoples. Social science's tasks include adding to knowledge of social phenomena and interrogating empirically, methodologically, and politically the dominant "regimes of truth."

Our analysis of the discourse on Chinese copper mining in Zambia centers on a well-publicized 2011 report by HRW, whose stated purpose was to determine the human rights implications of Chinese investments in Africa. The report reviewed labor practices of the four CNMC firms in Zambia that operate Chambishi and Luanshya mines and two smelters. HRW said that the case study served as "a magnifying lens" that was intended to "begin to paint a picture of China's broader role in Africa."[44] Thus it used the activities of one Chinese mining company in Africa to represent the whole of Chinese investment in the continent. Even were HRW's contentions accurate, its "magnifying lens" claim would still amount to the classic fallacy of composition: that something is true of the whole from the fact that it is true of a part of the whole.

HRW conflated Chinese state-owned enterprises with the Chinese state by imputing CNMC's activities to the Chinese government or to the country itself. But the governmental body that oversees central-government-owned firms, China's State Asset Supervision and Administration Commission, does not administer subsidiaries of those firms abroad. Indeed, in 2011 the commission was still struggling to identify those assets, since China had considerably expanded its state-owned portfolio in the decade before the HRW report.[45] Moreover, many state-owned firms privatize their overseas operations, either for economic reasons or because of local law, such as the one in Zambia that disallows registration of state-owned firms.[46] The Chinese government thus has no legal responsibility for overseeing such operations.

Foreign firms accounted for 89 percent of Zambia copper output, but CNMC firms contributed only 6.6 percent of the 481,000 tonnes produced by foreign firms in 2010.[47] HRW's report focused only on Chinese-owned mining facilities even though other foreign concerns were more important to Zambian mining. Those companies include Vedanta, owned by the United Kingdom and India (Konkola copper mine); Glencore, of Switzerland (Mopani copper mines); First Quantum Minerals, of Canada (Kansanshi mine); and Equinox, owned by Australia and Canada (Lumwana mine).

The HRW report relied mainly on interviews with miners from Chinese and non-Chinese corporations, but its sampling was highly problematic. Miners who had worked at both non-Chinese and Chinese firms said safety conditions at the latter were worse. What HRW did not reveal was that the forty-eight interviewed miners from non-Chinese operations were permanent employees of these "other multinational mining operations" and not contract workers for the Zambian companies that supply contract workers.[48] The interviewees were thus not representative of these mines' workforces, which were (and are) composed of either a strong minority or a majority of contract laborers. Contract miners at non-Chinese-owned mines are paid much less and work in worse conditions than the permanent employees who were interviewed by HRW. If HRW had interviewed contract miners, the story would have been different.

HRW also failed to contextualize its own research: that it was conducted at a time of strong anti-Chinese sentiment in Zambia's politics. Such sentiment had been building since 2006, stirred by the Yellow Peril invective from Michael Sata, Guy Scott, and other leaders of the opposition Patriotic Front, which was intended to discredit the then putatively pro-Chinese ruling party.[49] The Patriotic Front took power in 2011, just before the HRW issued its report. The HRW interviewees' observations should have been placed in the context of the anti-Chinese racial animus that pervaded Zambia's Copper-Cobalt Belt, which was evident in our own interviews of miners.[50] Zambian miners were the stronghold of a party that rose to prominence on an anti-Chinese platform, and some of them displayed explicitly anti-Chinese slogans in strikes.[51] These miners' observations and perceptions about their situations in Chinese-owned mines were likely shaped by the anti-Chinese campaign, continuous since 2006.

HRW also argued that compared with Western-operated mining firms in Zambia, CNMC mines had the worst safety standards and pay. It averred that Chinese-owned firms trivialized safety concerns and imported low labor standards from the home country. These allegations poured fuel on the fire of anti-Chinese sentiment, contributing to the negative Western "China in Africa" discourse by echoing longstanding stereotypes about Chinese cruelty and disregard for human life.

The effects of the HRW report were immediate. Thousands of news sources around the world reported on the organization's findings, and millions of internet documents discussed them. While a few websites in Africa and Asia carried a debate spurred by our critique, the HRW report dominated public attention and perpetuated the negative discourse about Chinese mining in Africa.[52]

Safety at Chinese- and other foreign-owned copper mines in Zambia

In assessing safety, the HRW report relied on interviewees' perceptions, whereas we argue that the industry standard—that is, fatalities—should be used as a starting point. According to this standard, Chinese firms' safety record was the same or better than other foreign firms.[53] Data we gathered from Zambia's Mine Safety Department showed that CNMC mines' share of fatalities from 2001 to 2011 almost equaled CNMC's share of the total copper mining workforce in those years. From 2001 to 2008 CNMC had 7.2 percent of the total copper mining workforce and 8.2 percent of total copper mining deaths; from 2009 to 2011 the proportion of CNMC fatalities increased as the firm expanded its mines, mine construction often being more unsafe than mineral extraction.[54]

Safety comparisons must also consider factors beyond the fatality rate—namely, the characteristics of mines being compared, particularly whether they are underground or surface (open-pit) mines. The latter are much safer than the former. For example, in the United States the death rate for underground mines is three times higher than for surface mines; most of the deaths are caused by falling and bursting rocks.[55] The deeper the underground mine, moreover, the more dangerous it is apt to be. In the 2010–2011 period covered by HRW, CNMC was the one foreign company in Zambia that had only underground mines, and they were fairly deep. Other foreign firms had mines that were surface only or a combination of surface and underground. Despite the fact that Chinese firms had underground mines, they basically achieved a fatality rate as low as other companies.

The HRW report ignored other safety-related factors: Due to long-term closures under previous owners, Chinese-owned mines drew from a labor pool with less experience and thus less skill compared to other mines. Less-skilled, less-experienced workers are more likely to be involved in fatal accidents than their more-skilled, more-experienced counterparts.[56] HRW thus constructed a false Chinese versus non-Chinese dichotomy on safety. When we examine the factors mentioned above, including fatality rates and characteristics of mines, HRW's claim does not hold up. Overall safety at CNMC-owned mines in Zambia was not markedly worse than at other mines.

Pay at Chinese- and other foreign-owned copper mines in Zambia

HRW reported that CNMC firms paid as little as one-fourth of non-CNMC firms' wages, and attributed the gap to Chinese labor standards.[57] Table 3.1 compares 2012 wages negotiated between unions and mining companies in 2011, when HRW was carrying out its research.[58] It is based on collective bargaining agreements that we obtained from two miners' unions and the Ministry of Labour and Social Security, covering Mopani copper mine, owned by the Swiss company Glencore (one of the two largest non-Chinese mining companies), and

Table 3.1 Wages of Chinese- and Swiss-owned underground copper mines in Zambia, 2012 (in Zambian Kwacha)

	Lowest wage in underground mine	Ratio of Mopani wage	Highest wage in underground mine	Ratio of Mopani wage
Mopani	2,926	100%	4,508	100%
Chambishi	2,040	70%	2,903	64%
Luanshya	2,867	98%	5,426	120%

Chambishi and Luanshya copper mines, owned by CNMC. Mopani and the CNMC mines all employed underground miners, and therefore the wages paid can be compared.

Based on this research, we argue that the wage gap, rather than being a function of Chinese labor standards, was the product of structural conditions in CNMC's operations in Zambia, such as differences in productivity and profits between Chinese- and non-Chinese-owned mines. The smaller scale, mainly underground character, lower copper content, and less-skilled workforce at CNMC mines in Zambia led to lower productivity there as compared not only with other major Zambia copper mines, but also with copper mines in China. CNMC's Zambia mine productivity, at 7.1–7.7 tonnes per miner in 2011, was much lower than the 12.3 tonnes per miner among all Zambia copper miners (2010) and 13 tonnes per miner in China (2009), yet CNMC has paid comparable wages in Zambia. In considering its lower pay at the time of HRW's report, CNMC Zambia's low profit rate is also relevant. Most other firms had higher profits, and some had much higher profits: in 2010 Canadian-owned Kansanshi mine's profits were almost $1 billion a year, more than 25 times those of the Chinese-owned Chambishi mine, CNMC's principal property. United Kingdom/India-owned Konkola copper mine had 7.7 times Chambishi's profits, and Australian-owned Lumwana mine had 13.2 times.

Moreover, the actual wage gap between CNMC operations and the highest-paying foreign-owned mines in Zambia was likely narrower than it appeared, if the contract workers at non-Chinese-owned mines were taken into consideration.[59] Since 2008–2009 CNMC's Chambishi and Luanshya mines have paid almost all workers permanent worker-level salaries. That was not so at most non-Chinese-owned mines in 2010–2011, during HRW's study. Overall, about half of the non-Chinese firms' workers were contractors and earned much less than permanent employees, often one-fourth to one-half as much.[60] And the high proportion of contractors at non-Chinese-owned mines has grown since the HRW study. Kansanshi had about 2,000 permanent employees and 10,000 contractors in 2012.[61] About 11,000 of Konkola's 18,000 workers were contractors in 2014.[62] Lumwana then had about 4,000 permanent workers and 8,000 contractors.[63] In late 2015, Mopani had 4,300 permanent workers and 8,000–10,000 contractors.[64] If HRW had accounted for contract workers at

Table 3.2 Wages of CNMC and other foreign-owned operations, 2014

	n	Average wage (Zambian Kwacha)	Ratio of Konkola wage
Konkola	62	4,497	100%
Mopani	54	4,363	97%
CNMC (Chambishi and Luanshya)	59	4,252	94.5%

non-CNMC firms, the wage gap between Chinese and non-Chinese companies might well have disappeared.

Beginning in 2009, the Chinese-owned mines in Zambia, despite lower productivity and profits, worked to reduce the wage gap. This process, which began well before the HRW report, appears to be leading to rough pay parity. Table 3.2 shows our 2014 survey of Zambian copper miners in Chinese- and non-Chinese-owned mining operations.[65] Zambian workers at Chinese-owned mines were paid an average of 96 percent of the wages earned by local permanent workers at two big non-Chinese-owned mines.

No firm in Zambia pays miners enough for most to have savings, and many mining families remain impoverished.[66] Compared to miners' wages in developed countries, wages for local miners in Africa, irrespective of firm ownership, are exploitative: nowhere in Africa are local miners paid close to the average of $58,000 that Canadian copper miners earned in 2014.[67] African miners do, however, earn much more than the average for local workers. In 2014 the average monthly wage for Zambia's entire workforce was $200, whereas the average miner's salary was $800.[68]

Uniquely bad or neoliberally bad?

The discourse of Chinese mining in Africa exaggerates its scale and distorts the ways that Chinese practices diverge from those of other foreign investors in the industry and continent. The misinformation is typically attributed, even by astute observers, to gaps in the knowledge of journalists and researchers.[69] If true, ignorance is still not the whole story, because the discourse's claims accord with Yellow Peril-inflected Western depictions of Chinese investment in Africa and the alleged "China threat" in Chinese–African interactions portrayed in Western media.[70] The persistence of the demonizing discourse of Yellow Peril and its extension to "China in Africa" should be understood in the context that Western, especially U.S. elites, view China as a strategic competitor.[71]

HRW proclaimed CNMC a uniquely bad employer of miners in Zambia. With its focus on CNMC's practices as a source of woe for Zambian workers, the HRW report turned a blind eye to the impact of neoliberal privatization on mine workers in Zambia. Pressured by the World Bank and the International Monetary Fund, the Zambian government privatized 280 parastatals from the late 1990s to early 2000s. That change affected 85 percent of Zambia's formal economy.[72]

Zambia's conversion of state-owned copper mining operations to private hands between 1997 and 2000 was one of the most rapid cases of privatization the world has seen, and it earned praise from the World Bank, the International Monetary Fund, and wealthy donor countries. As Lungu and Fraser have pointed out, however, affected communities in Zambia expressed frustration "through strikes, protests and the ballot box" with the absence of benefits that they were promised with the privatization. In the years leading up to the global financial crisis, copper prices doubled and tripled, earning handsome profits for large multinational companies, yet by 2008 the quality and conditions of employment in the copper mining industry had seriously deteriorated:

> There has been a collapse in the quality of employment, with around 45% of those working in the mines now unable to access permanent, pensionable contracts. Most mining companies have shifted workers onto rolling, fixed-term contracts on significantly less beneficial terms and conditions, or the jobs have been "contracted-out" to companies that pay in many cases less than half the monthly wage offered permanent workers for the same work in the same mine, and in some cases, just one tenth of this figure.[73]

Using Konkola mine as a case study, a group of non-governmental organizations concluded in a report that the development of Zambia copper mining had not benefited Zambian society at large but rather had brought suffering and disadvantages.[74] In that report, Konkola mine, which is owned by the United Kingdom and India, served as a case study for the industry because of its sheer size, and not because of its ownership. A different study, which looked at the Mopani mine, fundamentally questioned "the link between development and mining in general" and pointed out that Mopani was "far from a stand-alone case" in its activities not contributing to Zambia's social development.[75]

In contrast to those studies, HRW specifically chose to examine CNMC by virtue of its ownership. Even though more than two thousand major privately owned Chinese firms were operating in Africa, HRW generalized about the behavior of Chinese firms in the continent based on only one example, yet scholars have shown that Chinese companies' behavior in Africa is highly variegated.[76] That is not only logically fallacious but also analogous to racial profiling, a form of policing that links criminality and other malpractices to specific races, ethnicities, or national origins. That approach was well refuted by Martin Luther King Jr. a half century ago. Confronted by claims that Jewish landlords and shopkeepers exploit African Americans, he said, "The Jewish landlord or shopkeeper is not operating on the basis of Jewish ethics; he is operating on the basis of a marginal businessman." The solution "is for all people to condemn injustice wherever it exists."[77] The same principle applies to the activities of mining firms in Africa.

There was one non-neoliberal practice, unique to CNMC, that HRW failed to mention: During the global financial crisis of 2008–2009, when 30 percent of all miners in Zambia and 50 percent of salaried miners lost their jobs, no layoffs occurred at CNMC firms.[78] CNMC adopted a "three no's" policy: no layoffs, no cutbacks on planned investment, and no hesitation to make new investments.

It hired new workers, including those who had been laid off by other firms.[79] This countercyclical practice was enabled in part by CNMC's being a state-owned enterprise that chose to prioritize stability over short-term maximization of profit.

To show that Chinese mining in Africa is not what HRW claims it to be is not to argue that all is well: fundamental problems exist, but they pertain to mining generally and to mining in Africa specifically. There are the general problems that attend extraction of largely irreplaceable wasting assets and that pose labor, environmental, and other challenges inherent in a sector in which corporate interests dominate the political economy. There are also specific problems that arise when foreign firms garner most of the profits made from African mining, with the continent's states and people only incidental beneficiaries.

The problems of Chinese- and other foreign-owned mining operations in Africa cannot be effectively resolved when Africans are merely secondary stakeholders. Our 2014 survey of mine workers in Zambia asked whether they would see national ownership of mining assets as beneficial. Of the 190 workers who responded, 180 hoped to see national ownership. Of 170 who specified reasons for their choice, 89 linked nationalization with improved working conditions and 81 focused on the benefit of nationalization for national development. Thus, Africans might no longer consider a modicum of tax revenues and employment opportunities an acceptable exchange for their mineral wealth, and they might revive the issue of natural resource ownership that international financial institutions took off the table late in the twentieth century. Such steps can only be realized, however, if a pan-African opposition to Western-dominated asymmetries of power arises not only in mining, but also across the spectrum of African relationships with the wider world.

Notes

1 David Dollar, "China's Engagement with Africa: from Natural Resources to Human Resources" (Washington, DC: Brookings Institution, 2016), ix, www.brookings.edu/wp-content/uploads/2016/07/Chinas-Engagement-with-Africa-David-Dollar-July-2016.pdf; Jamie Farrell, "How do Chinese Contractors Perform in Africa? Evidence from World Bank Projects," *Working Paper no. 3* (China Africa Research Initiative, 2016), https://static1.squarespace.com/static/5652847de4b033f56d2bdc29/t/573c970bf8baf3591b05253f/1463588620386/Working+Paper_Jamie+Farrell.pdf.
2 "but not exclusively": See, e.g., Richard Aidoo, "The Political Economy of Galamsey and Anti-Chinese Sentiment in Ghana," *African Studies Quarterly* 16: 3–4 (2016): 55–72.
3 Barry Sautman and Yan Hairong, "Differences at the Margins: Understanding the Chinese Presence in Africa" (draft work).
4 African Union (AU) and United Nations Economic Commission for Africa (UNECA), "Exploiting Natural Resources for Financing Infrastructure Development: Policy Options for Africa," (AU and UNECA, 2011).
5 Human Rights Watch (HRW), "'You'll Be Fired if You Refuse': Labor Abuses in Zambia's Chinese State-Owned Copper Mines," (HRW, November 2011), www.hrw.org/sites/default/files/reports/zambia1111ForWebUpload.pdf.
6 Peter Nolan, *Is China Buying the World?* (Cambridge, UK: Cambridge University Press, 2012). Martin Wolf, "Why China Will Not Buy the World," *Financial Times*, July 10, 2013.

7 United Nations Conference on Trade and Development (UNCTAD), "World Investment Report 2016," Annex Tables 4 and 8, http://unctad.org/en/Pages/DIAE/World%20Investment%20Report/Annex-Tables.aspx.
8 Miria Pigato and Tang Wenxia, "China and Africa: Expanding Economic Ties in an Evolving Global Context" (Investing in Africa Forum, 2015), http://documents.worldbank/org/curate/en/241321468024314010/Chinese-and-Africa-expanding-economic-ties-in-an-evolving-global-context.
9 UNCTAD, "Regional Fact Sheet: Africa" (World Investment Report, 2016), 35.
10 "Africa in 2014": UNCTAD, "Regional Fact Sheet: Africa"; "around the globe": Chen Jie, David Dollar, and Tang Weihai, "China's Direct Investment in Africa: Reality vs. Myth" (Washington, DC: Brookings Institution, September 3, 2015); Carrien du Plessis, "Is Chinese Investment in Africa 1% or Less of their Investments Globally?" *Africa Check*, December 4, 2015, https://africacheck.org/reports/is-chinese-investment-in-africa-1-or-less-of-their-investments-globally/.
11 David Pilling, "China's Testing Ground," *Financial Times*, June 14, 2017.
12 David Dollar, Chen Jie, and Tang Weihai, "Why is China Investing in Africa? Evidence from the Firm Level" (Washington, DC: Brookings Institution, 2016), 3, www.brookings.edu/wp-content/uploads/2016/06/Why-is-China-investing-in-Africa.pdf.
13 Miria Pigato and Tang Wenxia, "China and Africa," 30.
14 Ibid., 11.
15 Shen Xiaofang, "Chinese Manufacturers Moving to Africa—Who? What? Where? Does Africa Benefit?" (Washington, DC: Paper, China in the World Economy: Building a New Partnership with Africa, World Bank Conference, May 20, 2014).
16 Updated list of mining companies in Africa can be found at the Africa Mining iQ web site: www.projectsiq.co.za.
17 Karl Paul, "When Canadian Mining Companies Take over the World," *GJ Global Journalist*, October 12, 2013, http://globaljournalist.org/2013/10/when-canadian-mining-companies-take-over-the-world/.
18 Natural Resources Canada, "Canadian Mining Assets," (Natural Resources Canada, 2016), Table 1, www.nrcan.gc.ca/mining-materials/publications/19323#T1.
19 Colin Barnett, "Australian Mining in Africa" (Investing in Africa Mining Indaba, 2016), www.miningindaba.com/ehome/index.php?eventid=174097&&eventid=84507&tabid=186544.
20 Zhen Han, "China's Current Involvement in Mining in Africa," *Mining Journal* (January 19, 2016).
21 U.S. firms had $36 billion in investments in African "mining" in 2014. U.S. Department of Commerce, "U.S.–Sub-Saharan Africa Trade and Investment: An Economic Report by the International Trade Administration," (U.S. Department of Commerce, 2014), 18. Both the Chinese and U.S. figures might include oil and gas investments. Chinese OFDI in "minerals" in Africa was thus less than 20 percent that of the U.S.
22 "worldwide OFDI": Zhen Han, "China's Current Involvement"; "investment worldwide": Deloitte, "Greater China Mining M&A and Greenfield FDI investment Spotlight 2013," (Deloitte, 2013), 6, www2.deloitte.com/content/dam/Deloitte/cn/Documents/finance/deloitte-cn-fas-mining-ma-en-030913.pdf.
23 "production in 2011": AU and UNECA, "Exploiting Natural Resources," 5. "its coking coal": World Bank, "China 2030: Building a Modern, Harmonious, and Creative Society," (World Bank & Development Centre of the State Council, 2013), 385, www.worldbank.org/content/dam/Worldbank/document/China-2030-complete.pdf.
24 Vladimir Besov, "China is Burning through its Natural Resources," *Mining.com*, April 26, 2015, www.mining.com/china-burning-natural-resources/.
25 Masuma Farooki, "China and Mineral Demand – More Opportunities than Risks?" *Polinares Working Paper no. 75* (Raw Materials Group, 2012): 3, www.polinares.eu/docs/d5-1/polinares_wp5_chapter5_4.pdf.

26 UNECA, "Minerals and Africa's Development: The International Study Group Report on Africa's Mineral Regime," (UNECA, 2011), 36–37, www.africaminingvision.org/amv.../AMV/ISG%20Report_eng.pdf.
27 Vladimir Basov, "The Chinese Scramble to Mine Africa," *Mining.com*, December 15, 2015. See also, Ben Lampert and Giles Mohan, "Making Space for African Agency in China-Africa Engagements: Ghanaian and Nigerian Patrons Shaping Chinese Enterprise," in *Africa and China: How Africans and Their Governments are Shaping Relations with China*, A. W. Gadzala (ed.) (Lanham: Rowman & Littlefield, 2015), 109–126.
28 Steven Guo, "China's Investment in Africa: The African Perspective," *Forbes*, July 8, 2015.
29 Han Wei and Shen Hu, "China's Harsh Squeeze in Zambia's Copperbelt," (Caixin, November 10, 2011), https://wikileaks.org/gifiles/docs/47/4755664_-os-zambia-china-mining-china-s-harsh-squeeze-in-zambia-s.html.
30 "China Nonferrous Mining Corporation Limited Announces 2012 Annual Results," *ACN Newswire*, March 21, 2013; CNMC "Announcement of Annual Results...," (CNMC, 2013), www.hkexnews.hk/listedco/listconews/sehk/2014/ 0323/LTN2014 0323087.PDF; CNMC, Announcements of Annual Results..." (CNMC, 2014), www.hkexnews.hk/listedco/ listconews/SEHK/2015/0326/LTN20150326139.pdf; CNMC, "2015 Annual Report": (CNMC, 2015), 83, https://euroland. com/pdf/HK-1258/AR_ENG_2015.pdf/.
31 "$4,300 in 2016": "Chinese Companies and Africa Need each Other like Never Before," *Nikkei*, August 5, 2016. "at least 2015": "Going Home; Chinese Mining Companies Losing Billions of Dollars as they Retreat," *Forbes,* January 16, 2016.
32 Tom Wilson, "What Crisis? Undeterred, China Snaps Up Vital Congo Resource as Western Miners Bail Out," *Mail & Guardian*, March 7, 2016.
33 Stanislas Mwimbe, Mineworkers Union of Zambia Branch Chairman, Luanshya, interview, August 17, 2011; Gao Xiang, Executive Vice General Manager, CNMC International Trade Ltd., interview, Beijing. October 20, 2011.
34 Han Zhen, "China's Current Involvement."
35 Tom Wilson, "What Crisis?" See also Huang Hongxiang, Zander Rounds, and Zhang Xianshuang, "China's Africa Dream Isn't Dead," *Foreign Policy*, February 18, 2016, http://foreignpolicy.com/2016/02/18/africa-kenya-tanzania-china-business-economy-gdp-slowing-investment-chinese/.
36 "belief in the West": See, for example, the statement of David Mulroney, Canadian ambassador to China from 2009 to 2012, that in the developing world, "China" will "ship in its own workforce and run the project as a Chinese enclave." Dan Beeby, "Chinese Mining Companies Feel Misled by Canada, Report Says," *CBC News*, June 7, 2016. "86 percent of workforces": Hilary Joffe, "Research Busts China Myths," *Business Day* (South Africa), May 8, 2017.
37 "in that sector": Miria Pigato and Tang Wenxia, "China and Africa," 11. "mining job in Zambia": Greg Mills and Dickie Davis, "Wake Up, Zambia, and Smell the Roses," *Daily Maverick*, August 30, 2016.
38 China Nonferrous Mining Corp. Ltd. "2014 Interim Report" (CNMC, 2014), 30, reports/interim2014eng.pdf.
39 Elliot Sichone, Mineworkers Union of Zambia chairman, Chibuluma Mine, interview, Kitwe, 6 July, 2013.
40 Sifuniso Nyumbu, President, Gemstones and Allied Workers Union of Zambia, interview, Lusaka, August 3, 2012.
41 Richard Parker, "A Personal Journey to the Heart of Greece's Darkness," *Financial Times*, February 15, 2012.
42 "into two 'classes'": "European Unity on the Rocks," *Pew Global Attitudes Project*, May 29, 2012, www.pewglobal.org/2012/05/29/ european-unity-on-the-rocks/.
43 Italienische Fahrerflucht [Italian Hit-and-Run], *Der Spiegel Online Politik*, January 23, 2012, www.spiegel.de/ politik/deutschland/0,1518,811817,00.html.

44 HRW, "'You'll Be Fired if You Refuse,'" 1, 13.
45 "Guanyu jiaqiang zhongyang qiye jingwai guoyou zichan guanli youguan gongzuo de tongzhi" [Announcement on Strengthening Central State-Owned Enterprises' Overseas State Asset Management Related Work], *SASAC*, October 14, 2011,www.sasac.gov.cn/ n1180/n1566/n258222/n259188/13863071.html; "Zhongyang qiye jingwai guoyou chanquan guanli zanxing banfa" [Interim Measures for the Administration of Overseas State-Owned Property Rights of Central Enterprises] (*SASAC*, June 27, 2011), www. sasac.gov.cn/n1180/n1566/n11183/n11244/13624758.html.
46 Li Jinping, China Geo-Engineering, Zambia Branch, interview, Lusaka, September 11, 2008.
47 "2010 Mineral Production (1st Half of Year)" and "2010 Mineral Production (2nd Half of Year)" (Lusaka: Photocopies, Office of the Chief Mining Engineer, August 19, 2011); "Zambian Copper Production by Year," *Index Mundi*, www.indexmundi.com/minerals/ ?country=zm&product=copper.
48 "latter were worse": HRW, "'You'll be Fired if You Refuse,'" 1. "supply contract workers": ibid., 99.
49 Barry Sautman and Yan Hairong, "African Perspectives on China–Africa Links," *The China Quarterly* 199: 728–759.
50 Rohit Negi, Beyond the 'Chinese Scramble': The Political Economy of Anti-China Sentiment in Zambia," *African Geographical Review* 27 (2008): 41–63. Studies by psychologists and historians show that racial prejudice affects the ability to accurately evaluate or even report on events or persons. Paul Kellestedt, *The Mass Media and the Dynamics of American Racial Attitudes* (Cambridge, UK: Cambridge University Press, 2003); Spencer Piston, "How Explicit Racial Prejudice Hurt Obama in the 2008 Presidential Election," *Political Behavior* 32 (2010): 431–451; Marianne Bertrand and Sendhil Mullianathan, "Are Emily and Greg More Employable than Lakisha and Jamal? A Field Experiment on Labor Market Discrimination," *American Economic Review* 94 (2004): 991–1013.
51 Han Wei and Shen Hu, "Zambian Workers Return to Jobs at Chinese-Owned Mine," *Caixin*, October 23, 2011, http://english/caixin.cn/2011-10-23/100316622.html.
52 "spurred by our critique": Barry Sautman and Yan Hairong, "Gilded Outside, Shoddy Inside: The Human Rights Watch Report on Chinese Copper Mining in Zambia," *Asia-Pacific Journal* (December 26, 2011), http://apjjf.org/2011/9/52/Barry-Sautman/3668/ article.html; Barry Sautman and Yan Hairong, "The Wrong Answers to the Wrong Question," *Pambazuka News* 570 (February 3, 2012), www.pambazuka.org/governance/ debate-wrong-answers-wrong-question.
53 See Barry Sautman and Yan Hairong, *The Chinese are the Worst? Human Rights and Labor Practices in Zambian Mining*. Maryland Series in Contemporary Asian Studies No. 210 (Baltimore: University of Maryland School of Law).
54 Barry Sautman and Yan Hairong, "Beginning of a World Empire? Contextualizing Chinese Copper Mining in Zambia," *Modern China* 39 (2013): 131–164, 140, Table 2, Figure 1).
55 See table "Mining Fatalities: All U.S. Mines Accident/Injury Classes" in *Briefing Book for the Niosh Mining Program* (section 1.4 "Research Needs") (Atlanta: Center for Disease Control and Prevention, 2005). The fatality ratio between surface and underground mining is calculated by us for the period of 1990 to 2004, using employment figures for 1992 to 2002.
56 "Inexperienced Workers and Cost Pressures Factors in Mine Death Increase, Safety Inspector Says," *ABC News*, October 14, 2015, www.abc.net.au/news/2015-10-14/ inexperienced-workers-a-factor-in-increased-mine-deaths/6851872.
57 "non-CNMC firms' wages": HRW, "'You'll Be Fired if You Refuse,'" 24. "Chinese labor standards": ibid., 2.
58 Figures from collective bargaining agreements supplied to authors by Mineworkers Union of Zambia and National Union Miners and Allied Workers.
59 Barry Sautman and Yan Hairong, "Gilded Outside, Shoddy Within."

60 "workers were contractors": Barry Sautman and Yan Hairong, "Beginning", 137, Table 1. "one-half as much": Barry Sautman and Yan Hairong "Beginning", 141.
61 "CU Later: Zambia 2014," *The Business Year*, www.thebusinessyear.com/zambia-2014/cu-later/review.
62 "KCM Workers Fight Back," *Foil Vedanta*, July 7, 2014, www.foilvedanta.org/articles/kcm-workers-fight-back/.
63 "Nevers Promises to Build University for North Western Province," *The Post* (Zambia), December 29, 2014.
64 "Mopani to Lay Off 5,331 Miners," *The Post*, November 12, 2015.
65 With the facilitation of the local unions, the survey was administered by Chilayi Mayondi, lecturer at Zambia's Mulungushi University, and four university students in mining towns where mine workers concentrate.
66 Alister Fraser and Miles Larmer (eds.), *Zambia, Mining, and Neo-liberalism: Boom and Bust on the Globalized Copperbelt* (New York: Palgrave MacMillan, 2010), Chapters 6–7.
67 Jack Caldwell, "Mining Salaries," *Technomine*, 2014.
68 "entire workforce was $200": Danish Trade Union Council for International Development Cooperation, "Zambia Labour Market Profile 2014" (Danish Trade Union Council for International Development Cooperation, 2014), 15, www.ulandssekretariatet.dk/sites/default/files/uploads/public/PDF/LMP/lmp_zambia_2014_final_version_revised.pdf. "miner's salary was $800": "CU Later: Zambia 2014". The $800 average miner's wage might include not only pay, but also allowances.
69 Deborah Brautigam, "Chinese Mining Projects in Africa: Is it an Investment?" *China in Africa: The Real Story*, February 15, 2016, www.chinaafricarealstory.com/2016/02/chinese-mining-projects-in-africa-is-it.html.
70 "Chinese investment in Africa": Emma Mawdsley, "Fu Manchu versus Dr Livingstone in the Dark Continent? Representing China, Africa and the West in British Broadsheet Newspapers," *Political Geography* 27 (2008): 509–529; Stephen Chan, "Is China Good for Africa," *New Internationalist,* October 2012, 30–32. "portrayed in Western media": Carola Richter and Sebastien Gebauer, "Die China-Berichterstattung in den Deutschen Medien," [China's portrayal in German media], (Berlin: Heinrich Boll Stiftung, 2010), www.boell.de/de/content/die-china-berichterstattung-den-deutschen-medien.
71 "Bush Makes Clinton's China Policy an Issue," *Washington Post*, August 20, 1999; Gabriel Rachman, *Easternization: Asia's Rise and America's Decline from Obama to Trump and Beyond* (London: Other Press, 2017).
72 Barry Sautman and Yan Hairong, "Bashing the Chinese: Contextualizing Zambia's Collum Coal Mine Shooting." *Journal of Contemporary China* 23 (2014): 1073–1092.
73 John Lungu and Allastair Fraser, "For Whom the Windfalls: Winners and Losers in the Privatization of Zambia's Copper Mines" (Civil Society Trade Network of Zambia, 2008), 2. "There has been a collapse": ibid., 3.
74 Action for Southern Africa et al., "Undermining Development? Copper Mining in Zambia" (Action for Southern Africa, October 2007).
75 Counter Balance, "The Mopani Copper Mine, Zambia: How European Development Money has Fed a Mining Scandal" (Counter Balance, 2010), 16–17.
76 "privately owned": Jing Gu "China's Private Enterprises in Africa and the Implications for African Development," *European Journal of Development Research* 24:1 (2009): 570–587. "Africa is highly variegated": For scholarly examinations of Chinese activities in Africa, see Wu Di, "The Everyday Life of Chinese Migrants in Zambia: Emotion, Sociality and Moral Interaction" (PhD dissertation, London School of Economics and Political Science, 2014); Jamie Monson, J. and Stephanie Rupp, "Africa and China: New Engagements, New Research," *African Studies Review* 56:1: 21–44. Karsten Giese and Alena Thiel, "The Vulnerable Other: Distorted Equity in Chinese–Ghanaian Employment Relations," *Ethnic and Racial Studies* 37:6 (2014):1101–1120.

77 Cheryl Lynn Greenberg, *Troubling the Waters: Black-Jewish Relations in the American Century,* (Princeton: Princeton University Press, 2010), 223.
78 Crispin Radoka Matenga, "The Impact of the Global Financial and Economic Crisis on Job Losses and Conditions of Work in the Mining Sector in Zambia" (Lusaka: International Labour Organization, 2010); Charles Muchimba, *The Zambian Mining Industry: A Status Report Ten Years after Privatization* (Lusaka: Friedrich Ebert Stiftung, 2010); Jean-Christophe Servant, "Mined out in Zambia," *Le Monde Diplomatique,* May 9, 2009; "Zambian Copperbelt Reels from Global Crisis," *Washington Post,* March 25, 2009.
79 Judith Fessehaie, "Development and Knowledge Intensification in Industries Upstream of Zambia's Copper Mining Sector," *MMCP Discussion Paper No. 3* (MMCP, 2011), 26.

4 Field vignette

Moving from prescriptive to performance-based regulation: the case of waste management

Andy Fourie, Mwiya Songolo, and James McIntosh

There are naturally highly variable regulations across different countries in Africa. Countries where mining has taken place for longer periods of time will usually have more mature and comprehensive regulations in place. Problems with closed mines are also only now beginning to manifest themselves. An example in South Africa is the rising levels of acidic water in mined-out underground voids, with some of this water now beginning to discharge to surface waters, threatening aquatic species and potentially impacting on the very important historical and tourist sites around the underground karst caves near Sterkfontein in Gauteng province. Other countries in Africa, particularly where large-scale mining activities are relatively recent, have not necessarily experienced first-hand the impacts that mining waste can have on the environment and on people, and regulations are accordingly less mature.

A notable exception is The AKOBEN program (the Akoben is a horn used to sound a battle cry), which is an environmental performance rating and disclosure initiative of the Ghanaian Environmental Protection Agency which has potential for improving waste regulation performance. Under the AKOBEN initiative, the environmental performance of mining and manufacturing operations is assessed using a five-color rating scheme (gold, green, blue, orange, and red) indicating environmental performance ranging from excellent to poor. These ratings are annually disclosed to the public and the general media, and it aims to strengthen public awareness and participation. The AKOBEN program uses formal standards for Water Quality, but EPA guidelines for effluent quality, air pollution and dust discharge. It is unclear if these guidelines are applied uniformly across the sector, how often they are reviewed and updated, and how stringently they are applied outside of the AKOBEN process.

AKOBEN ratings are evaluated by analyzing more than a hundred performance indicators that include quantitative data as well as qualitative and visual information. These ratings measure the environmental performance of companies based on their day-to-day operations once they have successfully cleared their Environmental Impact Assessments (EIA) and obtained their environmental permit to operate. These ratings indicate how well companies have met the commitments they made in their EIAs at the planning stage. AKOBEN, therefore, complements the EIA process and serves as a monitoring and verification program to ensure that companies follow environmental regulations on a continual basis.

Public disclosure of AKOBEN's rating is the end point of months of effort on data collection and analysis. However, AKOBEN's disclosure is not a single event; it is a process that follows a three-step procedure before the final disclosure to the public and the media. As a first step, AKOBEN privately shares the results of the ratings with companies; this step is called internal disclosure. If there is a disputable issue, companies inform the AKOBEN team in writing or hold a meeting in person. Upon receiving the feedback, AKOBEN reviews the ratings and re-sends the final ratings to companies. Next the ratings are reviewed by the senior management of the EPA and approved for public disclosure. The final ratings are then disclosed to the public through a press conference and are also posted on the AKOBEN website. The process is thus very transparent and interactive.

There is significant disparity between the approaches to mine waste regulation in the five countries studied (and probably even greater variability would become apparent if the study countries were expanded). It is suggested that it would be extremely beneficial to countries in Africa that have active mining industries to develop a Pan-African approach to regulating these activities, particularly when it comes to mine waste. Drawing on current best practice, such as the Ghanaian AKOBEN approach, as well as the long experience of regulators in countries such as Zambia and Ghana, a more integrated way of managing mining waste might be achieved. This would also be highly beneficial for responsible mining companies considering investing in Africa as expectations would become more uniform and transparent. The more mercenary international mining companies, who might be looking for regions with lax regulations (and thus perceived less expensive obligations to be fulfilled), would be thwarted in their attempts to improve their own profitability at the expense of the environment and local communities living near to mining operations.

5 Field vignette

Ghana's policy on artisanal and small-scale mining

Benjamin Aryee

The legal framework under which mining takes place in Ghana requires that mining, and other minerals-related activities, can only be undertaken legitimately in Ghana under a licence granted by the minister responsible for mining. While the law broadly provides for what and how mining and minerals-related activities can take place, it makes a special provision for small-scale mining. Participation in the latter is restricted to Ghanaian citizens. Other key limitations include a maximum land size of 25.2 acres that must be granted in an area designated by the minister as a small-scale mineral operation area.

Throughout the more than a century history of mining in Ghana (known as the Gold Coast in colonial times) gold has been the predominant mineral mined, irrespective of the scale of operation. Ghana has truly become the "Gold Coast," as gold is found across, almost, its whole length and breadth. Therefore, managing gold mining, which is characterized by especially high value but low volume, presents peculiar challenges for policy making.

Broadly, mining sector policy is considered within the following three areas of government responsibility: (1) ensuring growth and socio-economic development; (2) facilitating the generation of livelihoods for citizens; and (3) ensuring the rule of law—clear definition of, compliance with and enforcement of laws—to ensure that the rights of all residents (citizens or otherwise) are protected.

To promote growth and development, the government of Ghana has consistently facilitated the attraction of investment into the mining sector through various policies and laws. These policies are premised on the country's comparative advantage of having significant deposits of gold and other minerals. Like other mineral economies, legal, fiscal, regulatory, and institutional frameworks in place are aimed not only at the attraction of investment into the sector but also retention of investments to promote sustainable development. Consequently, the mining sector has been the single largest foreign exchange earner for Ghana since 1999 and contributed significantly to government revenues. Part of the policy impact is also that, currently, Ghana is the eleventh largest gold producer worldwide and the second largest producer in Africa.

Small-scale mining has also developed from predominantly artisanal operations to semi and fully mechanized operations, which together currently contribute more than a third of Ghana's total gold production. However, over the years, some of the policies that have encouraged the growth of small-scale mining include provision of viable designated areas, establishment of district offices and provision of extension services to small-scale miners, simplification of the licensing regime, lower fee regime, and establishment of plant pools and promotion of processes to improve recovery rates.

To facilitate livelihood creation for its impoverished population the government has restricted small-scale mining only to Ghanaian citizens. Over a million Ghanaians are estimated to be engaged in small-scale mining, while more than 4 million depend on it for their livelihood. Despite contributing significantly to Ghana's gold production, the small-scale mining sector is highly challenged, including associations with illegal mining (e.g., encroaching on large scale concessions, involvement of foreigners, mining in prohibited areas) and smuggling.

Also, a number of challenges militate against government's mining-related livelihood promotion agenda. Notable among these are lack of viable lands, lack of funding, limited human and technological capacity, unsafe and environmentally unfriendly mining practices leading to accidents as well as loss of lives, and other negative impacts on society.

In practice, in acknowledgement of these challenges, government interventions have included, "carrot" approaches, like:

- Establishment of legal framework within which to regularize such operations in a safe and environmentally friendly setting, supported by geo-scientific investigations of grounds for small-scale mining; sensitization and extension services to enhance efficiency, health, safety and environmental practices; ensuring access to formal markets; and facilitating access to funding, especially for organized groupings, e.g., associations, cooperatives, etc.
- Introduction of non-mining alternative livelihood projects to engage excess labor and diversify the local economy.

Operators who go beyond the confines of the law have to contend with "stick" approaches of law enforcement, including prosecution, confiscation of equipment and minerals, and other punitive sanctions. Apart from regular compliance enforcement, task forces have been instituted to stem periods of extreme infractions.

In addition to the above, the Ghana government has tried to enforce the rule of law in a number of ways. Legal, regulatory and institutional frameworks prescribe the rights and obligations of all key stakeholders—notably investors, communities, as well as others—and require compliance, failing which stipulated sanctions/ penalties are applicable. Investors are assured of sanctity of and right to enjoyment of contracts; communities are assured peaceful enjoyment of their socio-economic as well as other human rights; and every other person is granted appropriate rights within the society. It is noted that these rights go with concomitant obligations. Accordingly, in addition to the general enforcement of laws, the following measures seek to assure enjoyment of rights and also participation of all stakeholders in the governance of the sector:

- Mining sector institutions like the Minerals Commission and the Environmental Protection Agency have the responsibility of ensuring compliance with technical legislation, standards, requirements, etc.

- These, and other akin institutions also provide for administrative-type resolution of disputes among large scale and small-scale operators, communities and other stakeholders, aside from the formal adjudication of cases accessible in appropriate judicial fora.
- Additional to other initiatives to address the challenge of illegal mining, Ghana is exploring ways to recategorize mining operations to bridge the gap between large and small-scale mining operations.
- Ghana has also signed on to a number of international conventions and initiatives, such as the Extractive Industries Transparency Initiative, the Voluntary Principles on Security and Human Rights and the Kimberly Process Certification, for diamonds.

During 2017–2018, the government decided to upscale the management of the ASM sub-sector in Ghana through an integrated, multi-stakeholder approach termed the Multi-sectoral Mining Integrated Project (MMIP). The project is planned to span five years as the primary instrument for the management of ASM. It is expected to review and enforce the legal and regulatory regime, ensure reclamation of the degraded environment, implement social interventions to facilitate alternative livelihood creation, employing the requisite technology, notably tracking instruments and drones for monitoring operations and adopting the relevant human resource development and management methodologies as well as communication techniques.

Policy interventions in Ghana therefore aimed at and span using mining as a catalyst for sustainable development; ensuring collaborative, transparent stakeholder engagement over the mine life cycle, facilitating progressive capacity building and knowledge sharing with respect to benefits, costs, rights and responsibilities and assuring prompt and comprehensive dispute management among all stakeholders, within the rule of law.

Part II
Data and models
Supporting strategic planning for Africa's minerals

6 Developing accurate and accessible geoscientific data for sustainable mining in Africa

Judith A. Kinnaird and Raymond J. Durrheim

In brief

- Africa has vast resources of mineral wealth, but much of the continent remains underexplored.
- To attract investment, governments must provide digital geoscientific data that are accurate, logically presented and structured, comprehensive, and easy to use.
- Geological surveys and mining ministries are important institutions for the collection and dissemination of geoscientific data. The capacity of geological surveys to perform their functions should be regularly monitored and improved as necessary.
- Namibia provides a good example of sound practices in Africa and might serve as a role model for other developing countries.

Introduction

Africa is richly endowed with a number of metals and materials.[1] As shown in Table 6.1, 79 percent of the world's platinum and 50 percent of its chromium comes from South Africa and Zimbabwe, 53 percent of cobalt and 10 percent of copper is from the Central African Copper-Cobalt Belt in the Democratic Republic of the Congo (DRC) and Zambia, 25 percent of manganese is from South Africa, 40 percent of phosphate rock is produced by Morocco and Algeria, and just under 10 percent of gold is produced in Africa, mainly by Ghana and South Africa. Other critical metals produced on the continent include niobium-tantalum from the DRC and Rwanda, and rare earth elements from Malawi and South Africa.[2] Furthermore, African countries have the potential to develop or expand production of economically important metals whose supply is at risk, such as antimony, graphite, and tungsten. There are numerous factors that can endanger supply, ranging from limitations on exports and poor governance, to armed conflict.

Table 6.1 Selected African countries by commodities mined including oil and gas, proportion of mining to GDP, existence of online portals for license applications, and policy and investment attractiveness for exploration

Note: Percentages in parentheses indicate proportion of world supply; thus, "tantalum (17%)" means that the country produces 17 percent of the world's supply of tantalum. Ordinals in parentheses indicate ranking in world supply; thus, "diamonds (12th)" means that the country is the twelfth largest producer of diamonds in the world. The list of commodities mined is derived largely from United States Geological Survey country files updated in 2016; data are generally for 2013–2014. Where the percentage of GDP represented by mining is given, this is generally for 2013. Policy rank indicates the extent to which a country's policies encourage mining investment, as compared with other jurisdictions; investment rank reflects both the country's policy rank and its mineral potential, as compared with other jurisdictions. Both rankings are provided by the Fraser Institute in its 2016 based on 104 jurisdictions around the world, 18 of which are in Africa.

Country	Commodities	Proportion of GDP	Online portal	Policy rank	Investment rank
Botswana	Diamonds (24% by value), coal, cobalt, copper, gold, nickel, platinum-group metals, salt, aggregate, semiprecious gemstones, silver	23%	Yes	12	19
	Contact details: www.mines.gov.bw/Forms/ml%20mp%20applications%20guide.pdf				
Burkina Faso	Gold, zinc, phosphate, manganese, copper	Low	In prep.	51	48
	Contact details: www.burkina-emine.com/?p=3747				
Cameroon	Aluminum (from alumina mainly imported from Guinea), cement-grade limestone, petroleum, and sand and gravel; artisanal mining of diamonds and gold	Low	Launched in 2016		
	FlexiCadastre being implemented; see www.spatialdimension.com/News				
Côte d'Ivoire	Petroleum, gas, gold, manganese, silver, diamonds, industrial minerals.	2.3%	Launched in 2016	40	17
	FlexiCadastre being implemented; see www.spatialdimension.com/News				
Democratic Republic of Congo	Cobalt (< 50%), tantalum (17%), diamonds (12%) copper (5%), tin (1%), gold	20%	Yes	70	29
	Contact details: http://portals.flexicadastre.com/drc/en/				

Country	Minerals	Mining cadastre	% GDP		
Eritrea	Copper, gold, silver, cement-grade limestone, gypsum, lime, kaolin, marble, pumice, quartz, salt	No	1.7%	49	33
	Contact details: http://eritrean-embassy.se/invest-in-eritrea/mining-in-eritrea/				
Ethiopia	Tantalite (1%), gold, gemstones, cement-grade limestone, soda ash, kaolin, dimension stone	Yes	< 1%	79	68
	Contact details: www.mom.gov.et				
Ghana	Gold (one of the top ten producers), petroleum, diamonds, bauxite, cement-grade limestone, lead, manganese, salt, silver, construction materials	Launched in 2016	< 10%	31	22
	Contact details: www.ghana-mining.org				
Guinea	Bauxite (7%), diamonds (12h), alumina, cement-grade limestone, gold, iron ore, graphite, manganese, nickel	Launched 2015	25%		
	FlexiCadastre being implemented; see www.spatialdimension.com/News				
Kenya	Soda ash (4%), colored gemstones, gold, columbite, titanium, zircon	Yes	< 1%	76	86
	Contact details: https://portal.miningcadastre.go.ke/				
Liberia	Iron ore, barite, cement-grade limestone, diamonds and gold, aggregate	Yes	10%		
	Contact details: http://portals.flexicadastre.com/liberia/				
Malawi	Uranium (2% in 2013, but on hold in 2017 until price rises), cement-grade limestone, coal, aggregate, limestone, colored gemstones.	Launched 2016	5.2%		
	FlexiCadastre being implemented; see www.spatialdimension.com/News				

(continued)

Table 6.1 (continued)

Country	Commodities	Online portal	Proportion of GDP	Policy rank	Investment rank
Mali	Gold, phosphate, diamonds, iron ore, salt, sand and gravel, silver (as a by-product of gold mining), colored gemstones. Gold a major contributor to GDP.	No		61	42
	Contact details: https://revenuedevelopmentdotorg.wordpress.com/mali/				
Mozambique	Ilmenite (6%), zircon (3%), aluminum (1%), beryl, tantalum, cement-grade limestone, coal, natural gas	Yes	1.7%	72	95
	Contact details: http://portals.flexicadastre.com/mozambique/en/				
Namibia	Uranium (7%), diamonds (produced < 2% of world rough by weight in 2013, but ranked 1st in value per carat), arsenic, copper, gold, lead, manganese, silver, zinc, cement-grade limestone, dolomite, fluorspar, granite, marble, salt, semiprecious stones, wollastonite	Yes	13%	38	53
	Contact details: http://portals.flexicadastre.com/Namibia/				
Nigeria	Oil and gas (3%), tin, coltan, iron ore, coal, limestone, barite, gemstones, gold, salt, lead, zinc	Yes	< 1% (oil and gas, 13%)		
	Contact details: www.miningcadastre.gov.ng/				
Rwanda	Tantalum (50%), tin (1%), tungsten (1%), gas, niobium, gold	Yes	2%		
	Contact details: http://portals.flexicadastre.com/rwanda/				
Sierra Leone	Diamonds (10%), rutile (10%) bauxite, iron ore, cement-grade limestone, gold, ilmenite and zircon. The mineral sector accounted for > 90% of export earnings in 2013.	Yes		80	87
	Contact details: www.nma.gov.sl/home/licence-application-process/				

Country	Minerals	Online cadastre	Mining contribution to GDP		
South Africa	Platinum (72%), rhodium (56%), chromium (48%), kyanite/andalusite (47%), palladium (37%), vermiculite (36%), vanadium (27%), manganese (25%), gold (9%), coal (4%), fluorspar (3%), cobalt (3%), nickel (2%), iron ore, diamonds, uranium, ilmenite, rutile and zircon	limited	8.3%	84	74
	Contact details: www.gov.za/services/mining-and-water/apply-mining-right				
South Sudan	Oil, gold	Yes		97	80
	Contact details: http://portals.flexicadastre.com/southsudan/				
Tanzania	Gold (1.6%), cement-grade limestone, colored gemstones, tanzanite (100%), coal, diamonds	Yes	3% (mining and quarrying)	59	64
	Contact details: http://portal.mem.go.tz/map				
Uganda	Pumice (4%), aggregates, cement-grade limestone, cobalt, gold, iron ore, kaolin, lead, coltan, tin, tungsten, salt, vermiculite	Yes	1.5%	60	70
	Contact details: http://portals.flexicadastre.com/uganda/				
Zambia	Copper (8th), cobalt (6th), colored gemstones	Yes	8%	43	30
	Contact details: http://portals.flexicadastre.com/zambia/				
Zimbabwe	Diamonds (9%), platinum (7%), palladium (5%), gold, nickel, copper, chromite, cobalt, graphite, coal	Launched 2016	8%	102	96
	FlexiCadastre being implemented; see www.spatialdimension.com/News				

Worldwide demand for common metals such as aluminum, copper, zinc, and lead is increasing at 5 percent per year.[3] Large quantities of previously little-used metals are needed as well for cell phones, personal computers with touch screens, renewable energy technologies, and hybrid cars. Cell phone manufacturing, for example, requires a wide range of elements, such as bismuth, cobalt, copper, gallium, indium, lead, lithium, nickel, niobium, platinum group elements, rare earth elements, tantalum, tungsten, gold, and silver, in addition to metals for the casings and petrochemicals for plastics. Demand for these metals is likely to increase until around 2050, when it is expected to slacken off because of recycling.[4] It is projected that a large proportion of the world's population (especially in India and China) would have graduated to the middle class and acquired goods such as domestic appliances and motor cars. Thus, most purchases will become replacements rather than first-time acquisitions, making it possible to recycle the metals in the discarded goods.

The rising demand for metals has raised concerns over future supply. In 2010 a European Union report identified fourteen materials as critical based on economic importance and supply risk: the rare earth elements, the platinum group metals, germanium, magnesium, gallium, antimony, indium, beryllium, cobalt, tantalum, niobium, tungsten, fluorspar, and graphite; the report also listed metals whose availability was essential for high-technology, green, and defense applications but whose supply was vulnerable to fluctuations driven by political or economic considerations.[5] Materials considered critical vary by country, industry, perceived risk of future supply, and perceived demand, and these parameters can change over time.[6] In 2013, an EU review of the critical metals and materials identified in 2010, found that the supply risk had declined for many of those commodities, probably because of increased geological exploration and revisiting re-evaluation and revival of historical projects.[7]

In 2013 a European Union review of the critical metals included thirteen of the fourteen materials identified in the previous report, with only tantalum (due to a lower supply risk) moving out of the European Union critical material list.

Table 6.2 lists the 2015 criticality rankings for minerals linked to African production or reserve centers. The 3rd column, showing the relative supply risk index (criticality) rankings, ranges from 4.8 for aluminum to 8.1 for cobalt. (By comparison, the criticality ranking for rare earth minerals, in which China dominates production and reserves, tops the index at 9.5.) Each year the index is adjusted, and Africa's profile in the criticality rankings for minerals sometimes shifts. For example, the platinum group elements were at 8.6 in 2011 and 7.6 in 2015. Conversely, cobalt's rating has risen in recent years, and the DRC remains the dominant production and reserve center.

As the world population increases, the demand for more energy has led to the development of technologies for alternative sources of energy, which require their own supply of metals and materials. A typical wind turbine requires about 1,500 tonnes of steel, and its permanent magnets use about half a ton of rare earth elements; photovoltaic cells for solar power require copper, aluminum, glass, and steel; and hydropower requires concrete.[8] In 2050 the total amount of steel, concrete, aluminum, copper, and glass needed for wind and solar facilities is expected to be two to eight times the 2010 world production, and the aluminum needed

Table 6.2 Critical minerals for African production and reserves, 2015

Element or element group	Symbol	Relative supply risk index	Leading producer	Top reserve holder
Cobalt	Co	8.1	Democratic Republic of the Congo	Democratic Republic of the Congo
Platinum group elements	PGE	7.6	South Africa	South Africa
Tantalum	Ta	7.1	Rwanda	Australia
Fluorine	F	6.9	China	South Africa
Chromium	Cr	6.2	South Africa	Kazakhstan
Manganese	Mn	5.7	China	South Africa
Aluminum	Al	4.8	Australia	Guinea

Source: British Geological Survey, "Centre for Sustainable Mineral Development, Risk List 2015," www.bgs.ac.uk/mineralsuk/statistics/risklist.html, accessed March 15, 201

Note: "Relative Supply Risk Index" refers to the relative risk of disruption to the supply of forty-one economically important elements or element groups, seven of which are shown here

for such facilities will equal that used by all industrial sectors between 1970 and 2000.[9] This increase is even greater when it comes to critical metals: estimates of global demand for gallium, indium, selenium, tellurium, and rare earth elements in 2050 range from 10 times to more than 200 times the 2010 world supply. As ever-decreasing grades of metal are mined to meet the demand, additional energy is necessary for mining and processing the metals. Thus, the diminishing returns spiral, because treatment of ever-lower-grade ore requires more energy, while production of renewable energy requires more metals.

One country that is mindful of this situation is Japan, which is a world leader in electronic manufacturing and car production yet has a relatively small land area and few operating mines or resources. In 2013 Japan joined the United States and the European Union in a trilateral workshop on critical raw materials held in Brussels under the auspices of the European Commission. At that workshop, a representative of the Japanese Oil, Gas and Metals National Corporation, which is tasked with ensuring the supply of natural resources to Japan, described three criteria used to identify critical resources in Japan: "the importance of mineral resources for Japanese industries, the possibility of a failure in mineral resources supply, and the feasibility of securing supply resources."[10] To secure the necessary metal supplies for Japan, the Japanese Oil, Gas and Metals National Corporation is partnering with, and investing in, companies in Africa.

Given the inherent challenges in finding and extracting resources, mining and exploration companies consider a number of factors when deciding whether to invest in a country.[11] These include existence of favorable legislation for commodity exploration and extraction; a stable, competitive, and transparent taxation framework; adequate exploration investment, which, in Africa, is likely to be from outside the country because of the shortage of local venture capital; acceptance of international practices for capital markets and foreign trade; availability of suitable techniques for mineral exploration; clear environmental and social safeguards in line with international standards to ensure that best practices are adhered to; an easily searchable database that shows the availability of ground for exploration and mining; and prompt issuance of licenses for exploration and mining. Freedom to trade the minerals produced might also be desirable. In this chapter, we focus on the factors that directly involve geoscientific knowledge and skills. The political, economic, and environmental factors are covered in other chapters.

How can more metals and minerals be found?

Prerequisites for exploration activity

Two primary criteria are used to select an area for mineral exploration. The first is a favorable geological environment—that is, a region that is considered likely to contain ore deposits, either because mineral occurrences are known to exist or because the geology has similar characteristics to regions elsewhere in the world that are known to host deposits of target minerals. The second criterion is government policies and practices that make it attractive for companies to explore and mine.

In well-explored parts of the world, such as western Europe, north America, and Australia, most easy-to-find near-surface deposits have been discovered decades or even centuries ago. Consequently, the average depth beneath the surface of new discoveries in well-explored regions has increased, particularly since the development of geological, geochemical, and geophysical methods that can "see" hundreds of yards or even miles beneath the earth's surface.[12] In these regions, new mineral deposits are likely to be found only at greater depths or in areas where the geology demands further development in technology and expertise, legislative changes have made exploration more attractive, or new mineral treatment methods allow metals to be extracted from rocks that were not previously considered as ore.

Large parts of the world, notably in Africa and Asia, have geological potential but remain underexplored because of security concerns, economic policies, unfavorable legal frameworks, the lack of modern mining codes, or simply difficulties of reaching sites that are distant from roads. In these regions, current exploration technology could produce new discoveries, and advanced technologies could reduce costs and improve the success rate.

Exploration

To establish whether a province or district has mineral potential, exploration companies require modern geological and structural maps, regional geochemical analyses and geophysical data (e.g., aeromagnetics and radiometrics), and reports on past exploration and mining activity. This information is usually collected and made available by a government agency such as a geological survey or ministry of mines as part of a program to attract investment in exploration and mining. The data should be available at a reasonable cost and in a digital format that allows it to be reprocessed and interpreted with modern tools and concepts.

Mining companies assess national- and regional-scale data before seeking licenses to conduct exploration surveys, which can be hundreds of miles in extent. The techniques that are used to locate drilling targets depend on the mineral that is sought, the cover rocks, and the scale of the survey. Airborne geophysical surveys, geological and structural mapping, and regional geochemical surveys are usually conducted first, followed by detailed geological, geochemical, and ground geophysical surveys. Targets are ranked and drilled, and the borehole core is logged and assayed. The borehole may be surveyed using a range of geophysical tools.

A wide range of exploration technologies is used to map the surface and subsurface in great detail and depth, including global positioning systems, airborne gravimetry, and three-dimensional reflection seismology. These technologies are often adapted from the oil and gas sector, telecommunications, and the military. Advances in earth observation and mineral exploration are documented in the proceedings of the decennial International Conference on Mineral Exploration.[13]

Mining

Once an ore deposit is found, the viability of mining it is investigated. Advances in mining and metallurgical technology have made it possible to mine more deeply and more cost effectively, to treat ores with lower grades, and to extract metals and minerals from materials that were not previously amenable to processing.[14] The viability of a project also depends on many factors that are outside the direct control of the mining company, such as the existence of infrastructure (e.g., railways, ports, and power supply) and commodity prices.

Deep mines are prone to high rock stresses and temperatures.[15] Technologies have been developed to mitigate the risks posed by rock bursting, to cool and ventilate mine workings, and to operate machinery remotely. Equipment used in open pit mines has increased in size, bringing economies of scale, while Doppler radar systems have been developed to continually monitor the stability of the pit wall.[16]

Environmental and human impacts

Mining can have negative impacts on the environment. For example, noise, dust, and vibrations can disturb communities and wildlife; groundwater can be polluted by acid mine drainage; and the agricultural potential of a district is reduced if arable land is used for mining. Deeper mines and lower-grade ore require more energy to mine and increase competition for scarce water resources. Geoscientific data are required to monitor and mitigate these impacts.

The environmental and human impacts of mining in many countries have been mitigated by advances in technology and the development of guidelines, but there is considerable room to do better.[17] Increasing mechanization and automation of mining processes can improve worker safety and reduce costs, but also reduce employment, particularly among unskilled and semiskilled workers. "Green" mining, the notion that new mines will, by necessity, have less environmental impact, is gaining traction.

Skilled human resources

Mineral-producing countries legitimately expect that their citizens will benefit from employment opportunities in the extractives industry, and these benefits will ultimately raise citizens' standard of living and level of education, stimulate the economy, and contribute to the tax base. Skilled people are required to develop, operate, and maintain advanced technology. There is an increased expectation that mining companies should play an active role in the development of skills by supporting colleges and universities, and through on-the-job training.

Challenges to the development of mineral resources in Africa

Despite its rich mineral potential, Africa received only 14 percent of global exploration expenditure in 2017.[18] To facilitate mining investment, a number of African governments have created government mining cadastre portals for companies wanting online access to information about exploration and mining licenses (see Table 6.1). These portals allow registered parties to submit new applications for mineral exploration, make online payments, renew permits, and upload statutory reports, thereby improving access to information and transparency and reducing the potential for corruption.

Yet some countries—including South Africa, where mining has always been regarded as the backbone of the country's economy—offer limited electronic access to mining cadastre data. This might be one reason why South Africa's overall attractiveness as a destination for mining investment has dropped, from 66th out of 109 mining jurisdictions in 2015 to 74th out of 104 in 2016, and why its attractiveness in terms of mining policies is ranked 84th out of 104 jurisdictions in 2016, the 3rd worst of 18 African countries evaluated by the Fraser Institute.[19] The South African government's national development plan, adopted in September 2012, noted that whereas the top 20 mining exporting countries had averaged growth of 5 percent a year between 2001 and 2008, South Africa's mining industry had shrunk by 1 percent a year.[20]

South Africa's neighbor to the north, Botswana, is the highest-ranked African country based on policy factors, having risen from 14th of 109 jurisdictions in 2015 to 12th of 104 in 2016.[21] This rise reflects decreased concerns over labor regulations, increased availability of skilled labor, and improved infrastructure. Ghana has also risen from 52nd of 109 in 2015 to 31st of 104 in 2016 based on policy, and was 22nd of 104 in 2016 in terms of investment attractiveness.[22]

Namibia, like South Africa, has a long history of mining, but it has seen a steep decline in policy rankings, from the 19th most attractive jurisdiction out of 122 in 2014 to 29th of 109 in 2015 and 38th of 104 in 2016. Various factors have contributed to this downward trend, including uncertainty about the administration, interpretation and enforcement of existing regulations, the taxation regime, and trade barriers. Yet despite these challenges, Namibia has set the leading example in developing institutional capacity for mineral exploration in Africa.[23]

Namibia: an African example of the successful development of mineral resources

Mining has been the backbone of the Namibian economy for more than a century. It is responsible for about 50 percent of Namibia's exports and, at 12.2 percent (in 2017), is the economic sector that makes the largest contribution to the GDP.[24] Namibia's extensive mineral deposits, which include diamonds, uranium, gold, copper, lead, and zinc, and potential for discovery of more resources make it attractive to exploration companies. The government and the Geological Survey of Namibia (GSN) also play important roles in Namibia's success.

The role of Government

Realizing the importance of mining to its economy, the Namibian government has promoted mineral exploration and mining investment through sound legislation governing exploration and mining licenses, environmental clearance, labor issues, and financial transactions; a stable and competitive taxation framework; environmental and social demands that are balanced, progressive, and clear; and the acceptance of international practices for capital markets and foreign trade.[25] However, government policy is not static, and in 2016 the Namibian government announced that it intended to compel mining firms to reserve a stake of at least 20 percent for black Namibians. The state allows sufficient time to explore and develop and permits mineral rights to be used as collateral, with freedom to trade the minerals produced. Furthermore, Namibia has an excellent road and telecommunication infrastructure and is currently upgrading its harbors. Namibia also offers a good standard of living for employees in the exploration and mining industry.

The role of the geological survey

The mission of the GSN is twofold: to enhance knowledge and awareness of Namibia's geological resources through scientific investigation, and to encourage the application and dissemination of reliable research data, which will facilitate exploration, land use planning, and environmentally responsible mining and sustainable development. The GSN and the Directorate of Mines have successfully promoted investment in the minerals sector through a number of activities:

- *Obtaining and archiving historical data.* Companies are required to deposit data and core with the GSN. Information is moved into the public domain as soon as the license lapses.
- *Making information about exploration licenses available online.*
- *Building human capacity.* In 2016 the GSN employed sixty people, of which more than half are professional geoscientists and all but one are from Namibia. Professional staff members usually obtain their first degree in Namibia and post-graduate degree abroad.
- *Building infrastructural capacity.* The GSN has well-equipped laboratories, which provide analytical services to Namibians with limited means, such as small-scale miners, at a nominal cost.
- *Generating new data*
    o The GSN compiles digital geological maps at various scales from historical and recent data.

- The GSN regularly undertakes geochemical surveys. Every year one map sheet is sampled, XRF and ICP-MS analyses are performed, and the data are digitally collated.
- The GSN undertakes high-resolution geophysical surveys. In 2015, 98 percent of the country had been surveyed.

- *Marketing products locally and internationally.* The GSN and the Directorate of Mines are housed in the same building in the capital city of Windhoek, providing a one-stop shop where new and historic data can be studied, archived core examined, and licenses applied for. Metadata are available through the web and products are easily obtained at a reasonable cost.
- *Assisting other African countries in developing their own geological surveys.*

These services have encouraged company expenditure on exploration, identified numerous exploration targets, and led to the discovery of the world-class Husab uranium deposit with 137,700 tU in 0.039 percent ore.[26]

The critical role of national geological surveys

Much of the exploration in Africa is conducted by companies based in Australia, Brazil, Canada, China, India, and South Africa. Most of these companies are focused on only a few commodities such as gold, uranium, copper, and rare earths. Before any exploration takes place in a specific country, exploration firms will first study available data extensively to identify potential prospective ground, and then investigate whether the prospective ground is available for exploration. Some countries in Africa, such as Kenya and Namibia, make these steps very simple and allow a firm to purchase a license online quickly, cost-effectively, and easily. In other countries, firms must visit offices in person, and the process of obtaining an exploration license can take more than a year. These obstacles act as a deterrent to many companies, who then look at other options, be they in Africa or elsewhere. To encourage exploration, the first step is to emulate countries like Kenya and Namibia and instigate a user-friendly customer-oriented website that can promote the opportunities a country has to offer. This must be followed by efficient and effective customer service in the provision of exploration licenses, as well as providing the new license holder with any previous exploration data acquired in the license area.

Once a firm has obtained an exploration license, it must conduct sampling and surveys on site. The poor road network in many parts of Africa makes this a difficult task, particularly in the equatorial region of the continent. If a helicopter is needed to transport the rig, it could render exploration too costly. In this situation, data that the country's geological survey can supply from previous exploration attempts could eliminate the need for new analyses and surveys and thus save time and money.

If exploration is successful and a viable mineral deposit is discovered, work will begin on a mining feasibility study. This research seeks a host of non-geological data, such as availability and cost of electricity and water, as well as availability of a skilled workforce. The collection of geoscientific data becomes important once a mine opens, to ensure that an ore body is optimally exploited, and continues long after the mine closes, to ensure that negative impacts on the environment and population are mitigated.

Exploitation of Africa's undisputed mineral wealth could be an important driver of economic growth and human well-being. But mineral exploration and mining are risky ventures that require large sums of capital. It does not matter to an industrialist where a particular mineral or metal is mined; what matters most is the competitiveness of the price and the security of supply.

Many African countries lack the geoscientific data and governance infrastructure to attract mineral exploration. Furthermore, they often lack the technical capacity to ensure that ore bodies are mined in a way that optimally benefits the country. Consequently, it is imperative that African governments develop both the scientific and management skills of their staff, together with improving the analytical and digital infrastructure of their national geological surveys, so that they can provide a better service to potential exploration companies. The resulting investment will mean that citizens are able to benefit from their country's mineral endowment.

Notes

1 Information about mineral production is obtained from the United States Geological Survey's National Minerals Information Center. The country reports are available at https://minerals.usgs.gov/minerals/pubs/country/africa.html (accessed March 15, 2017). Ranking by policy and investment comes from Taylor Jackson and Kenneth P. Green, "Fraser Institute Annual Survey of Mining Companies 2016" (Fraser Institute, 2017); the full report is available at www.fraserinstitute.org/sites/default/files/survey-of-mining-companies-2016.pdf (accessed March 15, 2017).
2 "Other critical metals": Critical metals are those whose availability is essential for high-technology, green, and defense applications but are vulnerable to politically or economically driven fluctuations in supply. This designation applies particularly to the rare earth elements, tantalum, niobium, lithium, molybdenum, and indium (http://criticalmetalsmeeting.com, accessed March 15, 2017). The definition depends on the country and the technology industries of that country, the perceived risk of supply disruption, and the perceived demand. All these factors are subject to change.
3 Olivier Vidal, Bruno Goffé, and Nicholas Arndt. "Metals for a Low-Carbon Society," *Nature Geoscience* 6:11 (2013): 894–896.
4 Saleem H. Ali, Damien Giurco, Nicholas Arndt, Edmund Nickless, Graham Brown, Alecos Demetriades, Ray Durrheim, Maria Amélia Enriquez, Judith Kinnaird, Anna Littleboy, Lawrence D. Meinert, Roland Oberhänsli, Janet Salem, Richard Schodde, Gabi Schneider, Olivier Vidal and Natalia Yakovleva, "Mineral Supply for Sustainable Development Requires Resource Governance," *Nature* 543 (2017): 367–372.
5 European Commission, "Critical Raw Materials for the EU, Report of the Ad-hoc Working Group on Defining Critical Raw Materials" (European Commission, June, 2010), www.euromines.org/files/what-we-do/sustainable-development-issues/2010-report-critical-raw-materials-eu.pdf (accessed March 15, 2017).
6 In 2011, antimony, rhenium, and germanium were identified as the most critical metals for German industry by the Institut für Zukunftsstudien und Technologiebewertung, which researched the issue for the Federal Institute for Geosciences and Natural Resources and the German Mineral Resource Agency. A second, less critical grouping of materials included rare earth elements, palladium, gallium, tungsten, silver, tin, indium, niobium, chromium, and bismuth.

7 European Commission, "Report on Critical Raw Materials for the EU, Report of the Ad hoc Working Group on Defining Critical Raw Materials" (European Commission, May 2014).
8 E. Alonso et al., "Evaluating Rare Earth Element Availability: A Case with Revolutionary Demand from Clean Technologies," *Environ. Sci. Technol.* 46:6 (2012): 3406–3414. Estimates of the exact amount of rare earth minerals in wind turbines vary, but the numbers are staggering. According to the *Bulletin of Atomic Sciences*, a 2-megawatt (MW) wind turbine contains about 800 pounds of neodymium and 130 pounds of dysprosium. The MIT study cited above estimates that a 2 MW wind turbine contains about 752 pounds of rare earth minerals. Taken from "Big Wind's Dirty Little Secret: Toxic Lakes and Radioactive Waste" (Institute of Energy Research, October 23, 2013).
9 Vidal, Olivier Bruno Goffé, and Nicholas Arndt. "Metals for a Low-Carbon Society," *Nature Geoscience* 6:11 (2013): 894–896.
10 European Commission, "Minutes of the US-Japan-EU Trilateral Workshop on Critical Raw Materials, (Brussels: European Commission, December 2, 2013), 2, http://ec.europa.eu/DocsRoom/documents/5663/attachments/1/translations/en/renditions/native (accessed June 16, 2017).
11 C. J. Moon, M. K. G. Whateley, and A. M. Evans (eds.), *Introduction to Mineral Exploration*, 2nd edition (Oxford: Blackwell, 2006).
12 See Richard Schodde, "Uncovering Exploration Trends and the Future: Where's Exploration Going?" (Melbourne: Presentation to the International Mining and Resources Conference, September 22, 2014), www.asx.com.au/asxpdf/20141114/pdf/42tq79kchgfhww.pdf, accessed June 16, 2017.
13 The International Conference on Mineral Exploration is held in Canada every decade, first in 1967 and most recently in 2017. The proceedings of the decennial conferences can be downloaded at www.dmec.ca/Resources.aspx (accessed March 15, 2017).
14 For example, Ray Durrheim was manager of the DeepMine collaborative research program (1998–2002). Its objective was to gain knowledge and developing technology to mine at ultra-depth, i.e., 3.5–5 km. Some of the outputs are currently being implemented. See, for example Ray Durrheim, "The DeepMine and FutureMine Research Programmes—Knowledge and Technology for Deep Gold Mining in South Africa. Challenges in Deep and High Stress Mining" (Australian Centre for Geomechanics, 2007), 131–141; Ray Durrheim, "Keynote Address: Has Research and Development Contributed to Improvements in Safety and Profitability of Deep South African Mines?" M. Hudyma and Y. Potvin (eds.) (Sudbury, Canada: Proceedings of 7th International Congress on Deep and High Stress Mining, 16–18 September 2014, Australian Centre for Geomechanics), 23–40, http://technology.infomine.com/openpitmine/articles.aspx.
15 Yves Potvin, John Hadjigeorgiou, and Dick Stacey (eds.), *Challenges in Deep and High Stress Mining* (Perth: Australian Centre for Geomechanics, 2007).
16 G. Herrera, R. Tomás, F. Vicente, J. M. Lopez-Sanchez, J. J. Mallorquí, and J. Mulas, "Mapping Ground Movements in Open Pit Mining Areas Using Differential SAR Interferometry," *International Journal of Rock Mechanics and Mining Sciences* 47:7 (2010), 1114–1125; W. A. Hustrulid, M. Kuchta, and R. K. Martin, *Open Pit Mine Planning and Design Two Volume Set & CD-ROM Pack* (Boca Raton, Florida: CRC Press, 2013).
17 See, for example, the guidelines published by The International Council on Mining and Metals (ICMM), www.icmm.com/en-gb. See also ICMM, "Approaches to Understanding Development Outcomes from Mining" (ICMM, 2013); ICMM, "Land acquisition and resettlement: Lessons learned" (ICMM, 2015); ICMM, "Shared Water, Shared Responsibility, Shared Approach: Water in the Mining Sector" (ICMM, 2017).
18 www.miningweekly.com/article/africa-attracts-14-of-worlds-2017-exploration-budget---sps-2018-03-05 (accessed June 12, 2018).
19 Taylor Jackson and Kenneth P. Green, "Fraser Institute Annual Survey of Mining Companies 2016," 11, 15.

20 John Kane-Berman, "The Slow Destruction of our Mining Industry," *Politicsweb*, 5 March 2017, www.politicsweb.co.za/opinion/the-slow-destruction-of-our-mining-industry (accessed March 15, 2017).
21 Taylor Jackson and Kenneth P. Green, "Fraser Institute Annual Survey of Mining Companies 2016," 15.
22 Ibid., 11, 15.
23 The previous Director of the Geological Survey of Namibia, Dr Gaby Schneider, hosted the Resourcing Future Generations workshop at Goche Ganas in July 2015 and participated in the workshop, as did the Namibian Minister of Mines and the President of the Namibian Chamber of Mines, https://voices.nationalgeographic.org/2015/08/27/resourcing-future-generations/. Kinnaird and Durrheim have present collaborations with the Geological Survey of Namibia. Durrheim is Co-Director of the AfricaArray research and capacity-building network. Since February 2015, AfricaArray and the NGS have been jointly running of broadband seismograph network.
24 www.chamberofmines.org.na/index.php/home-menu/mining-industry-performance-2015/ (Accessed June 12, 2018).
25 See for example, http://ewn.co.za/2016/02/11/Namibia-aims-to-pass-new-mining-empowerment-laws-by-April-2017; www.reuters.com/article/africa-mining-namibia-idUSL3N15Q175; and www.chamberofmines.org.na/files/5014/6979/6192/MiningCharterFINAL19September2014.pdf.
26 See World Nuclear Association, "Uranium in Namibia," www.world-nuclear.org/information-library/country-profiles/countries-g-n/namibia.aspx (accessed March 15, 2017).

7 Challenges in measuring the local and regional contributions of mining

Lessons from case studies in Rwanda, Zambia, and Ghana

Julia Horsley, Shabbir Ahmad, and Matthew Tonts

In brief

- This chapter focuses on the challenges in quantifying the local and regional contributions of mining sector development across time and space.
- One of several ways to analyze drivers of inclusive approach is to analyze artisanal and small-scale mining sector case studies.
- Quantitative methods can be used for benchmarking local performance of mining companies and supply chain actors.
- National systems of innovation, skills development, production, employment, and income linkages could lead to overcoming current limiting factors to linkages development.
- Increases in mining sector output is expected to stimulate other sectors of local and regional economies.

Introduction

Since 2010 academic literature on mining activities in Africa has increasingly focused on China and other Asian countries.[1] The influx of foreign investment starting in the mid-2000s has led to an unprecedented rise in sub-Saharan mineral production and an increased dependence on the extractives industry in the region. For example, in Burkina Faso the mining sector accounted for 41 percent of exports in 2010, up from 2 percent 5 years earlier; in Somalia, more than one-third of export revenues were generated by the mining industry in 2012, up from 5.4 percent in 2005.[2] Even with unstable commodity prices, the long-term access to large reserves of metals and minerals granted to multinational corporations ensures that mining activity will remain the dominant sector of many African economies.

New frameworks have been developed to assess the macroeconomic impact of this investment on sub-Saharan Africa's trade, investment, and governance, among other areas.[3] The far-reaching structural reforms facilitated by the World Bank have been credited with transforming the resource "curse" into a blessing in countries such as Rwanda, Zambia, and Ghana, at least in terms of national economic indicators. But if resource wealth is to achieve sustainable and inclusive development objectives, there must be sound, reliable methods for measuring and monitoring its impacts across space and time. The impacts of mining usually reported pertain to production, investment, employment, taxation and royalties, and direct effects on the economy. These indicators are important in understanding the contribution that the mining industry can make to local, regional, and national economies.[4] But there is a growing demand for more-grounded and more-comprehensive analyses at the local and regional levels, as demonstrated by the multitude of so-called gray literature on the socioeconomic impacts of mining and the plethora of academic literature on the resource curse. Public policy initiatives, private sector reporting, and humanitarian concerns raised by international agencies and community interest groups have served to either highlight or reflect the limited availability of reliable local and regional data.

This chapter explores the challenges in measuring the local and regional contributions of mining that must be addressed when planning for inclusive growth in extractives development; it is based on case studies undertaken in Rwanda, Zambia, and Ghana. We begin by setting out traditional approaches to analyzing local and regional development impacts of mining, then present empirically based approaches to measuring linkage effects of operations in Africa. After reviewing the respective macroeconomic performance, mineral portfolios, and extractive industries structure in each of the three countries studied, we focus on supply chain value methodology and analysis in Rwanda. We then outline the findings of separate fieldwork using Q-sort methodology in all three countries, aimed primarily at developing an inclusive framework for gathering data at the local and regional scales based on the "five capitals" approach, discussed below.

The Rwandan case study, while also gathering community views, illustrates that the drive for more-diverse and more-inclusive approaches need not entail a complete departure from traditional methodologies. Methods such as input–output analysis, for example, still provide a sound underpinning and benchmark for comparative analysis that can be used in conjunction with newer approaches focused on economic data such as supply chain value analysis, as well as integrated into frameworks aimed at gathering broader sets of socioeconomic data relating to sustainable development, livelihoods, and community well-being. We conclude this chapter by discussing the key challenges and potential lessons illustrated by each of these case studies, and make suggestions for further research.

Traditional regional development perspectives on identifying and measuring the impacts of mining

The typical mineral project of around a hundred years ago, at least in developed countries, was a relatively simple and not very large enterprise intended to serve the industrial demand of nearby communities. With adequate capital and labor available within the region, the local employment effect of such projects was considerable. These conditions, together with a relatively low level of national taxation, allowed most of the output value to accrue to the local community. Measuring the impacts of mining on development was therefore a relatively straightforward task. The direct contribution to development consisted of the value added by the minerals venture and spent to compensate labor, capital, and the entrepreneurial effort, or to satisfy fiscal agents. Indirect or external development effects, then as now instructively organized around the concept of linkages, were more readily identifiable because of the scale and localized structure of operations.[5]

One of the primary methods of measuring linkages in regional development studies is input–output analysis, originally based on the pioneering work of Wassily Leontief.[6] An input–output table is a matrix showing the multiplier effects of the impact on other industries per unit change of output in another industry, as well as on labor use, imports, and final demand. That table provides the basis for calculating the impact of a change in the output of one industry on the supply of backward-linked producers and, simultaneously, on the production of forward-linked industries that use the originating outputs as inputs.[7]

In the extractive industries context, various elaborations on the notions of linkages and input–output analysis had their own strand of literature. For example, the staple thesis first developed by Harold Innis claimed development and economic growth would occur as a diversification process around staples, including minerals, through investment in the resulting downstream and upstream industrial linkages.[8] Innis and others emphasized, however, that such diversification and growth were not automatic.[9] Instead, they are the product of intertwining forces exerted by geography, institutions, and technology.[10]

In core industrialized economies and in some staple economies, backward linkages created by traditional mining ventures arise through various requirements for inputs, enhancing the prospects of both established and new industries to produce goods and services necessary for the ventures' operation. Forward linkages arise from the further processing of the raw material output, potentially involving a more extended chain than backward linkages. The establishment of infrastructure for mining can benefit other industrial or agricultural projects as well, a result sometimes referred to as a sideways linkage.[11] A simplified representation of potential impacts of traditional mining on 'development' at the local and regional scale can is summarized in Figure 7.1.

New perspectives on the relationship between mining and development: global trends and structural transformations

The technology, industry structure, and geographical scale of mineral resource exploitation underwent significant restructuring during the twentieth century and even more so in the early years of the twenty-first century. As changes continue to proliferate rapidly in both developed and developing countries, measurement of the developmental impacts of mining, particularly at the local and regional

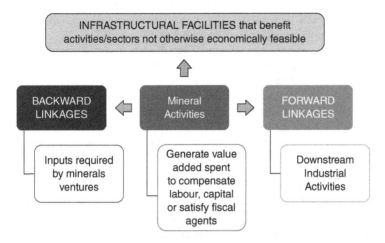

Figure 7.1 The development impacts of mining
Source: authors' summary of linkages described by Radetzki (1982)

levels, is becoming as complex as the plethora of conceptual shifts in the notion of development itself.

Parallel with the restructuring of the extractives industries, two other important changes in the global development landscape increased the challenges of its measurement. The first was a significant paradigm shift beyond narrow economic definitions of development. The second and related change is emerging recognition from within the mining industry of the need to demonstrate responsible social and environmental practices, which requires measurement of developmental impacts of mining at multiple spatial scales.[12]

The notion of development that rose to prominence after World War II was driven partly by the agitation for political independence in countries that had been colonized by Britain, France, Portugal, and others, including many countries in sub-Saharan Africa that would develop mineral-dependent economies.[13] The modernization approach, with its underlying assumption that the industrialization of developing countries should follow the trajectory of industrialization in the West, was implemented through encouragement of trade in commodities and through a series of interventions involving the transfer of technology, knowledge, resources, and organizational skills from the developed to the developing countries.[14]

When the expected prosperity did not materialize from the 1950s onward, and the poor became poorer in some cases, a theory of dependency arose as an alternative to the modernization approach. In essence this theory held that, although political colonization had been dismantled, effectively it had been replaced with a neocolonialism that kept exploitation of the core–periphery very much intact.[15] The development of natural resources as a way to emulate the economic growth of developed countries, which had been advocated by multilateral institutions such as the World Bank, the International Monetary Fund (IMF), and the United Nations, increasingly came under fire.[16] This pessimism culminated in the development of "resource curse" literature that asserted that natural resource endowment had a negative impact on economic growth rather than a positive one.[17]

Although dependency theory was influential, the modernization approach continued to dominate until at least the 1980s, when the advent of neoliberalism led to the introduction of structural adjustment programs to help countries deal with economic crises. Concurrently, issues such as the importance of participation, credit, and gender came to the fore.[18] Finally, the notion of sustainable development began to dominate the global agenda from the late 1980s; that notion, along with the introduction of the Human Development Index in 1990 and the ensuing bottom-up approaches to development such as the sustainable livelihoods framework, gave rise to a burgeoning literature on assessment frameworks and indicators for measuring development.[19]

Alongside these global trends in conceptualizing development were significant changes to the drivers and shapers of mining company activity that required innovative approaches to measuring the impacts of such activity. For most of the twentieth century, the mining industry focused almost exclusively on the need for profits.[20] Although companies sometimes provided social services such as housing, overall the main impacts were squarely within the economic pillar of development as governments generally let mining companies go about their business unimpeded as long as they generated foreign exchange, tax revenues, and employment.

Beginning in the 1970s, new environmental regulations required the industry to develop technologies to comply with emissions, pollutants, and tailings management. Decision making increasingly had to take into account the environmental pillar, leading companies to establish new environmental departments and to focus more attention on community relations. Social and cultural issues driven by global trends and changing public opinion caused the social pillar to become a sphere of strategic importance for the industry from the mid-1990s.[21] By the early 2000s, large mining companies widely adopted the practice of corporate social responsibility, which required them not only to provide evidence of direct community benefits but also to measure sustainable development outcomes over the long term.[22]

The view that sustainable development is the "ultimate culmination of development theories," is plausible, given the theory's potential for more-inclusive approaches to development outcomes, among other factors.[23] But, paradoxically, the very strength of acknowledging multiple perspectives also serves to create insurmountable challenges to sustainable development's consistent and coherent measurement. In the context of mining and regional development, inconsistencies in indicator selection across varying scales have frequently led to conflicting, or even directly opposing, assessments of the impacts of particular mining activities.[24] Furthermore, local communities most often were not directly asked by international agencies, national governments, or mining companies for their opinion on which indicators of development were the most important or relevant to measure.[25]

The rise of foreign direct investment (FDI) from China has led to further changes in the business model for the mining industry in Africa through its practice of offering large infrastructure investments in return for access to mineral deposits.[26] This paradigm shift has led to further scrutiny and questions on how mining can act as an engine of community and regional development, particularly in sub-Saharan Africa. In essence, the dominant strategic driver of the mining industry and the development priorities of governments has returned to regional economics but, as McMahon and van de Veen note, it is a much more complex type of economics than thirty years ago, encompassing local, regional, and national scales.[27]

This paradigm is not fundamentally different from the one that governed traditional regional development approaches, in which mining operations were expected to generate upstream and downstream linkages, rather than enclaves, that would produce new industrial clusters and economic diversification. While not discounting the complexity of and need for more-inclusive approaches in the current development context, the renewed focus on regional development linkages provides a useful thread to trace the vertical and horizontal linkages of a mining operation and its source of FDI, and provides needed empirical evidence on the extent to which the rapid expansion in mining revenues in many African countries has contributed to economic growth and socioeconomic development at local and regional levels.

Current approaches to measuring resource-sector investment linkages in Africa

The widespread restructuring of the mining industry in the 1980s has in some respects increased the potential for expanding linkages, at least for large-scale mining projects. For example, the shift from internal company structures to horizontally structured operations would seem to offer opportunities for more local and regional input (Figure 7.2).

As yet, there is limited empirical evidence on the full extent to which the rapid expansion in mining revenues in many African countries have contribute to economic growth and socioeconomic development at local and regional levels. Bloch and Owasu note that the linkage concept is, from some perspectives, dynamic

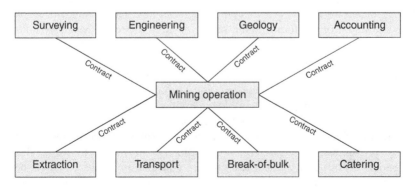

Figure 7.2 Transition from internalised firm structure to horizontally structured operations

rather than static, with linkages either decaying or becoming enhanced over time.[28] They quote Hirschman's definition of linkage effects:

> The linkage effects of a given product line [are] investment-generating forces that are set in motion, through input–output relations, when productive facilities that supply inputs to that line or utilize its outputs are inadequate or nonexistent. Backward linkages lead to new investment in input-supplying facilities and forward linkages to investment in output-using facilities.

Hirschman added two further types of linkages for commodity production: fiscal linkages, in the form of state taxation of the income streams associated with the commodity; and consumption linkages, represented by income (profits and wages) emanating from commodity production that was spent nationally and in the local vicinity on the outputs of domestic industries.[29]

While more optimistic about the prospects of linkages in other sectors, Hirschman, building on Singer, argued that the natural resource sector was an enclave in nature and had weak positive externalities, a view later supported in part by "resource curse" literature.[30] Thus, in contrast to the employment and industrial linkages emphasized in earlier regional development models, FDI became the most significant mining linkage or potential revenue stream.

The possibility that linkages could offer growth prospects for local firms was not one that enclave proponents embraced, according to Fessehaie and Morris. Such linkages and their governance in terms of large-scale operations, however, are a primary focus of global value chain (GVC) analysis. Upgrading in the GVC framework refers to firms' abilities to defend and sharpen their competitiveness by improving their method of production (process upgrading) and what they produce (product upgrading), and by either moving into new links in the value chain (functional upgrading) or into more remunerative chains (chain upgrading).[31]

In the context of GVC analysis, mineral-based value chains have been inadequately studied, according to Morris, Fessehaie, and others.[32] There is also a paucity of empirical work on the extent to which value chain dynamics and governance are shaped by firm ownership (specifically Chinese- and Indian-led firms) and social and institutional contexts.[33]

Fessehaie has investigated the role of public policy and value chain dynamics in shaping backward linkages to Zambia's copper mining sector. In Zambia the breadth and depth of the local mining supply chain was deeply shaped by policies adopted in the 1990s under the structural adjustment program. These policies attracted much-needed FDI in the mining sector, including Chinese and Indian FDI, but reduced the level of value addition undertaken by local suppliers. Forward linkages to buyers other than Chinese and Indian mining companies and backward linkages to parent companies and technology providers were found to be critical in supporting supply firms' success in the mining value chain.[34]

In Ghana gold mining has often been depicted as isolated from the rest of the economy. In contrast, the research findings of Bloch and Owasu demonstrate that after a period of strong investment and growth, gold mining could no longer be viewed as an enclave activity: rather, they believe that it is deeply woven into the Ghanaian economy through a set of as-yet under-researched but promising economic linkages that can potentially be strengthened by policy and support measures. These linkages take various forms, including principally backward linkages, final demand, and consumption linkages. Fiscal linkages, meanwhile, have also been strengthened. The linkages are also manifested spatially in the form of visible and differentiated clusters (i.e., geographic or sectoral agglomerations of enterprises) of mining activity, which appear to benefit in different ways from external economies of scale (agglomeration economies), notably the localization economies variant.[35]

Furthermore, extensive empirical research on linkages in and out of the commodities sector and six other sectors (copper, diamonds, gold, oil and gas, mining services, and timber) in eight sub-Saharan African countries indicates great potential for industrial development in the countries studied. Researchers identified a number of linkage industries related to resource extraction that developed in tandem with mining operations; these include processing firms, specialized machinery plants, makers of engineering products, providers of transport services and equipment, and software companies.[36]

Mining activities and trade in minerals by small companies and artisanal miners are also increasingly being recognized for their potential contribution to growth and income generation, particularly at local and regional levels. Reforms have been introduced in many African countries to formalize the artisanal and small-scale mining (ASM) sector in addition to exploring strategies that create additional linkages and diversify economic activity. The ASM sector employs more than 10 million people in sub-Saharan Africa, but its development potential has been mostly overlooked by regional and local development policies, which tend to favor large-scale mining operations.[37] The sector is characterized by complex labor hierarchies, unique forms of production, and informal systems of assistance, which have evolved in an ad hoc manner in a largely unregulated environment.[38] While the nature of ASM puts it outside the scope of GVC analysis, the basic tenets of supply chain value analysis are still applicable. Researchers are considering how to construct ethical mineral supply chains in sub-Saharan Africa through policies that focus more on increasing value and benefits to ASM miners and other suppliers along the chain—in line with the fair trade principles currently applied in agriculture that are aimed at facilitating market access—rather than on maintaining transparency and eliminating corruption in mineral revenue streams.[39]

The next section reviews empirical evidence on the value of ASM activity by examining a case study conducted in Rwanda at the district level; this study takes into account local and regional development multiplying effects. The discussion forms the basis of a comparative analysis of new approaches to evaluating the linkages and distributions of supply chain value in Rwanda, Zambia, and Ghana.

Macroeconomic performance, mineral portfolios, and the structure of extractive industries in Rwanda, Zambia, and Ghana

As shown in Table 7.1, from 2007 to 2014, the economies of Rwanda, Zambia, and Ghana grew as a result of investment in their extractive industry sectors and

Table 7.1 Selected macroeconomic indicators for Ghana, Rwanda, and Zambia

Ghana	2007	2008	2009	2010	2011	2012	2013	2014
GNI Per capita (US$)	800	1170	1210	1260	1410	1570	1740	1590
GDP growth	4.3	9.1	4.8	7.9	14	9.3	7.3	4.0
Agriculture, value added (% of GDP)	29.7	31.7	32.9	30.8	26	23.6	23.2	22.4
Industry, value added (% of GDP)	21.2	20.9	19.7	19.8	26.2	28.9	28.7	27.7
Services, value added (% of GDP)	49	47.3	47.4	49.4	47.7	47.5	48.1	49.9
Exports of goods and services (% of GDP)	24.5	25	29.3	29.5	36.9	40.4	34.2	39.5
Imports of goods and services (% of GDP)	40.8	44.5	42.3	45.9	49.4	52.8	47.5	48.9
FDI, Net flows (US$ Million)	1383	2715	2373	2527	3248	3295	3227	3363
Rwanda								
GNI Per capita (US$)	360	440	500	550	590	640	670	700
GDP growth (%)	7.6	11.2	6.3	7.3	7.9	8.8	4.7	7.0
Agriculture, value added (% of GDP)	35.1	32.7	33.9	32.6	32.3	33.4	33.4	33.1
Industry, value added (% of GDP)	12.4	12.6	12.3	12.9	14.4	14.4	14.9	14.4
Services, value added (% of GDP)	52.5	54.7	53.8	54.6	53.3	52.2	51.7	52.5
Exports of goods and services (% of GDP)	15.9	12.7	11.9	12.1	14.4	14.1	15.6	14.9
Imports of goods and services (% of GDP)	24.5	27.8	26.8	28	29.6	32	30.7	30.5
FDI, Net flows (US$ Million)	82.3	10.3	119	42.3	106	160	258	292
Zambia								
GNI Per capita (US$)	880	1160	1260	1310	1400	1650	1700	1680
GDP growth	8.4	7.8	9.2	10.3	6.3	6.7	6.7	6.0
Agriculture, value added (% of GDP)	13.2	12.5	12.4	10.5	10.2	10.3	9.6	---
Industry, value added (% of GDP)	34.9	33.9	32.4	35.5	35.9	34.4	33.9	---
Services, value added (% of GDP)	51.9	53.6	55.2	54	53.9	55.3	56.5	---
Exports of goods and services (% of GDP)	33.6	28.9	29.3	37	38.1	42.1	43.3	40.9
Imports of goods and services (% of GDP)	32.2	30.5	26.9	30.9	31.8	37.1	41.1	37.8
FDI, Net flows (US$ Million)	1324	939	695	1729	1109	1732	2100	1508

Source: World Bank, "World Development Indicators," 2015

revenues from mineral exports. Per capita gross national income (GNI) steadily increased, and FDI inflows were also significant, more than doubling in Rwanda and Ghana. These growth patterns, however, have since been tempered with commodity price decreases, fiscal and current deficits, inflationary pressure, and currency depreciation.

Table 7.2 shows the Human Development Index ratings of each of the three countries from 1980 to 2013 and provides regional and categorical comparisons. Rwanda made the most dramatic progress in human development according to this index, with a rating that nearly doubled from 0.291 in 1980 to 0.506 in 2013. Zambia showed significant improvement in human development starting in 2005, according to its rating. Ghana also grew during the thirty-three-year span, ranking the highest among the three countries with a rating of 0.573 in 2013.

Figure 7.3 shows the relationship between political risk and operating environments in selected African countries based on a report in 2013.[40] Rwanda, Zambia, and Ghana perform better than many other African countries in these rankings. Rwanda in particular shows the best operating environment vis-à-vis other

Table 7.2 Human Development Index for Rwanda, Zambia, and Ghana

Countries	1980	1990	2000	2005	2008	2010	2011	2012	2013
Rwanda	0.291	0.238	0.329	0.391	0.432	0.453	0.463	0.502	0.506
Zambia	0.422	0.407	0.423	0.471	0.505	0.53	0.543	0.554	0.561
Ghana	0.423	0.5.2	0.487	0.511	0.544	0.556	0.566	0.571	0.573
Low Human Development Countries	0.345	0.367	0.403	0.444	0.471	0.479	0.486	0.49	0.493
Sub-Saharan Africa	0.383	0.399	0.421	0.452	0.477	0.488	0.495	0.499	0.502

Source: "Human Development Report," 2014

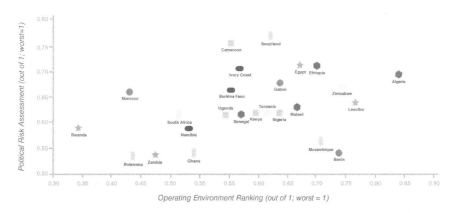

Figure 7.3 Relationship between political risk and operating environment Africa
Data sources for this graph: KPMG (2013) PRS, WB

countries in the region, while Benin, Botswana, Ghana, and Zambia have significantly less political risk compared with countries such as Swaziland and Cameroon.

Evaluating the supply chain value of ASM in Rwanda: input–output analysis and multiplier effects

In order to harness the full potential of production outcomes of the growing mineral sector in countries such as Rwanda, it is essential to measure the revenue distribution and multiplier effects among national, regional, and local supply chains, and carefully analyze future growth models. ASM is the dominant extraction and production mode in Rwanda, and the supply chain due diligence costs associated with recent developments in international regulations are significantly high. As a result, the significant trends in Rwanda's mineral sector raise the question of how revenues from mining operations and mineral trade are distributed along supply chains within the country and among diverse local and regional stakeholders.

As shown in Table 7.3, the mining sector's contribution to GDP in Rwanda more than doubled between 2009 and 2014.

Figure 7.4 shows annual exports of cassiterite (used for tin), wolframite (used for tungsten), and coltan (used for tantalum), or the 3Ts, compared to exports of traditional items such as coffee and tea. From 2001 to 2013, exports of the 3Ts increased from $50 million to $228 million, surpassing exports of coffee and tea.

Figure 7.4 shows the annual quantities of mineral exports from 1998 to 2013, which grew at an average annual rate of 2.6 percent. Coltan exports grew the fastest, at an average rate of 12.9 percent each year. By the end of 2013, exports of all three mineral commodities reached about 9,580 metric tonnes, and export revenues from the three commodities totaled $2,25.7 million. If these exports continue to grow at 2.6 percent per year, the total quantities exported will reach 10,881 metric

Table 7.3 Sectoral contribution to Rwanda's GDP (%) 2009 and 2014

	2009	2014
Agriculture, forestry, fishing, and hunting	36.2	35.0
of which fishing	0.4	0.4
Mining and quarrying	0.8	1.9
Manufacturing	5.6	5.1
Electricity gas and water	0.6	0.7
Construction	6.0	7.5
Wholesale and retail trade; repair of vehicles household goods; restaurants and hotels	16.3	15.5
of which hotels and restaurants	3.0	2.3
Transport, storage and communication	5.5	5.8
Finance, real estate and business services	18.0	14.8
Public administration and defense	2.8	3.6
Other services	8.2	10.1
Gross domestic product at basic prices/factor cost	100.0	100.0

Source: "Rwanda Establishment Census," 2011

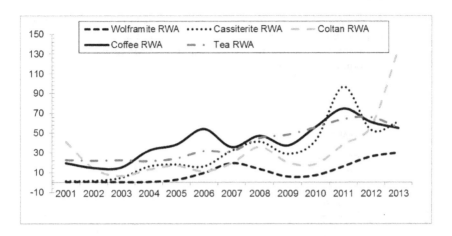

Figure 7.4 3Ts exports in Rwanda

Source: Adapted and modified from R. Cook and P. Mitchel, "Evaluation of Mining Streams and Due Diligence Implementation Costs Along Mineral Supply Chains in Rwanda" (Public Research Report for BGR and development partners, 2014)

Table 7.4 Projections of export quantity (metric tonnes)

Year	Cassiterite	Coltan	Wolframite	Total
1998	188.68	198.88	122.14	509.89
1999	529.00	330.00	84.00	943.20
2000	365.04	602.77	144.00	1111.81
2001	554.85	1540.21	161.56	2256.61
2002	672.07	1086.32	324.69	2083.08
2003	1458.00	732.00	120.00	2310.00
2004	3553.18	861.05	157.52	4571.74
2005	4531.83	1061.64	557.02	6150.49
2006	3835.33	724.25	1435.57	5995.15
2007	4565.91	968.96	2686.11	8220.98
2008	4193.29	1190.33	1708.04	7091.70
2009	4269.17	949.92	874.45	6093.57
2010	3874.20	748.72	843.42	5466.35
2011	6952.07	890.08	1006.24	8848.38
2012	4636.64	1144.68	1750.57	7531.89
2013	4895.27	2466.02	2217.93	9579.22
CAGR	2.61%	12.91%	4.45%	2.58%

Source: Authors calculation applying the formula to compute average growth of exports of three commodities $X_t = X_{t-1} \times \left(1 + \overline{dX_t / dt}\right)$, Where, X_t and X_{t-1}, represents the exports in current and preceding year and $\overline{dX_t / dt}$, denotes average growth rate for the entire period.

tonnes by the end of 2018. Table 7.4 also includes a projection from 2014 to 2018 based on this growth rate of the export quantity for each of the commodities based on a seven-year annual growth rate. By the end of 2013 the total export revenues from all commodities were recorded as $225.7 million dollars.

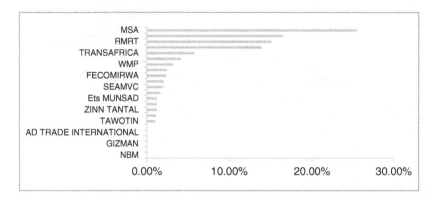

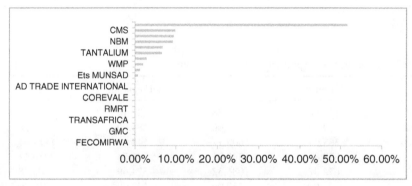

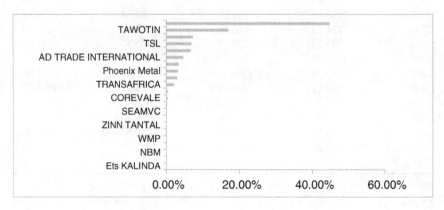

Figure 7.5a–c a) cassiterite exports of top twenty-three mining companies, 2013; b) coltan exports of top twenty-three mining companies, 2013; c) wolframite exports of top twenty-three mining companies, 2013

Source: Authors' calculations from RNRA data, 2013

 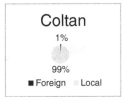 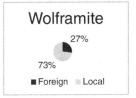

Figure 7.6 Share of foreign and local mining companies in 3Ts production source: authors calculations from company level data

Source: *RNRA*

The value of various companies' 3T exports in 2013 appears in Figure 7.5a–c. That year, the top 23 mining companies in Rwanda exported about 8.19 million kilograms of the 3Ts, with cassiterite comprising about half of the exports (4.17 million kilograms) and the remainder being divided between coltan (2.16 million kilograms) and wolframite (1.86 million kilograms). MSA was responsible for about 37 percent of the total 3T exports: 26 percent of the cassiterite, 52 percent of the wolframite, and 45 percent of the coltan. In comparison, Phoenix and Rutongo mines contributed 10 percent and 8 percent to mining exports, respectively. Figures 7.5a–c provide a more detailed analysis. Rutongo mines, for instance, contributed about 16 percent of cassiterite exports, and Tawotin made about 17 percent of the total coltan exports. Phoenix was responsible for 14 percent and 9 percent of the total cassiterite and wolframite exports, respectively.

Figure 7.6 compares the share of foreign and local companies in the production of each of these commodities. As shown, 83 percent of cassiterite (tin) is produced by local companies. Likewise, 73 percent of wolframite (tungsten) and 99 percent of coltan (tantalum) is produced by local companies. Of the 6,842 tonnes of coltan produced by 175 companies, Rutongo was responsible for about 11 percent (777 tons). Similarly, of the 2,251 tonnes of wolframite produced by 73 companies, EUROTRADE International was responsible for 11.3 percent (255 tonnes).

In 2013 all three mineral commodities generated approximately $366.7 million in revenue, based on the average export price of each commodity. The total monthly revenues based on production of all values are shown in Figure 7.7. We calculated the average prices based on monthly totals at four different stages including (1) opening price, (2) closing price, (3) highest price, and (4) lowest price for each respective commodity. According to these calculations, the average prices for the year 2013 were cassiterite at $22.17 per kilogram, coltan at $115.21 per kilogram, and wolframite at $43.46 per kilogram.

The total revenues derived from the production of all three commodities reached $366.7 million in 2013: cassiterite at $151.7 million, coltan at $117.6 million, and wolframite at $97.5 million. The revenue growth rate was 1.8 percent monthly as calculated using the compound growth rate formula $\sqrt[n]{R_t / R_0} - 1$, where n is the number of years, R is the revenue in the current period (t) and is the base period (0). Cassiterite and wolframite grew at an average rate of 2.8 percent

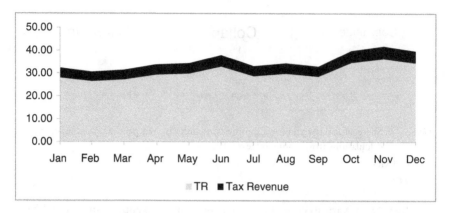

Figure 7.7 Estimates of monthly revenue and taxes for all commodities—2013
Source: "RNA Statistics," 2013

and 2.2 percent, respectively, while coltan showed a negative growth in revenue (–0.8 percent).

The case study was based on a concept developed by the Geology and Mines Department (GMD) of the Rwanda Natural Resources Authority (RNRA) in partnership with the German Federal Institute for Geosciences and Natural Resources (BGR). In addition to the original focus on supply chain analysis, and as requested by the Rwandan government, we identified and evaluated potential mineral supply chain efficiencies and solutions to contribute to a clearer understanding of the sector's growth and development potential.

The case study report findings are based on field research carried out from May to June 2014 in four districts, and includes twenty different 3T mine sites. We selected the mining companies and cooperatives, as well as their sites, as a representative sample of Rwanda's mining sector, covering the whole spectrum from small companies and cooperatives with manual operations to relatively large foreign companies running semi-industrial mines. The case study report was supplemented by research carried out at six mineral traders and exporters, four district offices, and with relevant national authorities. On-site interviews took place with 285 miners (in focus groups) and with mine management at each site. We interviewed a range of other interlocutors, including individuals from various other government agencies, international development partners, local government, local communities, and the private sector, such as the Rwanda Mining Association (RMA).

The study area and district profiles

We undertook a case study in Rwanda with the aim of considering the entire supply chain: (1) miners (both employees and contractors) at the extraction site, (2) internal traders, (3) the mining company/cooperatives, (4) upgrading facilities for the mineral concentrates, (5) aggregation at the export stage, and (6) downstream buyers beyond

Rwanda's borders, with a focus on local and regional multiplier effects primarily through using a form of input–output analysis. The qualitative and quantitative data used in this analysis were collected from four districts: Rutsiro, Muhanga, Rulindo, and Kayonza (see Figure 7.8).

Rutsiro: Rutsiro District has a population of 326,000, representing about 13.7 percent of the western province population and 3 percent of Rwanda's population. The population is 52.4 percent female and 47.8 percent male. Census statistics show that half of the district population is poor, and it ranked as the third-poorest district in the province. The statistics also reveal that only 60 percent population of the district has access to improved drinking water, of which half use protected spring water. The overall employment of the population aged 16 years and above in the district is 86 percent; 63 percent of the population has major source of income from farm activities. On the other hand, only 0.4 percent of the population has access to electricity for lighting purposes, which is well below that of the rural areas generally (4.8 percent) as well as the national level (10.8 percent).

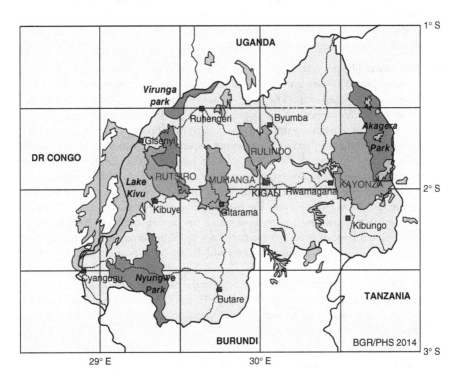

Figure 7.8 Rwanda study area map

Source: R. Cook and P. Mitchell, "Evaluation of Mining Revenue Streams and Due Diligence Implementation Costs along Mineral Supply Chains in Rwanda" (Public Research Report for BGR and development partners, 2014)

Muhanga: Muhanga District has a population of approximately 319,141, representing about 12.3 percent of the southern province population and 3 percent of Rwanda's population. The population is 52 percent female and 48 percent male. The majority of the district population lives in rural areas (67.7 percent). The district has higher labor participation rate in rural areas (80.8 percent) as compared to urban areas (71.2 percent). The unemployment rate in rural areas was much lower (1.4 percent) than urban areas (6.5 percent). Agriculture activities are main contributors to economic activities of the local economy. Other economic activities include trade (7.3 percent), construction (2.1 percent), and mining (1.2 percent). Health and education facilities are sufficient and most of the population (91.1 percent) is covered for health insurance. Primary and secondary school enrolment rate is 90.7 percent and 20.8 percent, respectively. The overall literacy rate is 71.1 percent in the district.

Rulindo: The population of the Rulindo District is approximately 288,452, representing about 16.7 percent of the southern province population and 2.7 percent of Rwanda's population. The district employment rate is higher than the national employment rate: About 87 percent of the population above 16 years is employed, compared with 80 percent at the national level. Rulindo has very low unemployment rate (0.4 percent) as compared to national unemployment figures (2.4 percent). Moreover, about 75 percent of the district population has access to improved drinking water; Rulindo district ranks fourth in safe water access among Rwandan districts.

Kayonza: The population of Kayonza District is approximately 332,000, representing about 13.3 percent of the southern province population and 3.3 percent of Rwanda's population. The population is 52 percent female and 48 percent male. About 55 percent of the population is aged 19 years or younger. Almost 83 percent of females are under 40 years. Twenty-three percent of the population is poor and the percent live in extreme poverty. Statistics on basic facilities indicate that 72 percent of the population have access to improved drinking water, obtained from springs, piped water, boreholes, or public standpoints. However, only 7.5 percent of households use electricity as a source of lighting in the district, which is above the rural average compared with other districts. The urban population, on the other hand, has higher access (46 percent) than the national level of access to electricity for lighting purposes. The unemployment rate is also high when compared with many other districts, at 2.4 percent.

Figure 7.9 sets out a comparison of the percentage of population living in poverty across each Rwandan district. Kayonza and Rulindo have almost same level of population living under the poverty line (42 percent) whereas Rutsiro District has a higher level (52 percent).

Analysis of survey data, productivity, and income of local mining activities

Table 7.5 describes household characteristics involved in 3T mining activities across different value chain operations. The age of miners varies between 19 to 68 years,

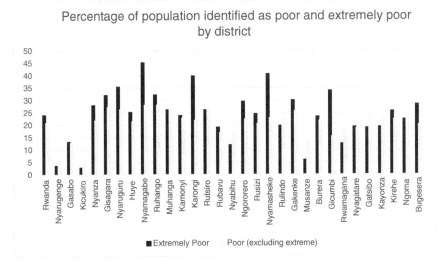

Figure 7.9 District level comparison of poverty

Source: "National Institute of Statistics Rwanda," 2011, EICV3 District Profile (Kigali–Gasabo)

with a median of 30 years. Sample statistics show that miners and other workers' experience vary between 0.1 to 36 years, with an average of 4.3 years. Average monthly production of all three minerals was approximately 37.9 kilograms with maximum quantity of 116.7 kilograms. The median income from mining activities was RWF 60,000, which constituted about 96 percent of the total income of households on average.

Table 7.6 provides comparative data on the median earnings of mine employees and subcontractors corresponding to their respective occupations within the mining operation. We computed median income based on different categories of mining activities and note that (1) miners earned a median per capita of RWF 79,000; (2) subcontractors earned approximately RWF 30,208; (3) washers earned RWF 50,000 per capita on average; (4) coop miners earned RWF 216,700 monthly; and (5) team leaders earned RWF 55,000 monthly per capita.

We also noted gender differentiation. We found male per capita earnings (approximately RWF 50,000) were on average five times more than their female counter parts (approximately RWF 10,500). Admittedly, our sample included only 9 female respondents, equating to only 1 percent of the total sample; however, general employment statistics disclose that the mining sector employment in Rwanda is dominated by male workers (84 percent) compared to female workers (16 percent). According to Rwanda Natural Resources Authority statistics, 33,638 workers were employed in the mining sector, of which 28,245 were male and 5,393 were female.

Table 7.5 Descriptive statistics of some household characteristics

Characteristic	Minimum	Maximum	Average	Median
Age (years)	19	68	33	30
Years working at site	0.1	36	4.3	2.5
Average production (kilogram/month) (as noted by miners)	0	116.7	37.9	32
Household size	1	11	4.7	4
Number of dependents	0	11	4.2	4
Family members doing mining work	0	10	0.3	0
Family members doing agricultural work	0	1	0.7	1
Income from mining (RWF/month)	0	450,000	81,300	60,000
Income from other sources (RWF/month)	0	240,000	23,100	25,00
% of total monthly income from mining	0	100	79	96
Typical expenditure on food (RWF/month)	0	240,000	28,600	20,000
Typical expenditure on rent (RWF/month)	0	20,000	235	0
Typical other expenditure (RWF/month)	0	214,000	19,400	0
Typical savings (RWF/month)	0	250,000	28,100	10,500

Source: Authors calculations from survey data

Table 7.6 Earning differentials of mine employees and sub-contractors

	Miners	Subcontractors	Team Leader	Washer
Mean	100058.78	35471.73	58428.57	86861.76
Median	79000.00	30208.33	55000.00	50000.00
Standard deviation	85868.75	32587.93	14729.30	79779.21
Minimum	0.00	4000.00	38000.00	22000.00
Maximum	480000.00	110000.00	84000.00	363000.00
Count	245	8	7	17

Source: Authors calculations from survey data

Data on household savings set out in Table 7.6 does not distinguish between those earned from mining activities and those derived from other non-mining business activity. Moreover, there seem to be some discrepancies in disclosure of mineral production because miners were not able to exactly quantify the production. For instance, monthly production disclosed by miners was higher than that calculated based on their income. Table 7.7 represents the monthly per capita income and expenditures of workers engaged in mining activities across different mining companies.

Looking at firm revenue and productivity maximization across different commodities, ideally we would have access to information on output, inputs, and output prices of respective commodities. However, in the case study, we only had access to commodities output and labor input for mining companies. Nevertheless, this allows us to derive some partial revenue and productivity measures at the firm level. For the purposes of this analysis, we made

Table 7.7 Averages income, expenditures, and age of miners working in mining companies

Company	Income	Expenditures	Age	Per capita income	Per capita expenditures
Alfa minerals—Nyanga	53090.91	33181.82	32.73	21762.99	12943.72
Cominya—Kinogi	30750.00	7000.00	41.00	5593.24	1312.88
Eurotrade—Nyakabingo	169909.09	79818.18	32.91	34445.89	16215.37
Excellent mining—Gituntu	46400.00	21000.00	30.07	16350.16	7287.70
Kalinda—Bahozi	22230.77	22230.77	41.62	5364.47	5364.47
Kuaka—Giseke	95866.67	24066.67	28.93	36628.10	9642.38
Multi-serve consult—Bugambira	51726.73	20733.33	33.93	27181.92	10530.16
Mushonyi mining—Biruyi	234466.67	87333.33	42.80	46858.71	16097.47
Rap—Kagano	62538.46	6615.38	35.77	11571.43	1177.38
Sindambiwe—Rongi	100000.00	45000.00	27.09	33257.58	14965.91
WMP—Gahengeri—B	106326.67	39533.33	37.67	23952.78	8677.78
Rutongo—No. 16	28200.00	28200.00	29.00	14453.97	14453.97
Comar	168080.00	77000.00	34.80	43758.48	16490.48
Cominyabu	216250.00	72437.50	35.81	49996.73	15297.02
Jasper minerals	80240.00	32666.67	30.53	19090.00	7553.65
Kamico	227200.00	130933.33	37.20	41975.40	22305.98
Nrd	115000.00	86705.88	30.12	23678.29	17904.48
Rap—Krimbali	76665.63	27187.50	28.00	34969.27	11692.71
Rwanda Rudniki	24941.18	22823.53	28.18	5398.32	5018.49
Vimico—Muntambara	120176.47	58294.12	29.82	28243.14	15340.69

Source: Authors calculations from survey data

the following assumptions: (1) All firms used the same technology within each mineral commodity. (2) Commodity prices remain constant for the year of calculation (2013). Total employment for each firm remains the same across all three commodities.

We considered data on output, output prices, and labor input in order to derive partial measures of productivity and revenues at firm level. We used the average export price to compute total revenue given the price variability in commodities. Since we are using the average price for all firms, it will not change the ranking of firm in terms of revenue efficiency (compared with productivity) because price effect will be cancelled using the formula given in the methodology. Using the revenue maximization formula as

$$Max\ Revenue = \frac{\bar{P} \times q}{\bar{P} \times q^{max}}$$

where $\bar{P}$, q, and q^{max} represent average mineral price, quantity produced and maximum possible quantity of minerals that can be produced using the same input resources, respectively. Therefore, we end up with the ratio of average productivity to maximum productivity of each firm. In other words, we compute the technical efficiency in relative terms. Dividing the observed revenue (productivity) with the maximum revenue (productivity) yields estimates within the range of 0 to 1. A value closer to 0 implies that the firm is least efficient or not maximizing the revenue, whereas value closer to 1 means that the firm is maximizing the revenue or productivity.[41] We use the linear programming to derive these measures of efficiency.[42] The linear program is written as the ratio of output to input(s) and we obtain optimal weights by solving the mathematical programming as

$$\max\nolimits_{u,v}(\mathbf{u'q}_i / \mathbf{v'x}_i) \text{ st } (\mathbf{u'q}_j / \mathbf{v'x}_j) \leq 1, \text{ where, } \mathbf{u, v} \geq 0,$$

where, u and v represent output and input weights for their respective output (q) and input (x) quantities. We used the single output and single input (labor) to estimate productivity/revenue maximizing of firms.

Based on our findings, Rwanda Mineral Africa was the most efficient with annual revenue of $17.22 million, followed by Ushahidi with total annual revenue of $6.92 million. Most of the mineral companies are not maximizing revenue given the available labor input. Average production and revenue of the sample of companies producing cassiterite with different employment levels of these companies are presented in Table 7.8.

Only small firms (with fewer than 100 employees) were able to achieve maximum efficiency in revenue and productivity (e.g., Havilla, Quincallerie Piano). We also found similar trends in coltan and wolframite minerals as companies are small (e.g., MINECA). The total revenue for cassiterite, coltan, and wolframite was noted as $85.4 million, $56.2 million, and $30.5 million, respectively. While the monthly average growth in the revenue remained positive in cassiterite (2.2 percent) and wolframite (2.9 percent), coltan showed a negative monthly growth (−4.4 percent) in 2013.

Mining companies with 100 or fewer employees earned the highest average annual revenue of $981,831.44, which was followed by mining firms with 101–300 employees and revenue of $889,470.21. Mining companies with 101–300 employees exhibited an average productivity of 235.46 kilograms. However, large firms with 301 employees and more earned the lowest annual average revenue (i.e., $697,258.35) and average productivity (i.e., 69.68 kilograms). However, these findings should be interpreted cautiously because in many months mining companies did not record any mineral output production. Those missing values might have simply gone unreported rather than mining companies not producing any minerals in that period.

We have also calculated the annual average productivity of three mineral commodities (Table 7.9). Estimates show that cassiterite exhibits highest labor productivity with an average of 16.95 kilograms per laborer, followed by wolframite with an average of 5.58 kilograms, and coltan with an average of 2.55 kilograms. Table 7.9 presents monthly average productivity and revenue of the 3Ts for all mining companies. While cassiterite and wolframite showed a positive growth in

Table 7.8 Annual average revenue (RWF) and productivity (cassiterite)

	Employees < 100		Employees 101–300		Employees > 300	
	Revenue	Productivity	Revenue	Productivity	Revenue	Productivity
Mean	981831.4475	1530.08	889470.212	235.46	697258.35	69.88
Median	303959.7275	284.88	450082.49	76.97	349453.39	37.27
SD	2263987.813	4984.80	1189149.633	363.15	970609.042	102.59
Kurtosis	39.52250364	40.08	1.41408E+12	10.11	3.56411499	7.55
Skewness	5.792303441	6.08	2.943541331	2.92	1.98488827	2.53
Minimum	664.95	1.50	398.97	0.07	664.95	0.07
Maximum	17220719.95	36996.81	4576822.036	1948.01	3584989.27	447.65
Sum	68728201.33	107105.89	50699802.08	13421.40	16734200.4	1677.01
Sample Size	70		57		26	

Source: Author estimates using firm level data, 2013

Table 7.9 Monthly average revenue and labour productivity

	Cassiterite		Coltan		Wolframite	
	Revenue (Million $)	Productivity	Revenue (Million $)	Productivity	Revenue (Million $)	Productivity
January	12.06	15.77	7.91	2.85	8.11	5.08
February	11.07	14.47	7.83	2.55	7.62	4.77
March	11.59	14.88	8.41	2.42	7.24	4.5
April	12.51	16.69	8.80	2.38	8.07	4.97
May	10.94	15.41	12.28	3.17	6.67	4.53
June	10.61	15.95	15.85	3.99	6.54	4.95
July	11.80	17.26	10.01	2.43	6.98	5.29
August	12.31	17.89	11.13	2.62	6.47	4.9
September	12.08	16.46	8.54	2.03	8.02	6.07
October	13.80	17.7	10.71	2.5	10.19	7.28
November	16.48	20.32	8.90	2.04	11.00	7.43
December	16.83	20.59	7.18	1.63	10.57	7.14
Average	12.67	16.95	9.80	2.55	8.12	5.58
Total Annual	151.65		117.55		97.48	
CAGR (%)		2.2		−4.4		2.89

Source: Author estimates using firm level data, 2013

productivity and revenue (i.e., 2.2 and 2.89 percent, respectively), coltan displayed a negative trend in monthly revenues and productivity.

Modelling revenue differentials across the supply chain

To investigate how income varies across different supply chains, various segments in the supply chain and socioeconomic characteristics in ASM production, a multivariate analysis can provide some insights to estimate the nature of the relationship between earnings and various socioeconomic factors. A simple multiple regression model can be used for this purpose:

$$Y_k = \beta_0 + \beta_l \sum_{l=1}^{L} x_{lk} + \gamma_m D_m + \epsilon_m$$

Where x_{lk} is the set of exogenous variables representing socioeconomic factors, and D_m is the vector of the dummy variables to capture the effects of different supply chain segments and $\epsilon \sim N(0, \sigma^2)$ Table 7.10 presents the ordinary least square (OLS) results for various specifications.

We found that the estimated model is overall significant. In particular, findings included the following: (1) Household size positively contributes to the income/revenues among different actors; (2) additional sources of income such as agricultural activities add to total earning individual well-being positively; (3) there are

Table 7.10 OLS estimates of revenue distribution differentials (income is a dependant variable)

	model 1	model 2	model 3	model 4	model 5	model 6	model 7
Lage	−0.067	0.386*	---	---	---	---	---
	(0.18)	(0.17)					
HH size	0.059*		0.084***	---	---	---	---
	(0.03)		(0.02)				
Agrwork	0.645***		---	0.867***	---	---	---
	(0.11)			(0.11)			
Oincdum	0.498***		---	---	0.695***	---	---
	(0.10)				(0.1)		
Exper	(0.025)		---	---	---	0.038	---
	(0.05)					(0.05)	
Sex	0.579*		---	---	---	---	0.42
	(0.26)						(0.3)
Type=2	0.46	0.459	0.48	0.514	0.507	0.49	---
	(0.31)	(0.3)	(0.28)	(0.28)	(0.31)	(0.31)	
Type=3	−0.286	−0.266	−0.494	−0.591	−0.3	−0.148	---
	(0.33)	(0.32)	(0.30)	(0.31)	(0.34)	(0.34)	

(continued)

Table 7.10 (continued)

	model 1	model 2	model 3	model 4	model 5	model 6	model 7
Type=4	−0.044	0.023	−0.262	0.04	−0.073	−0.086	----
	(0.21)	(0.21)	(0.20)	(0.20)	(0.22)	(0.22)	----
Type=5	0.775	0.801	0.521	0.771	0.762	0.747	----
	(0.43)	(0.42)	(0.39)	(0.40)	(0.43)	(0.43)	----
Type=6	1.071*	1.177*	0.904*	1.039*	1.077*	1.131*	----
	(0.5)	(0.49)	(0.45)	(0.46)	(0.51)	(0.50)	----
Constant	9.890***	9.868***	10.805***	10.576***	10.845***	11.170***	10.797***
	(0.64)	(0.6)	(0.13)	(0.09)	(0.07)	(0.07)	(0.30)
R^2	0.325	0.085	0.083	0.223	0.196	0.043	0.048
N	279	279	279	279	279	279	279
F	11.54	2.81	4.12	12.86	11.06	2.03	2.28
BIC							

Source: Author estimates from survey data

notable differences in earnings among different types of workers, for example, team leaders share a major part of the revenues compared with other groups; (4) there were significant income differentials between males and females; (5) experience and duration of employment does not seem to have any significant impact on the income levels; (6) in general, the results indicated that distribution of ASM revenues differ significantly across different groups as well as across activities along the supply chain.

Inclusive approaches to measuring development in Rwanda, Zambia, and Ghana: five capitals framework and Q-sort methodology

Separately and concurrently with the study evaluating the supply chain value in Rwanda, the authors took part in a separate project through case studies in Rwanda, Zambia, and Ghana titled "Rapid Assessment Frameworks for Mining and Regional Development."[43] A central aim was the formulation or collation of reliable indicators of development that can be applied in and adapted to a range of different local and regional contexts by public or private sectors. This project followed the general principles of inductive and action-oriented research, and was therefore based on an iterative methodology that involved a process of framework design, testing, refinement, implementation, and extension. The work was undertaken collaboratively by a research team from the University of West Alabama's School of Earth and Environment, the University of Queensland's Centre for Social Responsibility in Mining, and partner researchers in each of the case study countries, including Copperbelt University, the Ghana Institute of Public Management and Administration (GIMPA), and the University of Rwanda.

Following a literature review of the vast array of approaches used in measuring development generally, and in resource contexts in particular, we determined that the five capitals approach underlying the sustainable livelihoods framework (SLF) would provide a sound and consistent conceptual underpinning for determining an initial list of indicators to be tested in selected case study areas.[44] However, partly due to recognition of increasing proliferations in the conceptualization of development and differences in the rationale underlying the use of various sets of indicators, we determined that this initial set of indicators would need to be tested against the perspectives of stakeholders (including industry, government, and community) in mining-affected areas to gauge the validity and utility of such indicators in measuring development at local and regional scales across varying contexts. For this purpose, a survey instrument based on Q-sort methodology, which essentially invites multiple stakeholders to rank indicators in order of relevance and importance, was developed and implemented as part of the case study research.

For the case studies in Ghana and Zambia, we selected a list of 25 indicators based on known impacts of mining as most commonly described in resource literature, and organized them under the five capitals (five indicators per capital), placing an emphasis on indicators that would be relevant at local and regional levels. We presented participants with the list of indicators in a random order—without identifying the associated category of "capital"—as set out below in Table 7.11. In addition, we invited participants to advise researchers of any particular indicators or

Table 7.11 List of 25 indicators based on five capitals selected for Q-Sorting

	Indicator	Explanation
1	Drinking water quality	Measures the presence of heavy metals (type and quantity) present in drinking water sources.
2	Road access	Measures the number and quality of paved roads in mining communities.
3	Government spending on local education	Measures proportion of the national education budget that is spent locally.
4	Total arable land	Measures the amount of agricultural land lost, gained, or changed through mining activities.
5	Cost of living	Measures the change in cost of basic food and non-food items such as charcoal, soap, water, electricity, education, health, and transport.
6	Perceptions of government corruption	Measures how well the government and/or local leaders manage the costs and benefits of mining operations.
7	Infant mortality rate	Measures the number of children (per 1000) who die before the age of five.
8	Adult literacy rates	Measures the proportion of adults (over the age of 15) who can read and write in the official language of the country.
9	Cultural continuity	Measures transmission of local cultural knowledge, customs, events, etc.
10	School participation rates	Measures the number primary and high school enrolments as a proportion of the school-aged population size.
11	Unemployment rates	Measures the overall gain or loss of livelihoods in mining communities taking into account the number of farmers displaced from their land by mining.
12	Morbidity by major health category	Measures the proportion of the local population effected by the following health problems: HIV/AIDS; malaria; respiratory diseases; accidents; metals/minerals poisoning.
13	Forward linkages	Measures the number of secondary industries established to process mining output prior to export.
14	Perceptions of safety and crime	Measures perceptions whether communities near the mine are becoming more or less safe.
15	Household income	Measures the total income of each household including income from the local 'informal' economy.
16	Perceptions of mining company transparency	Measures the perceptions of people involved in, or affected by, mining about how ethical mining companies are in the way they operate within the country.
17	Post-school qualification	Measures the prevalence, type and sector of training and education received outside of primary and high school, amongst the community.
18	Displaced peoples	Measures the number of people forced to relocate because of mining activity.

19	Land tenure security	Measures perceptions of local community about how confident they feel about their rights to be self-determining with regard to their land.
20	Access to and quality of energy supply	Measures the proportion of households in the mining community with access to a reliable and affordable source of energy in the form of electricity, coal, or firewood.
21	Biodiversity/access to natural resources	Measures levels of access to, and health of, natural resources such as forests, fish, and other food sources.
22	Government spending on local health	Measures proportion of the national health budget that is spent locally.
23	Health access	Measures the number of doctors per 1000 people.
24	Air quality	Measures the amount and nature of emissions and poisonous gases present in the air within proximity of a mine-site.
25	Backward linkages	Measures the number of local businesses used by mining companies to support their operations.

Source: Horsley et al., 2015

issues related to development that they considered were either missing completely or not adequately covered by the range included in this list of 25 indicators, based on their own experience from their respective stakeholder perspectives.

The Rwanda case study also used Q-sort methodology, investigating quantitatively analyzed perspectives regarding impacts of mining-led development at a district level in Rwanda. However, in this case we used a series of statements, rather than a set of indicators, as the domain or concourse for the Q-sort exercise. We selected forty-six statements across all categories of the five capitals, with some statements representing two to three categories. For example, Statements 25 and 35 represent social, financial, and human capitals. Statements were constructed to balance positive and negative views of similar topics and the degree of extremity of views expressed so there was minimal bias when sorting.[45]

For the purposes of this chapter, the relevant findings of these case studies applicable across all three countries can be summarized as follows: First, among all the views expressed through the quantitative and qualitative data obtained in this research, there was a general consensus that mining activity had an important role to play in contributing to development at local and regional scales. However, there was also a general consensus that this activity is not currently reaching its potential in producing the benefits to development at local and regional scales; this was the case regardless of how development was conceptualized across the varying perspectives.

Second, on the whole most participants indicated that the twenty-five development indicators covered the range of their concerns regarding mining and development and were appropriate items for measurement. Among the few participants who responded otherwise, the predominant concern was the lack of indicators to address equity concerns, specifically gender inequality, disparity in income among foreign nationals and local employees, and perceived bias in recruitment practices (favoring internal/external immigrants over locals).

Third, criticisms of the current governance structure in each country included lack of community engagement at the local and regional levels in areas where mining activity is occurring, weak political accountability in the resource sector generally, and hindrances to transparency in both government and mining company policy and practice. Finally, the range of impacts that the economic activity of extractive industries potentially produces, and the various groups that it affects, can make policy and investment planning in this area difficult, particularly given the potential of competing agendas of stakeholders and power differentials between these parties. Q-sort methodology, by taking numerous individual views and revealing areas that are shared among many, takes a step towards overcoming obstacles in this regard. Where areas of shared concern are revealed, policy can be directed to address and prioritize these issues. Where areas of shared consensus on the benefits of a particular activity or impact emerge, they can provide evidence to support the continuance of current practices and perhaps demonstrate ways in which other issues of concern can be addressed.

Conclusion

Initially, the theory of development underlying the positions of the World Bank and the United Nations, among others, that mining would serve as a viable route to national development in developing countries seemed to be based on an analogue of modernization theory and an optimistic version of staple theory, partly based on the historical experience of resource-rich countries such as Australia, Canada, and the United States.[46] This view envisaged resource economies moving through three stages: (1) an export-oriented staples economy dependent on foreign capital and skills; (2) a diversified second stage when multipliers and linkages and externalities encourage the growth of other sectors, and that transforms the extractive base toward resource upgrading and manufacturing; and (3) the final stage of economic independence when the local economy has acquired its own indigenous capital and skills.[47] In theory, if not in practice, the subsequent trends toward sustainable development and neoliberalist strategy are not inherently inconsistent with this original position and could effectively be absorbed into the model in terms of selection of relevant indicators.

Recent approaches to measuring the linkage effects of mining across multiple scales such as GVC or supply chain value analysis, however, are more empirically based rather than theoretically driven by conceptualizations of development stages. When combined with traditional methods in measuring regional development linkages, important employment and multiplier effects can be identified and monitored at different links along the chain, as the Rwandan case study demonstrates. The possible challenge to this approach, which uses methods such as input–output analysis, is the risk of focusing on financial indicators that, while providing reliable data and analysis on regional economic growth, do not fully encapsulate the breadth of development priorities of affected communities, as demonstrated in the Q-sort case studies. On the other hand, approaches such as the five capitals framework that are more conceptually based and that emphasize inclusion both in terms of the objectives and measurement priorities of multiple stakeholders, could face challenges in providing the precision and empirical depth of the more-tested and more-consistent parameters of economistic methods that produce quantitative data more readily used by decision-makers. Accordingly, an approach that combines the breadth and depth of each of these approaches is worth exploring further, as some of the key lessons from our case studies demonstrate.

For example, while the Q-sort methodology case studies uncovered qualitative data in the form of community concerns that the range of development indicators should include those that monitor gender inequality, the data obtained in the Rwandan input–output analysis case study provided clear quantitative evidence that such inequities were in fact occurring and that they are a sound method for measuring their increase or decrease overtime. In addition, while the Rwandan case study confirmed linkages to employment and income generation along the supply chain, suggesting increases in mining sector output would stimulate other sectors of the local and regional economy, the findings from the Q-sort analyses on stakeholder perspectives provided a strong reminder that economic impacts should not focus exclusively on the exclusion of other important indicators. In particular, in areas where water contamination and air quality were an issue due to mining activity, the importance and relevance of natural capital indicators ranked consistently higher than financial capital indicators as preferred measures of development impacts. The qualitative analysis of the Q-sorts also highlighted important community expectations about the roles and responsibilities of the government versus the private sector.

The differences of opinions in regard to the positive and negative impacts of mining and the roles of different actors revealed in the qualitative case studies highlight the need to ensure that economic planning and investment in the community contributes to realizing the goals of poverty reduction and economic development at local and regional scales. As both the supply chain analysis in Rwanda and the Q-sort methodology studies in Rwanda, Zambia, Ghana demonstrated, one component of this could be support for skills training and technical education, because this was also found among most groups to be a long-term benefit from the mining activity. Another lesson learned was the importance of using revenues from the extractive industries for sustainable social development programs such as creation of multiplier effects in other, more-durable sectors.

The findings of these case studies further support a primary conclusion of the linkage research of case studies conducted in other African countries: that national systems of innovation, skills development, and production linkages in promoting industrialization and knowledge intensification could play a positive role in overcoming current limiting factors to linkages development.[48]

In line with increasingly global trends toward the development of knowledge economies, fostering employment and local business multipliers in technology-related sectors, in particular, might serve to bolster the multiplier effects of mining operations during boom times, while facilitating the diversification of economic structure at local, regional, and national scales. To build this resilience to fluctuations (and eventual decline) in mineral sector activity, and to ensure a trajectory of inclusive growth, requires policy decisions in both public and private realms that not only promote linkage effects, but also directly address the many challenges in their ongoing monitoring and measurement across multiple scales and dimensions of development.

Notes

1 G. Mohan, "Beyond the Enclave: Towards a Critical Political Economy of China and Africa," *Development and Change* 44:6 (2013): 1255–1272; P. Carmody, *The New Scramble for Africa* (Cambridge, UK: Polity Press, 2011).
2 D. Haglund, "How Can African Economies Turn the Resource Curse into a Blessing?" *Bridges Africa: Trade and Sustainable Development News and Analysis on Africa* 1:3 (July 2012): 7–10.
3 D. Haglund, "In It for the Long Term? Governance and Learning Among Chinese Investors in Zambia's Copper Sector," *The China Quarterly* 199 (2009): 627–646; P. Carmody, "Exploring Africa' Economic Recovery," *Geography Compass* 2:1 (2008): 79–107.
4 D. Franks, C. Parra, and A. Schleger, *Approaches to Understanding Development Outcomes from Mining* (London: International Council on Mining and Metals, July 2013).
5 M. Radetzki, "Regional Development Benefits of Mineral Projects," *Resources Policy* 8:3 (1982): 193–200.
6 W. Leontief, *Input-Output Economics* (New York: Oxford University Press, 1966).
7 J. Cypher, *The Process of Economic Development* (London: Routledge, 2009).
8 H. Innis, *Problems of Staple Production in Toronto* (Toronto: Ryerson Press, 1933); T. Gunton, "Natural Resources and Regional Development: An Assessment of Dependency and Comparative Advantage Paradigms," *Economic Geography* 79 (2003): 67–94.
9 J. Horsley, "Conceptualising the State, Governance and Development in a Semi-Peripheral Resource Economy: The Evolution of State Agreements in Western Australia," *Australian Geographer* 44:3 (2013): 283–303; T. Barnes, R. Hayter, and E. Hay, "Stormy Weather: Cyclones, Harold Innis, and Port Alberni, BC," *Environment and Planning A* 33 (2001): 2127–2147.
10 T. Barnes and R. Hayter, "No 'Greek-Letter Writing': Local Models of Resource Economies." *Growth and Change* 36 (2005): 453–470; R. Hayter and T. Barnes, "Neoliberalisation and its Geographical Limits: Comparative Reflections from Forest Peripheries in the Global North," *Economic Geography* 88 (2012): 197–221.
11 M. Radetzki, "Regional Development Benefits of Mineral Projects," *Resources Policy* 8:3 (1982): 193–200.
12 J. Horsley, S. Prout, M. Tonts, and S. Ali, "Sustainable Livelihoods and Indicators for Regional Development in Mining Economies," *The Extractive Industries and Society* 2 (2015): 368–380.
13 E. M. Biggs, E. Bruce, B. Boruff, J. Duncan, J. Horsley, N. Pauli, K. McNeill, A. Neef, F. Van Ogtrope, J. Curnow, B. Haworth, S. Duce, and Y. Imanari, "Sustainable Development and the Water–Energy–Food Nexus: A Perspective on Livelihoods." *Environmental Science & Policy* 54 (2015): 389–397.
14 J. Cypher, *The Process of Economic Development*.
15 S. Morse, *Indices and Indicators in Development: An Unhealthy Obsession with Numbers* (London: Earthscan Publications Ltd, 2004).
16 S. Pegg, "Mining and poverty Reduction: Transforming Rhetoric into Reality," *Journal of Cleaner Production* 143 (2006): 376–387.
17 R. M. Auty, *Sustaining Development in Mineral Economies* (London: Routledge, 1993); J. D. Sachs and A. M. Warner, "The Curse of Natural Resources." *European Economic Review* 45 (1997): 827–838.
18 S. Morse, *Indices and Indicators in Development*.
19 E. M. Biggs, et al., "Sustainable Development and the Water–Energy–Food Nexus," 389–397.
20 G. McMahon, and P. Van der Heen, "Strategic Drivers of the Mining Industry: From Enclave Production to Integrated Development" (Milos: Proceedings from the Third International Conference on Sustainable Development Indicators in the Minerals Industry, 2007).

21 Ibid.
22 G. Hilson, "Corporate Social Responsibility in the Extractive Industries: Experiences from Developing Countries," *Resources Policy* 37:2 (2012): 131–137.
23 S. Morse, *Indices and Indicators in Development*, 28.
24 J. Horsley, S. Prout, M. Tonts, and S. Ali, "Sustainable Livelihoods and Indicators for Regional Development," 368–380.
25 Ibid.; F. Weldegiorgis, and S. Ali, "Mineral Resources and Localised Development: Q-Methodology for Rapid Assessment of Socioeconomic Impacts in Rwanda," *Resources Policy* 49 (2016): 1–11.
26 J. Fessehaie, "What determines the Breadth and Depth of Zambia's Backward Linkages to Copper Mining? The Role of Public Policy and Value Chain Dynamics," *Resources Policy* 37:44 (2012): 443–451.
27 G. McMahon, and P. Van der Heen, "Strategic Drivers of The Mining Industry,"
28 R. Bloch, and G. Owasu, "Linkages in Ghana's Gold Mining Industry: Challenging the Enclave Thesis," *Resources Policy* 37:4 (2012): 434–442.
29 A. Hirschman, *Essays in Trespassing: Economics to Politics and Beyond* (Cambridge, UK: Cambridge University Press, 1981), 65.
30 J. Cypher, *The Process of Economic Development*; A. O. Hirschman, *The Strategy of Economic Development* (New Haven, CT: Yale University Press, 1958); H. Singer, "The Distribution of Gains between Investing and Borrowing Countries," *American Economic Review* 40 (1950): 473–485.
31 J. Fessehaie, and M. Morris, "Value Chain Dynamics of Chinese Copper Mining in Zambia: Enclave or Linkage Development?" *European Journal of Development Research* 25: 4 (2013): 537–556.
32 D. Kaplan, R. Kaplinsky, and M. Morris, "'One Thing Leads to Another'–Commodities, Linkages and Industrial Development: A Conceptual Overview," *Discussion Paper No. 12* (University of Cape Town and Open University, Making the Most of Commodities Programme, June 2011); J. Fessehaie and M. Morris, "Value Chain Dynamics of Chinese Copper Mining in Zambia: Enclave or Linkage Development?" 537–556.
33 J. Fessehaie, "What Determines the Breadth and Depth of Zambia's Backward Linkages to Copper Mining?" 443–451.
34 Ibid.
35 R. Bloch and G. Owasu "Linkages in Ghana's gold Mining Industry: Challenging the Enclave Thesis," *Resources Policy* 37:4 (2012): 434–442.
36 D. Kaplan, R. Kaplinsky, and M. Morris, "One Thing Leads to Another." The countries studied were Angola, Botswana, Gabon, Ghana, Nigeria, South Africa, Tanzania, and Zambia.
37 G. Hilson and J. McQuilken, "Four Decades of Support for Artisanal and Small-Scale Mining in Sub-Saharan Africa: A Critical Review," *Extractive Industries and Society* 1:1 (2014): 104–118.
38 G. Hilson, C. J. Garforth, "Everyone Now is Concentrating on the Mining Drivers and Implications of Changing Agrarian Patterns in the Eastern Region of Ghana," *Journal of Development Studies* 49:3 (2013): 348–362.
39 G. Hilson, "Constructing Ethical Mineral Supply Chains in Sub-Saharan Africa: The Case of Malawian Fair Trade Rubies," *Development and Change* 45:1 (2014): 53–78.
40 KPMG, *Mining in Africa Towards 2020* (Survey Report, KPMG, 2013) www.kpmg.com/africa/en/issuesandinsights/articles-publications/pages/mining-in-africa-towards-2020.aspx.
41 M. J. Farrell, "The Measurement of Productive Efficiency," *Journal of the Royal Statistical Society* 120:3 (1957): 253–281.
42 T. J. Coelli, D. S. P. Rao, C. J. O'Donnell, and G. E. Battese, "An Introduction to Efficiency and Productivity Analysis, 2nd edition (New York: Springer, 1998).
43 J. Horsley, M. Tonts, S. Prout, F. Weldegiorgis, and S. Ali, "Rapid Assessment Frameworks for Mining and Regional Development" (IM4DC Action Research Report Publication,

June 2015), https://im4dc.org/wp-content/uploads/2013/07/Action-research-summary-to-September-30-2015.pdf.
44 J. Horsley, S. Prout, M. Tonts, and S. Ali, "Sustainable Livelihoods and Indicators for Regional Development," 368–380.
45 F. Weldegiorgis, and S. Ali, "Mineral Resources and Localised Development."
46 S. Pegg, "Mining and Poverty Reduction: Transforming Rhetoric into Reality," *Journal of Cleaner Production* 143 (2006): 376–387.
47 J. Horsley, "Conceptualising the State, Governance and Development in a Semi-Peripheral Resource Economy," 283–303.
48 M. Morris, R. Kaplinsky, and D. Kaplan, "One Thing Leads to Another."

8 Measuring transformative development from mining
A case study of Madagascar

Fitsum Weldegiorgis and Cristian Parra

In brief

- Madagascar is used as a case study to examine the long-term impacts of large-scale mining to economic development and social progress.
- With efficient resources management, large-scale mining could be a source of long-term impacts (inputs) and transformative development at local, regional, and national levels.
- Mining could account for up to 14 percent of GDP by 2025 while dominating exports.
- In 2025 about 51 percent of the projects' operating costs are expected to be spent domestically.
- Direct employment, indirect employment, and linked employment are estimated to rise between 2015 and 2025.
- Policymakers, non-governmental organizations, civil society, and mining companies need to coordinate efforts to develop complementary policies, strategies, and actions to transform mining opportunities into effective results.

Introduction

Large-scale mining development could contribute to the economic development and social progress in developing countries by generating significant monetary flows through capital investments, exports, local expenditures, salaries, and fiscal and non-fiscal payments. Together, these flows could transform the social baseline, improve fiscal budgets, and strengthen the domestic economy. Positive transformation has already occurred across different mining-intensive economies in Africa, Latin America, and Asia, where the resource boom of the past twenty-five years has generated unprecedented opportunities. Chile, Botswana, and India, among other countries, have been transformed to varying degrees.

On the other hand, the impressive progress in mining's contribution since the beginning of the twenty-first century has caused a new set of societal concerns,

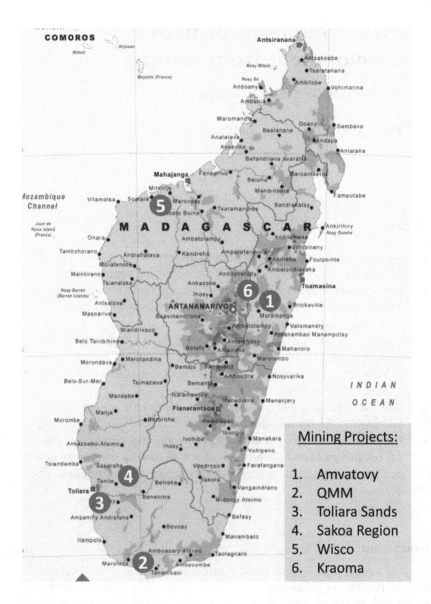

Figure 8.1 Map of Madagascar

demands, and expectations regarding how mining activities have been carried out and how the sector must conduct itself going forward. The sector has faced complex challenges associated with distribution of mining revenues and benefits, the relationship between mining operations and directly affected people (e.g., traditional land owners and indigenous communities), and operational performance regarding transparency, legitimacy of the mining legal frameworks, and effects on the environment. Because of these growing and complex concerns, mining projects have been delayed or even shut down in response to public opposition. The outcomes in terms of development and social progress will depend on how the mining sector and its interested groups address growing societal demands and expectations, and engage various actors to transform economic benefits into real social progress.

The contribution of large-scale mining development to a country's economy has been at the center of various analytical perspectives from academics, development institutions, research centers, and think-tanks, among others. In theory, these important intellectual and conceptual contributions could be used to understand the flows of benefits into the main economy and the ultimate effects on society. In practice, however, these analyses have had limited contributions toward understanding large-scale mining development in a context characterized by very low levels of socioeconomic conditions and structural constraints, and weak institutional capacities (or will) to foster sustainable and equitable economic growth. In particular, these studies have not sufficiently considered poverty and human development levels of a country as indicators in discerning the level of mining influence and the magnitude of mining contribution to absorb and redress negative conditions.

This chapter uses a case study of Madagascar to demonstrate a more comprehensive approach to analyzing the potential of mining to contribute to transformative development in Africa. It is based on a study conducted by the Centre for Social Responsibility in Mining (CSRM) at the University of Queensland.[1] The core research and analysis of the study conducted by the authors of this chapter provides the main source of data and information. In addition, the authors refer to the different training and capacity-building activities they undertook in Madagascar during the years 2014 and 2015.[2] Madagascar is useful as a case study because it provides a typical scenario where large-scale mining investment dominates an economy that has growth potential; in addition, a political-economic transition in Madagascar could prove critical in policy and regulatory decision-making regarding the mining sector.

Madagascar and mining: an overview

Madagascar is characterized by a complex political situation and extremely difficult socioeconomic conditions with no clear positive trend of progress. Despite the new mining ventures that have begun to influence the economy, Madagascar remains one of the world's least developed countries, ranking 158 out of 188 on the United Nations Development Programme's (UNDP's) Human Development Index. Between 2008 and 2012 a protracted political crisis dramatically reduced the country's development indicators and led to a notable increase in poverty: people living below $1.25 a day has topped 87.7 percent of the population.[3] Madagascar's poor socioeconomic condition is characterized by high levels of illiteracy, low levels of agricultural productivity, poor infrastructure, and substantial levels of informal employment and underemployment.[4]

With a GDP of $9.17 billion, the country's economy is overwhelmingly dominated by the informal sector, which accounted for an estimated 61.5 percent of total employment in 2010.[5] The country suffers from large-scale corruption, as indicated by the Ibrahim Index for African Governance,[6] which ranked Madagascar at 29 out of 54 African countries in 2015, and the Transparency International index, which ranked it at 133 out of 175 countries in 2014.[7] Madagascar's democratic institutions are not yet developed, leading to socioeconomic structural problems and political and social unrest. The return to an elected government in 2014 has raised hopes that these unfavorable trends could be reversed, however.

The objective of this study was to review the current and future direct economic impacts of mining development in Madagascar, as a first step to a better understanding of the final potential contribution of the sector to societal well-being and development in general. In developing the analysis, authors have kept in mind the following two points: (1) Mining development exists within a very complex international scenario characterized by growing societal expectations and demands for better benefit-sharing mechanisms, as well as by highly volatile commodity prices; (2) Madagascar remains in a very vulnerable and fragile socioeconomic context.

Mining in Madagascar and the legal framework on which it is based are still gaining maturity. Until the 2000s chromite was the only commodity exploited industrially in Madagascar: In 1968 a French company, Compagnie Minière d'Andriamena (COMINA), began exploiting this mineral ore. The company was nationalized in 1975 and is now known as Kraomita Malagasy, or Kraoma SA. This medium-size mining company exported 47,200 tonnes of concentrate chrome and 38,222 tonnes of rock chrome in 2011, and had an accumulated initial investment of $2.83 million as of 2013.[8] Large-scale mining markedly increased in 2005–2010 with the development of two large industrial mining projects that are now in their exploitation phase: (1) QIT Madagascar Minerals (QMM), an ilmenite, rutile, and zircon mine in Tolagnaro that started production in 2009; (2) and the Ambatovy Project, a nickel, cobalt, and sulphate of ammonia (as a refining by-product) mine near Moramanga, and a processing plant at Toamasina that started production in 2012.

QMM is 80 percent owned by Anglo-Australian mining giant Rio Tinto and 20 percent by the Malagasy government. The site accommodates a mine, a separation floating plant, and port facilities. The extracted ilmenite is exported to be enriched within the metallurgical complex of Sorel-Tracy of Rio Tinto in Canada. The project represents a global investment (in Canada and Madagascar) of $1.1 billion, of which $930 million had been invested in Madagascar as of 2008. QMM has a potential production at full capacity of 496,000 tonnes per year of ilmenite and 26,710 tonnes per year of ZIRSILL (a mixture of zircon and sillimanite).

The Ambatovy Project was launched at the end of 2012 by a consortium of Sherritt International Corporation of Canada (40 percent), Sumitomo Corporation of Japan (32.5 percent), and Korea Resources Corporation of the Republic of Korea (27.5 percent). The ore slurry is conveyed by the help of gravity to the processing plant in Toamasina through a 220-kilometer-long pipeline. Processed product is shipped to international markets from the port of Toamasina. With an accumulated investment of $7.2 billion as of 2013, the project has a production capacity of 60,000 tonnes per year of refined nickel and 5,600 tonnes per year of cobalt.

The political crisis between 2008 and 2012 and Madagascar's political transition has slowed mining exploration and degraded the country's attractiveness as an investment destination. In light of these factors and the history of mining development in Madagascar, it is likely that QMM, Ambatovy, and Kraoma will remain the only large-scale mining projects operating in Madagascar for the foreseeable future. Three deposits have been the focus of much attention and could come to life in the medium term, however:

1. Ilmenite extraction on the east coast and near Toliara on the west coast. The Madagascan coast is rich in mineral sands. The coast near Toliara has been explored, and a deposit there could contain resources of 959 million tonnes, including ilmenite and zircon. On the west coast Toliara Sands has an exploitation license (EL), and by the end of 2013 had an environmental impact study (EIS) approved and under way. Depending on the results of the feasibility study and according to preliminary studies, that potential mine could produce 407,000 tonnes per year of ilmenite and 44,000 tonnes per year of zircon concentrate.[9] On the east coast around Toamasina, a Chinese company already extracts ilmenite on a small scale.
2. Coal extraction at Sakoa. Several companies in parallel explore coal deposits in this region.[10] Initial estimates have identified resources of 1,100 million tonnes to 1,200 million tonnes of coal.[11] If the feasibility studies confirm the initial estimates, the Greater Sakoa Basin could develop coal mines with a capacity of 5 million to 10 million tonnes/year.
3. Iron ore extraction at Soalala. According to initial estimates, this deposit could contain more than 800 million tonnes of reserves available for exploitation.[12] Wuhan Iron and Steel Corporation (WISCO), China's third-largest steelmaker, started exploration of the deposit in 2011. Limited information is available about this project; despite the ambitious plans, as of this writing the project has not advanced beyond exploration.

Other mineral deposits have some potential for development, although feasibility of some has not been demonstrated. These include:

- Gold in Betsiaka, Maevatanana, and Dabolava, where mining permits were issued to several operators in 2008 to transform artisanal exploitation into industrial production.
- Bauxite in Manantenina, where several companies, including Rio Tinto–Alcan, hold prospecting permits.
- Rare earths in Ampasindava and Fotadrevo. Occasional price peaks have renewed interest for rare earths, and several projects are active.

Artisanal and small-scale mining

Until recently, mining in Madagascar was dominated by the artisanal and small-scale extraction of gold, and of precious and semiprecious stones such as sapphire, ruby, aquamarine, tourmaline, topaz, amethyst, and emerald. At the end of the 1990s a rush for sapphire and ruby led to the sudden development of new mining towns, Ilakaka and Sakaraha, and turned Madagascar into one of the world's largest producers of these two gemstones. The artisanal extraction of alluvial gold has been a livelihood activity in many parts of the country, and in some years produces several tonnes of gold.

The year 2012 was one of the peaks of gold and gemstone production in Madagascar. The production of gold increased by an estimated 900 percent from 2011 to 2012, and that of ruby by an estimated 344 percent.[13] Data on exports of precious stones from Madagascar give a good indication of the importance of small-scale mining: Based on UN Commodities Trade Statistics, approximately $250 million worth of gold and stones were exported to other countries (mainly the United Arab Emirates) from Madagascar in 2011. The dramatic increase in exports since 2008 has accompanied the increase in gold price during that period.

Extraction of gold and precious stones is almost exclusively informal. There are an estimated 500,000 artisanal miners across the country, making the artisanal mining sector one of the three largest providers of employment, behind agriculture and ahead of the textile and clothing industry.[14] While its contribution to the economy of Madagascar and livelihoods is significant, artisanal mining has also been associated with considerable adverse impacts on health, safety, social harmony, the environment, and taxation revenue; it also is associated with corruption and illicit trade. To some extent, the development of large-scale mining has shifted attention away from investments and initiatives aimed at formalizing the artisanal and small-scale mining sector and the government's capacity to regulate the sector.

Quarrying

Madagascar produces a range of industrial minerals and other ornamental stones: graphite, gypsum, kaolin, mica, agate, quartz, labradorite, salt, granite, limestone, and marble. The most material example is cement: Holcim Madagascar SA was the second contributor of government payments after Ambatovy, according to data released by the Extractive Industries Transparency Initiative (EITI) in 2013.[15] Holcim employed 280 people in 2011, and consisted of 2 production centers: an integrated production capacity of 150,000 tonnes per year in Antsirabé (Ibity) and a bagging and silo facility with a capacity of 180,000 tonnes per year in Toamasina. Another company, Madagascar Long Cimenterie (Maloci) of China, also owns a cement plant with capacity of 360,000 tonnes per year.

Petroleum

Production of oil has started on a pilot basis at the Tsimiroro site, but exploration is expected to expand significantly. The Tsimiroro oil field is the most advanced on shore, with an estimated deposit of 1 billion to 3.5 billion barrels.[16] The company is owned by Madagascar Oil and is in a partnership with Total for the exploration of the neighboring field of Bemolanga. Off-shore oil potential is not proven but the proximity to fields in Mozambique suggests potential.

Methodology

The study provided a prospective view about future mining activities by building a simplified economic or cash-flow model to analyze potential scenarios of direct and indirect contributions of mining in Madagascar. This model considers projections of direct and indirect mining economic impacts using macroeconomic variables to evaluate the relative weight of impacts.

The data collection method included field research within Madagascar with visits to mining sites, semi-structured interviews, and workshops with local stakeholders.

Three possible scenarios for the future of large-scale mining activities in Madagascar were developed to forecast the potential impacts of the sector for the period of 2014–2035. Scenarios of large-scale mining development in Madagascar were constructed based on the history of mining in the country, a prospective analysis of operational and planned mining projects in various stages of development, and consultations with key actors.

- Mining Activities Scenario 1: Scenario 1 consists of the existing large-scale mining projects of Ambatovy, QMM, and Kraoma.
- Mining Activities Scenario 2: Scenario 2 considers Scenario 1 and two additional mining ventures: Toliara Sands and Sakoa Region (Greater Sakoa Basin).
- Mining Activities Scenario 3: Scenario 3 considers the first two scenarios and the development of the Soalala Project.

The study developed a simplified cash-flow model for individual mining projects based on empirical data, estimates, and assumptions of the cost structure of the projects; actual and forecast production; market conditions and prices; and an analysis of the fiscal regime for mining in Madagascar. The models for each individual company were then aggregated to estimate the monetary effect of the large-scale mining sector in Madagascar, including national costs, salaries, taxation, and royalties.[17] The final component of the model uses an analysis of the projected macroeconomic conditions in Madagascar to determine the impact of large-scale mining development on key economic variables, including exports, GDP, employment levels, fiscal revenues, GDP per capita, and fiscal revenue per capita. The study performed a sensitivity analysis to map the impact of variations in production and commodity price.

The analysis of scenarios estimates the time taken (in years) for key phases of planning and construction before projects operate at full capacity (see Table 8.1). The phases analyzed are feasibility studies; financial structure, legal approval, and permits; engineering and procurement; construction; start up; and production at full capacity. Analysis of the three scenarios of mining development in Madagascar indicates that the projects studied under Scenarios 2 and 3 are not likely to reach the construction phase before 2019–2020, with Scenario 2 and Scenario 3 achieving full capacity no sooner than 2023–2024 and 2025, respectively. The projects will not transition automatically from one stage to another but will do so only after significant effort on the part of government and other relevant actors.

Large-scale mining companies, in particular Ambatovy and QMM, invested around $8.13 billion in the period 2005–2013. In the following period

Table 8.1 Mining scenarios in Madagascar; production based on phases of mining development

Commodity	Projects	09	10	11	12	13	14	15	16	17	18	19	20	21	22	23	24	25	26	27	28	29	30	31	32	33	34	35
Scenario 1																												
Chromium	Kraoma	141	253	256	183	201	130	150	150	150	150	150	150	150	150	0	0	0										
Nickel	Ambatovy				5.7	25	37	60	60	60	60	60	60	60	60	60	60	60	60	60	60	60	60	60	60	60	60	60
Cobalt	Ambatovy				0.5	2	2.7	5.6	5.6	5.6	5.6	5.6	5.6	5.6	5.6	5.6	5.6	5.6	5.6	5.6	5.6	5.6	5.6	5.6	5.6	5.6	5.6	5.6
Ilmenite	QMM	160	287	470	562	562	562	475	396	480	496	496	496	496	496	496	496	496	496	496	496	496	496	496	496	496	496	496
Zirsill	QMM	5.3	12.6	17	30	30	30	24	23	24	26	27	27	27	27	27	27	27	27	27	27	27	27	27	27	27	27	27
Scenario 2 (additional projects)																												
Coal	Sakoa																	5,000	5,000	5,000	5,000	5,000	5,000	5,000	5,000	5,000	5,000	5,000
Ilmenite	Toliara															407	407	407	407	407	407	407	407	407	407	407	407	407
Zircon	Toliara															44	44	44	44	44	44	44	44	44	44	44	44	44
Scenario 3 (additional projects)																												
Iron Ore	WISCO																	14,000	14,000	14,000	14,000	14,000	14,000	14,000	14,000	14,000	14,000	14,000

Notes

Feasibility studies	Technical, economic and financial evaluation
Financial structure, legal approval, and engineering and procurement	Capital structure, lenders, shareholder structure, Environmental Impact Assessment, permits, agreements planning, programming, estimation, design, purchasing
Construction	Including construction of mine facilities, pre-stripping, complementary infrastructure
Start-up	First period of production phase
Production at full capacity	Two years after start-up

(2014–2024) the sector is forecast to invest an additional $1.98 billion, of which 89 percent would consist of investment by WISCO. The potential investments of the additional mining projects (Toliara Sands, Sakoa Region, and WISCO) are highly uncertain. The investments by QMM and Ambatovy represented a historical inflection point for the economy of Madagascar, accounting for 39 percent of total investment during 2005–2009. Ambatovy's investment continued to represent a significant addition to the total investment until 2013, accounting for 65 percent of total investment during 2010–2013. In contrast, foreign investment in all sectors did not exceed $256 million during each five-year period from 1970 to 2004.

Macroeconomic effects

Large-scale mining projects under analysis contributed to $126 million in total monetary flows to the Madagascan economy in 2012 (Table 8.2). Monetary flows are expected to increase as all three operating mines reach full production capacity. Large-scale mining operating at full capacity could generate monetary flows amounting to $460 million in 2015, $1,210 million in 2025, and $1,500 million in 2035, according to the model's most optimistic predictions. Domestic expenditure would dominate this flow with fiscal contributions increasing by proportion in the last seven years of the time period of analysis (2028–2035). The fiscal contribution (discussed further in the next section of this chapter) is forecast to represent 25 percent of total monetary flows in 2035 as compared to 9 percent in 2015 and 7.2 percent in 2025.

The total contribution from salaries is forecast to rise significantly in 2023–2024 when Toliara Sands, Sakoa Region, and WISCO enter the exploitation phase during Scenario 3. Salaries are forecast to remain steady in the final ten years of the analysis, hence their proportion of the overall monetary flow decreases over this period. A forecast increase in monetary flows over the final ten years of study period (2025–2035) is due to flows from Scenarios 2 and 3 that are slated to begin between 2023–2024. Over the period of analysis and beyond, income before corporate tax will progressively increase as accumulated losses decline, affecting corporate tax and withholding tax on dividends.

The final effect in terms of monetary flows will depend on how many companies ultimately operate in Madagascar (Scenarios 1, 2, or 3), the level of production, and commodity prices. International evolution of commodity prices could have an important effect on fiscal contributions (i.e., it could reduce the net profit and consequently the level of corporate tax) and could have a reduced effect on salaries and national procurement (i.e., because of the relatively fixed cost structure of mining).

Large-scale mining development in Madagascar could have a significant impact on the national GDP (i.e., the country's capacity to generate added value), and on future opportunities for development. In 2012 Scenario 1 companies accounted for $96 million of GDP, of which $54 million was direct and $42 million was indirect mining GDP. The higher direct-to-indirect GDP ratio (1 to 0.78) could be attributed to the domestic costs Ambatovy and QMM incurred during the early phases of mining development.

Table 8.2 Total monetary flows of large scale mining at full capacity (1, 2) (US$ million)

	2012		2015		2025				2035			
	Scenarios	%	Scenarios	%	Scenarios			%	Scenarios			%
	S1		S1		S1	S2	S3		S1	S2	S3	
Salaries	24	19%	99	22%	96	128	198	16.4%	96	128	198	13%
Domestic expenditure (3)	84	67%	318	69%	310	469	924	76.4%	310	469	924	62%
Fiscal contribution (4)	18	14%	43	9%	40	52	88	7.2%	38	89	378	25%
Total monetary flows	126	100%	460	100%	446	649	1,210	100%	444	686	1,500	100%

Note

1 Analysis based on mining companies' information and market data. Projections are based on full capacity and price forecasts.
2 The scenarios include the following mining activities:

   Scenario 1: QMM, Ambatovy and Kroama;
   Scenario 2: S1 + Toliara Sands and Sakoa Region;
   Scenario 3: S2 + WISCO.

3 Domestic procurement includes the potential expenditure at local or national level during operational phases.
4 Fiscal contribution includes: royalties, corporate taxes, withholding taxes on dividends, customs duties and import taxes, professional taxes, non-refundable VAT, foreign transfer taxes, and minimum corporate taxes. This fiscal contribution does not include indirect effects of taxes and personal taxes.

Mining development could generate a direct mining GDP of $554 million in 2015 and $1,898 million each year from 2025 to 2035. It is estimated that indirect GDP (the added value generated in other sectors because of mining salaries and local procurement) could amount to $159 million in 2015 and $462 million each year from 2025 to 2035. This means that for each dollar of GDP generated by mining, mining-related activities will indirectly generate $0.29 in 2015 and are expected to generate $0.24 each year from 2025 to 2035.

The significant contribution to GDP by large-scale mining development in Madagascar can be understood in relative terms to other sectors. Total mining GDP is forecast to be 110 percent of industry GDP, 46 percent of agriculture GDP, and 35 percent of services GDP in 2025. The contribution of mining to GDP emanates directly from the added value and productivity of large-scale mining investments in Madagascar over the study period (2014–2015). While this contribution is relatively stable given the realization of planned investments and productivity, it will not inevitably contribute to economic development and improved livelihoods. It is instead an opportunity that can be translated into sustainable human development, depending on how benefits are generated, managed, and distributed at the local level.

As shown in Figure 8.2, large-scale mining in Madagascar started to make a notable contribution to GDP after 2012. Mining accounted for only 1 percent of the total GDP in 2012. If Madagascar maintains the economic trend of the past five years, it could increase its GDP without mining from $10 billion in 2012 to $10.7 billion in 2015 and $14.2 billion in 2025.

With Scenario 1 mining companies operating at full capacity, the GDP of Madagascar could grow to $11.4 billion in 2015, with mining representing 6

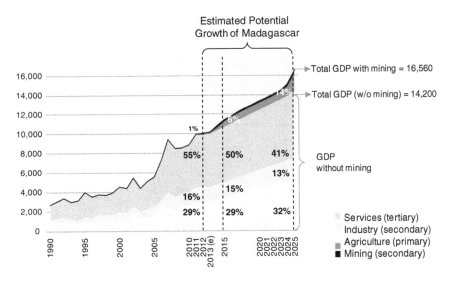

Figure 8.2 Gross Domestic Product by sector in Madagascar and the contribution of mining development (US$ million)

Note

1 GDP forecast (Primary, Secondary and Tertiary) calculated based on the 2001–2012 growth trend (World Bank data accessed 22 July 2014). Mining GDP based on scenario 3.

percent of the total GDP. By 2025 all mines under the three scenarios are forecast to contribute to 14 percent of the $16.6 billion overall GDP. The industry sector (including mining projects at full capacity) could represent up to 27 percent of the Madagascan economy by 2025. This potential macroeconomic contribution of mining development represents an opportunity to transform the economy through new capital investments and to reinforce Madagascar's economic growth.

The large-scale mining projects under analysis operating at full capacity could generate export earnings of $1,139 million in 2015 and $3,696 million each year from 2025 to 2035. As shown in Figure 8.3, mining exports are expected to account for 54 percent of total exports by 2025, a significant increase from 30 percent in 2015 and 8 percent in 2012; that increase is mainly due to the expected shift to full capacity of Ambatovy's production and forecast price increase for ilmenite and chromium in the cases of QMM and Kraoma. The rise in mining share of total exports in 2023 and onward is mainly due to the full-scale production from the additional mining activities in Scenarios 2 and 3, which accounted for 66 percent of total mining export (the remaining 34 percent being that of Scenario 1).

Fiscal contribution

The analysis presented is modeled using estimates of the cost structure of the large-scale mining companies; average realized ore prices for the companies currently in operation until 2014; price forecasts for the period 2015–2035 based on different forecasting methods, including the World Bank Group's forecasts; and the final or net effects of royalties and income taxes, indirect taxes, and direct payments.[18] The accuracy of the modeling depends on the stability of the tax regime and legal framework for mining, corporate policies such as accounting and investment

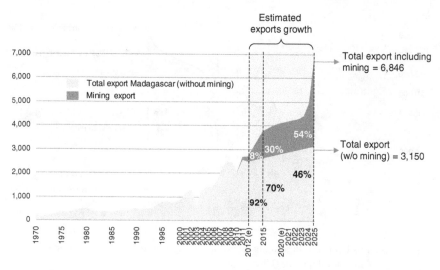

Figure 8.3 Total exports and mining exports in Madagascar (US$ million–current–FOB)

Note

1 1970–2012 = World Bank Data

policies such as depreciation, fiscal discounts, and dividends, commodity prices, production levels, and cost structure.

The model was benchmarked against data of payments disclosed by the 2011 EITI report for Madagascar.[19] While the model accounted for the tax and royalty payments that we determined to be fiscal contributions, EITI data represent both tax and royalty contributions, as well as payments (and taxes) that are not considered fiscal contributions in this study. Penalties, administration fees, personal income tax of employees, pension fund contributions of employees, environmental impact assessment (EIA) evaluation fees, visa and identity card fees, and other one-off payments are not included in the model but represent a substantial component of the payments to government represented by EITI data. Many of these payments are for services provided by government or for contributions by employees, which should not be considered as fiscal contributions of the mining sector. The magnitude of these payments, however, indicates an indirect value providing resources for services delivered by government or resulting from the labor of employees. It should be noted that although royalties are grouped under the term *fiscal income* in this study, they are in effect a payment for the state's ownership of mineral resources.

The model does not include all mining and mineral exploration companies operating in Madagascar. A number of small companies excluded from analysis do make a fiscal contribution, calculated to be around $2 million in 2011. The modeling presented here, does, however, account for state participation, which is not disclosed in EITI data. The state's participation is particularly significant in the case of Kraoma, where 97.2 percent of net profit goes to the state.

The large-scale mining projects currently in production (Scenario 1) contributed a total fiscal income of $18.2 million in 2012 (Table 8.3). This income consisted of state withdrawal from Kraoma (37 percent), royalties (19 percent), corporate taxes (16 percent), professional taxes (13 percent), and foreign transfer taxes (11 percent). In 2015 Scenario 1 projects were expected to produce at full capacity and the total fiscal income is forecast to increase to $43 million. All the large-scale mining projects under analysis (Scenario 3) expected to operate at full capacity in 2025 could generate a total fiscal income of $88 million in 2025. The majority ($52 million) of this total is from Scenario 2, which excludes WISCO.

In the period beyond 2022, state withdrawal is modeled to cease with the projected ending of Kraoma's operations. QMM, in which the state has 20 percent participation, is not projected to result in state withdrawal for the period of analysis due to carry-forward of forecast losses. By 2035 the sector could generate $378 million under Scenario 3. This is more than a fourfold increase from forecasted contributions of $88 million in 2025. This substantial increase in fiscal contribution begins in 2028 (Figure 8.4). During the period 2025–2035, corporate taxes make up a large portion of the total fiscal income, and are expected to contribute $212 million (56 percent of the total) in 2035. The next significant income comes from withholding taxes on dividends and royalties (21 percent and 17 percent of total in 2035, respectively). Royalties are the predominant fiscal income between 2014 and 2027 for all scenarios.

Corporate income taxes do not generate a majority of fiscal income for Scenario 1 at any point. Corporate taxes are forecast to generate a majority of fiscal income for Scenario 2 in 2035 and for Scenario 3 in 2028 (Figure 8.4). This is largely due to fiscal incentives such as investment depreciation, carry-forward of accumulated

Table 8.3 Total fiscal income from large-scale mining

	2012		2015		2025					2035				
	Scenarios	%	Scenarios	%	Scenarios			%		Scenarios			%	
	S1		S1		S1	S2	S3			S1	S2	S3		
Royalties	3.4	19%	14.4	33%	15.7	27.8	64	73%		15.7	27.8	64	17%	
State withdrawal	6.7	37%	8.6	20%	0	0	0	0%		0	0	0	0%	
Corporate taxes (3)	2.9	16%	2.2	5%	0	0	0	0%		0	18.3	212	56%	
Minimum corporate taxes	0	0%	0	0%	2.5	2.5	2.5	3%		2.8	2.8	2.8	1%	
Withholding taxes on dividends	0.7	4%	0	0%	0	0	0	0%		0	20.8	78.9	21%	
Non-refundable VAT (food and mobile fuels)	0	0%	10.2	24%	9.2	9.2	9.2	10%		8.2	8.2	8.2	2%	
Professional taxes	2.4	13%	3.4	8%	8	8	8	9%		8	8	8	2%	
Customs duties and import Taxes	0	0%	1.8	4%	1.8	1.8	1.8	2%		1	1	1	0%	
Foreign transfer taxes	2	11%	2.4	6%	2.4	2.4	2.4	3%		2.4	2.4	2.4	1%	
Total fiscal income	**18.2**	**100%**	**43**	**100%**	**39.5**	**51.7**	**88**	**100%**		**38**	**89.4**	**377.6**	**100%**	

Note

1 *Prepared based on mining companies' information and market data. Projections are based on full capacity and price forecasts.*

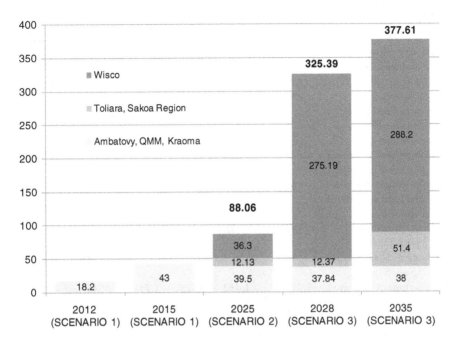

Figure 8.4 Total fiscal income from large-scale mining

losses, and investment tax credits. The total fiscal income from the currently operating large-scale mines is forecast to be $38 million in 2035. Royalties, state withdrawal (in the case of Kroama), non-refundable VAT, minimum corporate tax, and professional tax are the largest contributors to fiscal income from currently operating mines during the period of analysis.

Ambatovy and QMM are not forecast to pay corporate tax above the threshold of minimum corporate tax (0.5 percent of income). The benefits of these investments are realistically derived from the provision of employment, procurement, and infrastructure, as well as the strengthening of the investment climate for future growth of the large-scale mining sector. These two companies are expected to generate a greater degree of fiscal income after the period of analysis when fiscal incentives decline. For the period of analysis, and under all scenarios, corporate taxes (beyond the mandated minimum corporate tax) and withholding taxes on dividends are estimated to be generated almost exclusively by WISCO and Toliara Sands, with the coal projects of the Sakoa Region contributing in the final year of analysis (2035; see Figure 8.4). From 2028 WISCO is expected to account for 84 percent of yearly fiscal income, moderating to 76 percent in 2035 (averaging 81 percent in the period 2028–2035).

The impact of mining's contribution to government revenues in Madagascar is estimated in Table 8.4. Using the average fiscal revenue as a percent of GDP for 2008–2010, the fiscal revenue of Madagascar with mining as a percentage of GDP is assumed to be 14 percent each year for the period 2011–2035. This assumption

Table 8.4 Total estimated fiscal income of Madagascar

	2009	2010	2012	2015 (e)	2025 (e)	2035 (e)
Fiscal revenues in Madagascar						
Fiscal incomes (Internal)	526	536	0			
Fiscal incomes on international trade	385	386	0			
Other national incomes	41	125	0			
Donations and grants	98	81	0			
Government revenues (without mining)	**1,051**	**1,129**	**1,407**	**1,675**	**2,362**	**3,048**
Government revenues per capita (without mining)	**51**	**54**	**63**	**70**	**80**	**84**
Mining fiscal contribution (S1) (e)	1	23	18	43	40	38
Mining fiscal contribution (S3 less S1) (e)					48	340
Government revenues (with mining)	**1,052**	**1,152**	**1425**	**1,718**	**2,450**	**3,426**
Government revenues per capita (with mining)	**51**	**55**	**64**	**72**	**83**	**94**
Total government revenues with mining (% of GDP)	12%	13%	**14%**	**14%**	**14%**	**14%**
Mining contribution to government revenues	**0.13%**	**2%**	**1%**	**3%**	**4%**	**11%**

Note
1 Estimations based on "Annual Report of Central Bank of Madagascar: 2008–2010."
2 All numbers are in US$ million, except government revenue per capita.

might be affected by how fiscal revenues other than mining change over time. As shown in Table 8.4, government revenues without mining were $1,129 million in 2010 and $1,407 million in 2012. With mining, revenues total $1,152 million in 2010 (a 2 percent increase) and $1,425 million in 2012 (a 1 percent increase).

Future forecasts of government revenues without mining are estimated to be $1,675 million (2015), $2,362 million (2025), and $3,048 million (2035). With mining, total government revenues are expected to increase by 3 percent (to $1,718 million) in 2015 and 4 percent (to $2,450 million) in 2025. As discussed earlier, the increase is more significant starting in 2028, when the additional companies in Scenarios 2 and 3 are forecast to begin paying taxes, in particular corporate taxes and withholding taxes on dividends. The overall government revenues are forecast at $3,426 million in 2035, with mining accounting for 11 percent of the total. When expressed per capita, government revenues including mining are expected to increase from $64 in 2012 to US$72 in 2015, and are expected to continue rising to US$83 (2025) and $94 in 2035. Although a significant contribution of mining is expected from 2028, the change in government revenues per capita thereafter is insignificant due to forecast increase in population.

The large-scale mining projects currently in production (Scenario 1) generated $3.4 million in royalties in 2012. This is expected to increase to $14.4 million in 2015 when Scenario 1 mining projects operate at full capacity. Royalties are forecast to double in 2024, reaching $28 million as a result of additional royalties from mines in Toliara Sands and Sakoa Region ($11.85 million). From 2025 to 2035, Scenario 3 mining projects operating at full capacity are estimated to generate $64 million in royalties annually. WISCO's contribution is forecast to dominate accounting for 57 percent of the total royalties.

According to the mining code, an important part of the total royalties (70 percent) are managed by local governments, with the remaining 30 percent managed by the central government. The 70 percent of total royalties managed by local governments are further distributed to autonomous province (10 percent), region (30 percent), and commune (60 percent) levels.

Based on these royalty distributions, local governments received $2.4 million in royalties from currently operating (Scenario 1) projects in 2012, of which $1.4 million was allocated to communes. When Scenario 1 projects are expected to operate at full capacity in 2015, they are forecast to generate royalties amounting to $10.1 million for the local governments. Communes would receive $6.1 million of this total. The addition of Toliara Sands and Sakoa Region projects in Scenario 2 is forecast to raise the total royalties to $19.3 million in 2024 that is due to the local governments. Of this total, $11.6 million would be distributed to communes. From 2025 to 2035 Scenario 3 projects operating at full capacity are forecast to annually generate royalties totaling $44.8 million to local governments, of which communes would receive $26.9 million and regional governments $13.4 million.

The royalties accruing to local governments could have a significant impact on public policies at the local level given the current challenging fiscal situation in Madagascar that has reduced the capacity to develop new and tailored social programs. The analysis shows that royalties managed by all the relevant entities could have a direct impact on the provinces of Toamasina, Toliara Sands, and Mahajanga from the large-scale projects in their respective jurisdictions. Accordingly, Toliara Province received $1.2 million in royalties, while Toamasina and Mahajanga Provinces shared approximately equal amounts of the remaining $1.46 million in 2012. The three provinces are predicted to receive a total of $10.1 million in royalties, with Toamasina benefiting the most ($8 million) in 2015. These royalties are largely due to Ambatovy's contribution.

In 2024 Toamasina and Toliara are expected to be the only benefiting provinces, receiving royalties of $9.13 million and $10.14 million, respectively. These two amounts combined are almost double that of the total royalties the three provinces are expected to receive in 2015, which is US$10.1 million. Mahajanga Province is not forecast to receive royalties in 2024 because we estimate that Kraoma will have ceased mining and WISCO will not start operation before 2025. The total royalties received by the three provinces are forecast to more than double, reaching $44.95 million each year from 2025 to 2035.

The study identified that establishing a consensus formula for royalty distribution at the commune level has been an ongoing challenge. During the study period royalties were distributed to the communes where the mining area is located. There are two arguments for extending this distribution to other communes: (1) Mining activities make use of different industrial facilities and associated infrastructure beyond the communes where the mine is located; (2) the socioeconomic effects of mining activity extend beyond these communes. This was clarified by ministerial decree in February 2014, further breaking up royalties earmarked for communes into communes of extraction (60 percent), communes hosting processing plants (25 percent), and other affected communes (15 percent).[20]

The legal framework, tax, and royalty regimes for mining development in Madagascar is in three parts: (1) mining code–general regime; (2) Madagascar's law on large-scale mining (Loi sur les Grands Investissements Miniers, or LGIM regime); and (3) special agreements or QMM convention. They are a composite model based on different norms and rules according to different needs and state objectives in different time periods of Madagascar. The strength of the current LGIM regime is that it is legislated and not negotiated project by project. In the context of the political instability between 2008 and 2012 in Madagascar, the legal and tax framework for mining development has proved to be a competitive tool from the perspective of attracting foreign mining investment in the country. The regime has allowed the materialization of significant mining investment of $8.1 billion in a relatively short period (2005–2013). The magnitude of this investment is comparable to initial foreign mining investment in other mining countries in the 1990s (e.g., Botswana, Chile, or India). In addition, Madagascar's royalties (and indeed the tax rates) are competitive when compared to other African mining economies.[21]

Madagascar's fiscal regime has achieved its aim to attract mining investments despite the early stage of the mining industry, the recent political risk, and the unfavorable quality of infrastructure. Madagascar now needs to consider whether the current mining policy regime and fiscal settings are suited to seize the opportunities of the next generation of mining investment. The fiscal income forecasts in this study revealed that royalties are the dominant means of securing revenue in the early stages of investment, as rapid cost recovery and low rates of tax limit fiscal income. An increase of the royalty rates would appear justified, as would amendments to when incentives such as investment tax credits can be realized. Large capital investments in downstream processing can provide additional opportunities along the value chain, but they also attract significant fiscal incentives that limit the realization of fiscal income over long periods. Any revisions of the fiscal regime should respect existing stability clauses to maintain and enhance investor confidence.

Economic linkages: local procurement

This study estimated that the currently operating large-scale mining sector in Madagascar (Scenario 1) generated an operating cost of around $160 million in 2012.

This figure was forecast to rise to $585 million in 2015. With additional mining projects at Toliara Sands and the Sakoa Region (Scenario 2), mining could generate an operating cost of $888 million in 2024. This would double in 2025–2035 with the addition of the WISCO project (Scenario 3). Around 49 percent of operating costs is estimated to be spent on energy, power, and fuel until 2013. These costs are expected to decline in the period 2024–2035 when energy, power, and fuel costs would account for just 29 percent of total operating cost. Detailed disaggregation of costs was not made available by the companies covered in this study, and the study estimated the contribution to local procurement using the local proportion of cost that the companies revealed to the research team.

In 2015 operating costs of the currently operating mining projects were forecast to consist of 46 percent imports ($267 million) and 54 percent direct costs of local goods and services ($318 million). This local demand for goods and services will generate important business opportunities for large-, middle-, and small-size suppliers in the country. If we assume that 50 percent of the direct local cost could be spent procuring from large-size suppliers, 40 percent from medium-size suppliers, and 10 percent from small-size suppliers, the dollar amounts would be $159 million (large), $127 million (medium), and $32 million (small).

In 2025 direct cost in the form of local goods and services are forecast to increase to $924 million, which is more than double that of 2024. This represents 51 percent of the total operating cost with the remaining 49 percent ($874 million) representing imports. Based on the aforementioned assumption, $462 million of the total direct local cost in 2025 would be spent on goods and services provided by large-size suppliers. The remaining $370 million would be procured from medium-size suppliers and $92.4 million from small-size suppliers.

While the potential contribution is significant, it will depend on the final cost structure of mining in Madagascar. As with other mining activities worldwide, the cost structure of projects is a composite of different factors such as geology and mineral resources; availability and quality of local suppliers; skills, productivity, and cost of labor; mining and processing technology; financial and investment structure; and management capacities of the mining companies.

Ambatovy used more than seven hundred small- and medium-size enterprises across forty sectors in 2013. The company's database indicates that, as of 2014, 3,500 businesses, including more than 2,700 local companies, were used by Amabatovy and its subcontractors.[22] A report for the Chamber of Mines found that QMM has a database of nine hundred suppliers with three hundred active suppliers in the past three years.[23] The sectors with the most potential for linkages with mining include transportation, food services, miscellaneous services such as cleaning, technical assistance, insurance, and auditing.

In Madagascar, laws and regulations do not precisely define local procurement. The term *local* is used in different contexts as a reference to different levels, either the national level (local), or the regional level (local–local), but there is no consensus on the criteria used to identify these terms. Experts have suggested that a series of criteria, including the percentage of local staff, management, and ownership, would be appropriate to adequately capture truly local entities.[24] Large-scale mining companies often refer to local suppliers as any enterprise registered in Madagascar, and understand local purchases to mean any goods and services purchased from a local supplier. Statistics on local procurement have been provided by mining companies on this basis, but such data could obscure international suppliers registered locally in Madagascar.

The new large-scale mining investments in Madagascar have adhered to a number of international standards, regulations, and policies that promote local development. In some cases, these standards have been adopted following voluntary initiatives, but most of them have been included because of specific requirements and covenants of shareholders, investors, and mining lenders. The Malagasy law does not require companies to supply their goods and services locally, but QMM and Ambatovy have their own policies to enhance business relationships with the local economy, including (1) local preference for purchases and procurement; (2) training of small-size local suppliers to strengthen administrative and management capacities; and (3) specialized payment arrangements for small-size local suppliers.

These corporate efforts, sometimes with the collaboration of international organizations such as the GIZ (Deutsche Gesellschaft für Internationale Zusammenarbeit) and the World Bank Group, aim to establish long-term relationships with local suppliers and enhance inclusive development beyond the direct economic benefits at the national and regional levels. CARA (Anosy Regional Affairs Centre of QMM) and ALBI (Ambatovy Local Business Initiative) are two specific corporate initiatives to develop and support local suppliers. In 2012 QMM spent around $12 million in direct purchases linked to local suppliers and organized training programs for 335 local suppliers.

In addition to local procurement, large-scale mining development in Madagascar presents an opportunity for complementary development of public and private infrastructure. The QMM and Ambatovy projects are linked to the development of ports, roads, bridges, facilities for providing social services, a thermal power station, power lines, and a water distribution system, among other infrastructure projects (see Box 8.1).

Box 8.1 Local infrastructure investments

Examples of complementary infrastructure: QMM

Infrastructure	Description and public-interest goal pursued
Port Ehoala	Construction designed to support mining development and economic development of other sectors.
Toalognaro port	Rehabilitation to provide continuity for traffic during construction of the new port.
Roads, including main RN 13	Rehabilitation to provide access to landlocked region (90 km rehabilitated between 2005 and 2012).
Drinking water Treatment plant	Rehabilitation of water treatment plant, installation of new sewage system, construction of new plant to provide access to drinking water (urban population needs met).
Electricity	Purchase and installation of generator for the city to have a reliable source of electricity.

Roads	Built and/or enhanced 100 km of roads and bridges. Objective: Enhance safety, security, and access for local populations to transportation corridors and markets Linking plant to port.
Port	Upgrades to the port of Toamasina, Madagascar's largest seaport—over $70 million invested. Objective: Ensure a safer, cleaner transfer of energy products from tankers and bulk primary materials from cargo ships.
Railway	A 12 km railway line built in parallel to the existing line between the plant site and the port. Objective: create the capacity needed to move commodities efficiently between the port and the plant site.
Water pumps	Drilling of wells and installation of water pumps along the pipeline.
Anjoma market	In 2010, construction was completed on the new Anjoma Market, which was handed over to the Municipality of Toamasina. Municipality has developed a detailed management plan.
Training facilities	Project supported physical infrastructure renovation and investment in new equipment at institutions, such as the University of Toamasina and the Alarobia Polytechnic School in Antananarivo. The objective was to invest in local human capital.

Employment

The model on employment estimates that the current large-scale mining projects in Madagascar (Scenario 1) operating at full capacity will provide direct employment for 4,200 people during the period 2015–2022 (Figure 8.5). Employment is forecast to remain steady until Kroama has ceased production in 2023 as anticipated, with employment declining to 3,900 people that year (despite the potential for 200 additional jobs created by the start of Toliara Sands in Scenario 2). Direct employment associated with Scenario 2 is forecast to increase to 4,950 in 2024, largely due to the potential for the Sakoa Region projects to start operating at full capacity. Scenario 1 projects, however, are still expected to dominate employment during this period, contributing 100 percent (2015), 95 percent (2023), and 75 percent (2024) of the total jobs created.

Scenario 3 is forecast to raise direct employment to 10,950 jobs in 2025 (and each year until 2035). The number of jobs created by Scenario 1 and 2 projects is estimated to remain at 3,700 and 1,250, respectively, during this period. WISCO in Scenario 3 is forecast to generate a significant number of direct employees accounting for 55 percent of the total forecast of 10,950 jobs (between 2025–2035), with 34 percent of the total generated by Scenario 1 projects. This analysis does not include forecasts for job creation in the construction phase, during which more people normally are employed.

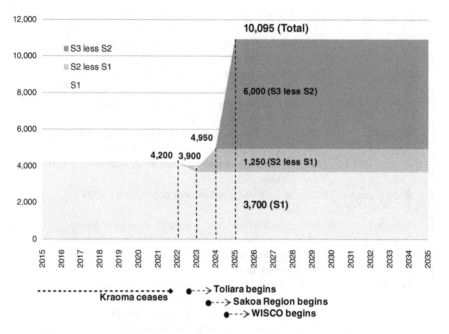

Figure 8.5 Total direct employment by year, with mining activities at full capacity

Large-scale mining is supported by a wide range of businesses in Madagascar. Modeling of indirect employment in this study assumed that three rounds of suppliers could benefit from the total domestic cost of the large-scale mining. Each round of supplier is expected to create its own jobs. To estimate jobs created by the mining sector and linked suppliers, we use the salaries generated by mining in relation to the base salary in Madagascar. The cost structure of the three rounds of suppliers is assumed to be 20 percent salaries, 50 percent direct cost, 10 percent taxes, and 20 percent net profit, with the average salary of mining sector and linked suppliers estimated to be five times the base salary in Madagascar (that is, $480 per month, or $5,754 per year). On the one hand, higher salaries associated with mining sector activities provide greater opportunities for mining suppliers and local spending. On the other hand, unusually high salary levels can cause inflation, inequality, social tensions, and economic disruptions, affecting the poorest most.

Using the assumptions above, it is estimated that the currently operating Scenario 1 projects are forecast to create 19,321 indirect and linked employment in 2015 (figure 8.6). In addition to the direct mining employment, a total of 23,521 people are beneficiaries from Scenario 1 mining employment.

As with the direct employment, the indirect and linked employment created by the additional Scenario 2 projects (Toliara Sands and Sakoa Region) are forecast to come into effect at different years depending on the start of their full capacity production. As a result, the total indirect and linked employment created by Scenario 2 projects is forecast to be 20,188 persons in 2023, rising to 28,552

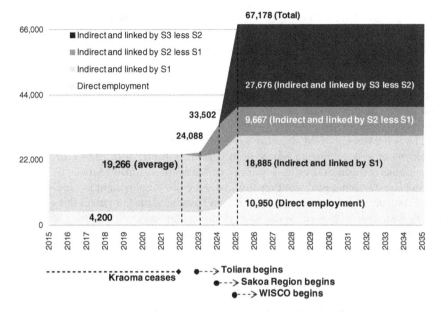

Figure 8.6 Total direct, indirect, and linked employment by year (mining activities at full capacity)

Notes:

1 The same values as 2025 for each year until 2035.

2 Indirect employment is considered here as the employment generated by Round 1 mining suppliers.

3 Linked employment is an overall estimation about the total labor force of Madagascar that has or will have a connection with mining activities.

Linked employment must be understood as:

a) 'existing employment that now is linked to mining local procurement', and

b) 'new employment positions created because of mining demand'.

Assumptions

Three Rounds of Suppliers: Round 1 supply to mining projects, Round 2 supply to Round 1, and Round 3 supply to Round 2

Cost Structure the three Rounds of Suppliers:

Salary	20%
Direct Cost	50%
Taxes	10%
Net Profit	20%

Average salary of mining suppliers and linked suppliers:

US$ 480 per month (if we assume it to be 5 times the legal minimum monthly salary in Madagascar)

Exchange rate: 1US$ = 2323 Ariary

in 2024. In addition to the direct jobs created, mining is therefore forecast to be responsible for a total of 24,088 direct, indirect, and linked jobs in 2023 and 33,502 jobs in 2024.

During the period 2025–2035 the potential full employment effect of all the large-scale mining projects studied (Scenario 3) is forecast to be realized. This means that, similar to the direct employment, the indirect and linked employment created could rise to 56,228 jobs in 2025. By the same proportion, the potential final employment effect of mining (direct, indirect, and linked) could be 67,178 people employed in Madagascar in 2025. This is projected to be the case each year from 2025 to 2035 (Figure 8.6).

In a context where 75–85 percent of the population in Madagascar is located in rural areas and only 15 percent and 3 percent of the labor force hold secondary and tertiary qualifications, respectively, training and skills development represent an important contribution to local development. For some people, the training initiatives offered by the mining industry represent their only opportunity for advanced professional training.

The government of Madagascar expects the mining sector to prioritize the employment and training of Madagascan citizens. Chapter 4 of the LGIM states that mining companies must give priority to Malagasy nationals (Article 103) and put in place a training plan that prepares Malagasies for employment at various levels of company activities (Article 104). Efforts have been made to meet this expectation, with over 84 percent of Ambatovy's direct employees and contract workers and 94 percent of QMM's direct employees (excluding Antananarivo office staff) Madagascar nationals.[25,26]

Transformative development: concluding remarks

The approach adopted in this study went a step farther to understand how mining outcomes are connected to country development goals. It acknowledges the need to account for the socioeconomic context in which mining exists as a starting point before we can assess the contribution of mining. As such, it enables identification of key considerations for policy and legal reform from the point of human development effect. Although progress can be made in achieving the direct and indirect positive impacts as well as in minimizing the negative social and environmental impacts, experience in recent years has brought to light the fact that those positive impacts have not been translated into societal well-being. Efforts to enhance mining's contribution through the various indicators analyzed in this study should be people centered, focusing on poverty reduction and improvement of social conditions. The Sustainable Development Goals (SDGs) which came into effect in January 2016 should present an opportunity for the mining sector to reengage with society using a new lens.

Realizing the forecasted financial flows and the potential contribution to sustainable development could be enormously difficult, as demonstrated by the significant disparity in what the studied projects in Madagascar are expected to contribute in the short and long runs. Discussion of the factors that explain this disparity could help channel appropriate policy reform:

- Status of projects: As described above, the projects profiled are in various stages of development, with Kraoma, QMM, and Ambatovy currently in operation and the rest in various stages of evaluation. QMM and Ambatovy were yet to reach full production capacities at the time of study, whereas Kraoma was phasing out with few years of production left. The rest were in the feasibility stage with the assumption that full production would not be reached until 2024–2025.

- Sizes of investments: Ambatovy and QMM represent significantly greater investments, with WISCO expected to take investment to a new high level. These large investments significantly affect the financial flows at construction and production phases.
- Mining legal regimes: The studied projects are administered by three different legal regimes (mining code–general regime, LGIM regime, and special agreements or QMM convention). As such, the fiscal contributions by the various projects vary depending on incentives and requirements they are subject to. For this reason, QMM and Ambatovy were not expected to make any payments on corporate tax, whereas WISCO, Toliara Sands, and Sakoa Region were expected to make such payments toward the end of the study period.
- Assumptions on commodity prices and production levels: The scenarios are based on forecasted assumptions regarding the start and continuity of production at full capacity, stability of commodity prices, and consistency of legal regimes. These factors are highly vulnerable to changes affecting the overall monetary flows estimated in this study. For example, the WISCO project, which caused the major inflection point in the amount of monetary flows toward the end of the study period, is included under the assumption that it will go ahead as planned. It is also highly likely, however, that it might not be realized due to a number of possible impediments.

While these factors indicate the uncertainties surrounding the realization of the forecasted monetary flow and other benefits, the study demonstrated that there are potential benefits, and that they can be harnessed with appropriate regulatory reforms and enforcements, and given stabilized commodity prices, barring any future major shocks. The transformation of the direct economic benefits into a real social progress will depend on the political stability of the country and how the main interested groups can manage this opportunity beyond their own particular interests. As witnessed in other resource-rich countries, the danger of a resource curse is very much present, and there is a risk that the potential of developing the mining sector could become another missed opportunity for the people of Madagascar.

The merit of revising the mining fiscal regime should be assessed in light of investment risks, commodity price volatility, mineral endowments, and competitiveness in terms of infrastructure, services, institutional capacity, and political stability. The fiscal regime for mining should be simple, predictable, transparent, and consistent. It should ensure a fair distribution of the economic benefits from mineral resources between mining companies and government, with fiscal rules complemented by an efficient and transparent tax administration.

In view of the cyclical nature of commodity prices, the government should consider adopting revenue management mechanisms, especially at the local level as mining revenues grow. These mechanisms could include establishing trust funds or foundations for strategic long-term investments, which would help avoid overreliance on cyclical mining revenues for recurrent spending. Governance of such mechanisms would have to be transparent, efficient, and inclusive. The capacity to administer and manage revenues at the national and local levels needs to be enhanced. It is imperative that internal capacities to collect, control, monitor, and evaluate mining taxes and royalties are developed. Local capacities, resources, and policies also need to be progressively enhanced to improve the collection, management, and investment of royalties by communes, regions, and provinces. Local decision makers can be trained on basic financial literacy, business administration, project management, communication, inclusive governance, and the design of trusts and funds for equitable and sustained social development.

Government should encourage local content, while industry should promote best practice that would foster economic linkages. Civil society, business organizations, and development donors should support and partner with industry to improve and develop good practices. Innovative programs for the incubation of mining service businesses can enhance the potential employment outcomes of mining and foster local–local procurement. Programs using enterprise facilitation; microfinancing, savings, and investment forums; and coaching on financial management and business development could be supported by local governments in mining regions with the assistance of development partners and the mining industry. Cooperatives have proven successful in Rwanda and Ethiopia, among other African countries, and can be replicated in grouping and supporting smaller businesses that could benefit from mining; they could also spearhead private sector development in a diversified economy.

Partnerships between the mining industry and government offer the best opportunity to simultaneously train a professional mining workforce. Strengthening access to and quality of secondary and tertiary education is important for the success of the mining industry in developing and utilizing local skilled labor. The Ministry of Education and the Chamber of Mines should identify synergies for further development of basic educational outcomes that can meet the future workforce demand of the mining sector. The Chamber of Mines and individual mining companies should continue working with government agencies at multiple levels as well as development partners to support the establishment of technical schools and targeted vocational programs (especially in areas of skills shortage) within existing institutions. The Chamber of Mines can be a forum for industry leadership and the promotion of best practice.

Notes

1 The authors would like to extend special thanks to the CSRM team, including Lynda Lawson, Daniel Franks, and Saleem Ali for their contributions in reviewing and editing the original report; the World Bank for commissioning the study and its team mainly Remi Pelon and Mylene Faure for their feedback; local partners including Lalalison Razafintsalama, Myora Henintsoa Razafindrakoto, and Rupert Cook for in-country support; mining companies particularly Ambatovy and QMM for supporting the study and providing data; and government, civil society and donor organizations that formed part of the study steering committee.
2 The findings, interpretations, and conclusions made in this study are those of the authors and should not be attributed in any way to the World Bank, its allied institutions, members of its board of directors, or the countries they represent.
3 UNDP, "Human Development Indicators: Madagascar" (UNDP, 2015), http://hdr.undp.org/en/countries/profiles/MDG (accessed January 17, 2016).

4 African Economic Outlook Madagascar, "World Bank. Madagascar: Three Years into the Crisis, an Assessment of Vulnerability and Social Policies and Prospects for the Future" (African Economic Outlook Madagascar, 2012).
5 IMF, "World Economic Outlook" (IMF, October 2015), http://knoema.com/IMF WEO2015Oct/imf-world-economic-outlook-weo-october-2015?country=1000980-madagascar (accessed January 17, 2016); J. Vaillant, M. Grimm, J. Lay, and F. Roubaud, "Informal Sector Dynamics in Times of Fragile Growth: The Case of Madagascar," *European Journal of Development Research* 26:4 (2014): 437–455.
6 IIAG, "Ibrahim Index of African Governance" (Mo Ibrahim Foundation, 2015).
7 Transparency International, "Corruption Perception Index" (Transparency International, 2014), www.Transparency.Org/Cpi2014/Results (accessed January 18, 2016).
8 Ernst and Young, "Rapport De Réconciliation des Paiements Effectués Par Les Industries Extractives À L'etat Malagasy Et Des Recettes Perçues Par L'etat Exercice" (Ernst and Young, 2011).
9 World Titanium Resources Ranobe, "Definitive Engineering Study Report" (World Titanium Resources Ranobe, August 2012), www.worldtitaniumresources.com/ (accessed August 21, 2014).
10 Those companies are MCM-SA Sakoa, PAM- SAKOA, and Lemur Resources.
11 Thomas R. Yager, "The Mineral Industry of Madagascar. Minerals Yearbook, 2012" (U.S. Department of the Interior, U.S. Geological Survey, 2014).
12 Madagascar Ministry of Strategic Resources, "Fiche Rela: Ve Au Projet De La Société Madagascar Wisco Guangxin Kam Wah Ressources" (S.A.U., 2014)
13 Thomas R. Yager, "The Mineral Industry of Madagascar."
14 Ibid.
15 Ernst and Young, "Rapport De Réconciliation des Paiements Effectués."
16 Nadia Ouedraogo, "Sub-Saharan Africa: Unconventional Oil Resources," *International Association for Energy Economics, Fourth Quarterly* (2014): 34–36.
17 Royalties are included in the analysis, acknowledging that these are payments for inputs owned by the state.
18 Including corporate tax, minimum corporate tax, withholding tax, professional tax, non-refundable VAT, foreign transfer tax, customs duties, import tax, and direct payments associated with state participation.
19 Ernst and Young, "Rapport De Réconciliation des Paiements Effectués."
20 République de Madagascar, "Arrêté Interministériel No 8887/2014 Définissant Les Modalités De Répartition Et D'u: Lisa: On Des Ristournes Minières Issues De Certains Projets Miniers" (République de Madagascar, 2014).
21 E. Sunley, "Madagascar's Fiscal Regimes for Mining: A Preliminary Assessment" (World Bank, May 2014).
22 Ambatovy, "Job Creation," (Ambatovy, 2014), www.ambatovy.com/docs/?p=433 (accessed October 30, 2014).
23 Georges Ramanoara, "Appui Au Groupe De Travail Sur Le Contenu Local (Gtcl) De La Chambre des Mines De Madagascar," (World Bank Group, 2013).
24 Ibid.
25 Ambatovy, "Job Creation."
26 QMM, "Employment at QMM," (QMM, 2013), www.Rio;Ntomadagascar.Com/French/Pubemp.Asp (accessed October 30, 2014).

9 Field vignette

The Extractives Dependence Index and its impact on Africa

Degol Hailu and Chinpihoi Kipgen

High and persistent dependence on the extraction of oil, gas, and minerals for export earnings and fiscal revenues are a concern for the sustainability of growth in resource-rich countries. Not so much a reliance on unexploited abundant resources in the ground, but dependency on extracted commodities for generating incomes. There are strong arguments for reducing resource-dependence—from declining terms of trade for commodity exports and volatility in their prices to potential weak governance and poor environmental and social safeguards associated with extraction.

Although initial reliance on the sector is inevitable to generate income, it is the eventual moving away from this dependency through diversification into other sectors (particularly towards manufacturing) that denotes successful resource-based development. Several researchers came up with indicators to measure resource-dependency. They look at shares of resource exports in GDP or in total exports; resource revenue as a share of total government revenue and extractive value-added in GDP. These traditional measures however do not take into account the contribution of other sectors to foreign income, taxes or to domestic value addition. Dependence on resource incomes is only problematic if other sectors are relatively weaker. This has unfortunately been the challenge for a number of resource-rich African countries.

This is what our new index tries to correct. We call it the Extractives Dependence Index (EDI). The EDI is a composite index consisting of six indicators. Three of them are well recognized in the literature: (1) the share of oil, gas and minerals in total export revenue; (2) the share of resources in total government revenue; and (3) oil, gas and mineral value-added in GDP.

However, we adjust each of the above three indicators to capture the productive environment under which the resource sector strives. To account for the productive capacity of an economy to generate additional foreign exchange (other than resources) we use a fourth indicator: "export revenue from high-skill and technology-intensive manufactured (HTM) exports as a share of total global HTM exports." This indicator takes into consideration the productivity and competitiveness of a country's export basket. The same productive capabilities and human capital can be used to diversify into a range of export products, hence lessening dependence on the resource sector.

To account for an economy's capacity to generate alternative sources of tax revenue, we use a fifth indicator: "total non-resource taxes from incomes, profits and capital gains as a share of GDP." This is a proxy for non-resource revenue base and also reflects higher institutional capacity required for tax collection and administration.

To adjust for a country's ability to domestically process its raw materials, we use a final indicator: "per capita manufacturing value added." Greater forward linkages from the resource sector imply higher levels of transferable skills that can increase technology transfer and employment mobility within and between sectors. In a nutshell, this indicator is a proxy for domestic industrial capability.

Using the above 6 indicators, we constructed an index that ranges between 0 and 100, with 100 being the highest dependence score. The index values are calculated for a total of 81 countries between 2000 and 2011, although not all countries have data for all years.

The comparison between Zambia and Norway illustrates the contribution of our index. In 2011, the two countries collected a similar share of revenues from the extractive sector, which contributed to 76 percent and 74.5 percent of total export revenues and 25 percent and 23.5 percent of fiscal revenues, respectively. Without accounting for Zambia's relatively lower domestic productive capacity, the traditional measures of resource-dependence would consider the two countries as being equally dependent on the sector. However, our EDI values for Norway and Zambia in 2011 were 34 and 45, and the countries were ranked 33rd and 41th, respectively.

Our index also implies a possible inverted-U shaped trajectory in countries with successful strategies that decrease resource-dependence over time. The cases of Mongolia, Nigeria, and Botswana demonstrate this point. Mongolia's EDI value in 2000 was 26. With the discovery of the Oyu copper and gold deposits, extraction intensified and by 2011, the EDI score increased to 63. Hence, Mongolia is becoming more dependent on its minerals, implying the country is may be in the rising part of the inverted-U.

Despite over 60 years of resource extraction, Nigeria has not undergone the structural transformation required to decrease its dependence on the sector and has maintained an EDI score greater than 80 in both 2000 and 2011. The economy continues to be dominated by the oil sector providing over 90 percent of foreign exchange earnings and financing 77 percent of total government revenues over the past decade. The country seems to be stuck in the flatter part of the inverted-U.

In the case of Botswana, the EDI score declined from 71 to 62 between 2000–2011. Although diamonds, nickel, copper, gold, and other resources continue to bring in an average of 85 percent of total export earnings, the country's efforts to promote downstream value addition, including diamond beneficiation, agriculture and tourism, is slowly moving it towards decreasing dependence on the mineral sector. The country may have begun the downward journey on the inverted-U.

It is well established that persistent dependence on resource incomes subjects an economy to volatile growth. And knowing where the dependence originates from, through the use of measures such as the index we introduced here, allows policy makers to design better diversification policies and strategies. The African Mining Vision and related programs may benefit from utilizing this index in their strategic planning efforts and setting targets for tangible progress.

10 Field vignette

The West African Exploration Initiative (WAXI) as a model for collaborative research and development

M. W. Jessell and the WAXI Team

Thee West African Craton (WAC) consists of two Archean nuclei in the northwestern and south-western parts of the craton juxtaposed against an array of Paleoproterozoic domains made up of greenstone belts, sedimentary basins, domains of extensive granitoid-TTG plutons and large shear zones, which are overlain by Meso- and Neoproterozoic and younger sedimentary basins. The region has a thousand-year history of gold mining, and the Malian Emperor Kankou Moussa in 1324 had gold assets estimated to be worth 400 Billion dollars at today's value. Not surprisingly the WAC is therefore typically referred to as a gold province, although it also hosts contains world class iron ore and bauxite deposits.

The AMIRA International P934 West African Exploration Initiative (WAXI) is a collaboration between the principal actors in the west African large-scale minerals sector which started in 2006 and the third phase of the project is due to end in 2018. The overall aim of WAXI is to enhance the exploration potential of the West African Craton through an integrated program of research and data gathering into its "anatomy," and to augment the capacity of local institutions to undertake this form of work. We believe that this project has resulted in positive outcomes for all partners, and the lessons learned in this project help provide an improved model for collaborative research in developing countries.

The West African Craton extends over 2.5 million square kilometers, and is hosted by 13 countries that use 3 administrative languages.[1] This has resulted in a fragmentation of effort and is a barrier to integrated research, as data are held by each country in different languages and formats, and with very heterogeneous levels of detail. Publicly available syntheses are principally limited to harmonized regional-scale maps. The Paleoproterozoic terranes represent one of the world's most important gold provinces, with output increasing from approximately 50 metric tonnes per year to 220 between 1993 and 2013 this period saw some countries, such as Burkina Faso, becoming significant gold exporters for the first time.[2]

The shared challenges facing geoscientists wishing to undertake scientific research in west Africa (from industry, governments and academia) include the limited flow of scarce government resources to STEM activities in general, and geoscience research in particular. This contrasts with the significant revenue flow that many west African countries receive as the result of mining activity, and the costs of managing the environmental impacts of mining (and in particular by variably regulated artisanal mining).

Research activities

The WAXI research program was designed following a gaps analysis and consists of a set of integrated research modules which are grouped into three themes: architecture and timing; mineralizing systems; and surface processes. The results of the research program have been progressively published over the last ten years (see www.tectonique.net/waxi3 for a complete list of papers and access to the thesis collection), but in particular as four special issues that were compiled following the end of confidentiality for stage two of the project.[3,4,5,6]

This research represents one form of support of economic development in the region, as it provides key datasets (in GIS-ready formats) that help to reduce geological risk in the minerals exploration decision-making process. The development of an integrated exploration GIS has proved to be a very efficient approach for transferring information to partners, all of whom have access to an exploration GIS which has now expanded into a more than 600 Gigabyte, 180-layer online and static GIS database of metadata and data related to west African exploration. More that eighty of these layers consist of new data or compilations developed by project partners.

Capacity Building Activities

According to UNESCO/Geological Society of Africa data, there are currently thirty-eight public higher education institutions that provide geoscience training in west Africa (including Schools of Mines), of which twenty-nine are in Nigeria. In the last ten years, the African Union via the African Mining Vision (AMV) and its associated action plan,[7] the World Bank and the African Development Bank have recognized the need for renewed investment in higher education to support the growing African population, which is predicted to have the world's largest labor force by the year 2040.[8] The African Development Bank, UNESCO's Earth Science Initiative in Africa, the AMV and African Network of Earth Sciences Institutions (ANESI) share many proposals in terms of enhanced regional collaboration, better integration of the private sector and the minerals industry, in particular via Public-Private Partnerships, and the use of ICT-based training programs as supports for Africa's education needs. Many of these same themes are also found in the UN's Sustainable Development Goals, and map directly to WAXI activities.

The different stages of the project, together with associated one-on-one industry projects and university-funded scholarships, have allowed us to support a total of thirty-five PhD projects, thirty-five masters and honors projects, and six postdoctoral fellowships, with more than half of these being African. Nevertheless, the demand for scholarships, and in particular projects which provide support for research costs, far outstrip the capacity of the WAXI project's resources, and partnering with other training schemes provides a partial solution to meeting this demand, as many outside scholarships do not provide research funds.

To date, the project has supported 23 3–5-day field and classroom training courses to over 350 industry, university and geological survey personnel. The courses have been held in Ghana, Burkina Faso, Côte d'Ivoire, and Senegal. The vast majority of the attendees of these courses were African geoscientists. In addition to these technical courses, we organized a five-day course on research management aimed at west African university and geological survey personnel. Finally, two international conferences (in Ougadougou, Burkina Faso in 2007, and Dakar, Senegal, in 2015) were held to stimulate new research ideas in the region and present project outcomes.

Conclusion

The success of the WAXI program comes from the commitment of the individuals and organizations that have been involved over the last ten years, but equally from the recognition that these different organizations have different drivers. The WAXI project forms part of a broad range of development geoscience initiatives in Africa that have overlapping aims that partially fulfil the aims of the African Mining Vision and the Sustainable Development Goals. This initiative demonstrates the significant achievements that become possible when the different stakeholders in the minerals sector (industry, academia, government, and nongovernment organizations) partner together to combine datasets and knowledge across national borders in order to achieve their diverse goals.

Acknowledgements

We wish to gratefully acknowledge AMIRA International and the thirty-four industry sponsors, as well as AusAid and the ARC Linkage Project LP110100667, for their support of the different stages of the WAXI P934 series of projects. We are also appreciative of the contribution of the twelve geological surveys/departments of mines in west Africa who joined as sponsors in kind, as without their ongoing logistical support and provision of a range of country-level datasets, this project would not have been possible. Finally, we wish to recognize our research colleagues from twenty institutions from around the world.

Notes

1 Algeria, Burkina Faso, Ghana, Guinea, Ivory Coast, Liberia, Mali, Mauritania, Niger, Sierra Leone, Senegal, Western Sahara/Morocco, and Togo.
2 R. J. Goldfarb, A.-S. André-Mayer, S. M. Jowitt, and G. M. Mudd, "West Africa: The World's Premier Paleoproterozoic Gold Province," *Economic Geology* 112 (2017): 123–143.
3 Goldfarb, R. J., A.-S. André-Mayer, "West African Gold," *Economic Geology* 112 (2017): 1–2.
4 K. A. A Hein, "West African Mineral Atlas Monograph," *Ore Geology Reviews* 78 (2016): 556–557.
5 M. W. Jessell, J.-P. Liegeois, "Editorial: 100 Years of Research on the West African Craton," *J. Afr. Earth Sci.* 112 (2015): 377–381.
6 M. W. Jessell, P. A. Cawood, and J. M. Miller, "Editorial: Craton to Regional-Scale Analysis of the Birimian of West Africa," *Precambrian Research* 274 (2016): 1–2.
7 African Union Commission, "Action Plan for Implementing the African Mining Vision, Building a Sustainable Future for Africa's Extractive Industry: From Vision to Action" (Addis Ababa: African Union Commission, 2011), 45.
8 African Development Bank, "The Bank's Human Capital Strategy for Africa (2014–2018)" (African Development Bank, 2014).

Part III
Environment, health, and innovation

11 Conservation priorities and extractive industries in Africa

Opportunities for conflict prevention

Mahlette Betre, Marielle Canter Weikel, Romy Chevallier, Janet Edmond, Rosimeiry Portela, Zachary Wells, and Jennifer Blaha

In brief

- Africa, rich in natural resources including tropical forests, wildlife, water resources, and fisheries, is also rich in mineral, oil, and gas resources. Conservation priority areas across the continent often coincide with areas of importance for mineral development; national and local governments will need to decide about trade-offs inherent in competing development pathways.
- Case studies from Mozambique and Madagascar highlight the pressures, complexities, and responses in balancing extractive development and conservation. The Mozambique case study takes a broad view of a country on the verge of a mineral boom, and the steps it is taking to address conflicting land uses. The Madagascar case study highlights methods that estimate nature's benefits, relying on basic natural science data within a social context of multiple beneficiaries and economic sectors, to inform decision-making and management choices, particularly as they relate to conflicts between development and conservation needs.
- Despite recent fluctuations in commodity prices, there is a consensus that Africa is approaching an unprecedented mining boom. Therefore, governments and the mining industry alike must use cutting-edge tools, information, policy frameworks, effective institutions, and thoughtful planning to balance these development pathways.

Introduction

Africa has enormous natural resource wealth, much of it largely unexploited.[1] A closer look at specific natural resources, however, shows diverging trajectories with potentially alarming implications. On the one hand, resources such as tropical forests, wildlife, water resources, and fisheries, while relatively abundant and the source of livelihoods for millions of Africans, are being degraded and are declining at unprecedented rates. On the other hand, global demand for resources such as minerals, oil, and gas—also significant on the continent—is increasing, driving the economies of many African nations. At the intersection of these trajectories is the challenge to both conserve and, if possible, enhance Africa's rich biodiversity and ecosystems, while also harnessing its mineral wealth to drive development for current and future generations. The coincidence of mineral and biodiversity wealth implies that national and local governments will need to make decisions about the inherent trade-offs in these competing development pathways. While extractive development can affect a country in positive ways—as an economic catalyst, a significant revenue stream, and a creator of jobs and infrastructure—it can also have direct and indirect effects on the biodiversity and ecosystem services that the country's people depend on. According to the United Nations Environment Programme (UNEP), at least 40 percent of all intrastate conflicts can be linked to natural resources, and are often precipitated by environmental issues.[2] Strict conservation, however, can mean leaving valuable mineral wealth beneath the soil. It falls to government and society at large to evaluate and make decisions regarding potential trade-offs between mining and conservation. Cutting-edge tools, information, policy frameworks, effective institutions, and thoughtful planning are all needed to balance conservation and development.

In this chapter we examine the complex relationship between extractive development (with an emphasis on mining) and biodiversity. We outline Africa's current conservation priorities; review trends in mining and the environment; and examine two case studies (one from Mozambique and one from Madagascar) illustrating the pressures, complexities, and responses in balancing extractive development and conservation; and conclude with recommendations for improving decision-making and reducing conflict between these two objectives.

Africa's current conservation priorities

The second-largest continent, with more than 30 million square kilometers, is Africa, a land mass endowed with rich and abundant biodiversity (see Box 11.1).[3] The African landscape is a dynamic mosaic of ecosystems that millions of people depend on for their livelihood: tropical forests, woodlands, open plains, deserts, and saltwater and freshwater coastal areas. According to UNEP's *Africa Atlas*, Africa is home to about 1,200 mammal species (one-quarter of the global population), 2,000 bird species (one-fifth of the global population), 2,000 fish species, and 950 amphibian species. The continent also has between 40,000 and 60,000 plant species and about 100,000 known species of insects and arachnids.[4] This biodiversity, with some exceptions, is in better condition than in many parts of the world.[5]

Africa's rich biodiversity underpins the diverse ecosystems and services that benefit millions of people on the continent, particularly those living in rural and remote communities whose livelihoods are heavily dependent on water, forest products, and wildlife.[6] Benefits also include more-complex services that regulate and mitigate climate, protect against natural disasters, and support cultural heritage of nature-based traditions.[7]

Unfortunately, over the past fifty years biodiversity, ecosystems, and the services they provide have experienced substantial loss, having changed more rapidly

> **Box 11.1 Definitions**
>
> Biological diversity, or biodiversity is defined as "the variability among living organisms from all sources including terrestrial, marine and other aquatic ecosystems and the ecological complexes of which they are part; this includes diversity within species, between species and of ecosystems."[8] It comprises all the millions of different species that live on our planet, as well as the genetic differences within species. It also refers to the multitude of different ecosystems in which species form unique communities, interacting with one another and the air, water, and soil.
>
> Ecosystem services are "the benefits people obtain from ecosystems. These include provisioning services such as food and water; regulating services such as regulation of floods, drought, land degradation, and disease; supporting services such as soil formation and nutrient cycling; and cultural services such as recreational, spiritual, religious and other non-material benefits."[9] Ecosystem services are critical to the societies and economies of African nations.

and more extensively than in any comparable period in human history, largely in response to rapidly growing populations and their demands for food, freshwater, timber, fiber, and fuel.[10] Often considered free and invisible, these services tend not to get factored into decision-making and planning.

The effects of the loss and unsustainable use of ecosystem services are complex. While some of these effects are relatively simple and have been well studied (e.g., over 80 percent of the world's marine fisheries are either fully exploited or overexploited), others are more complex.[11] For example, there can be a time lag between an impact on an ecosystem and the resulting effects. Also, the effects of an impact might be apparent only at some distance from where the ecosystem was changed (e.g., forest fires in one country can degrade air quality in adjacent countries). Degradation of ecosystem services has been linked to exacerbating poverty in developing countries, thereby contributing to outbreaks of conflicts.[12]

In order to address biodiversity and ecosystem services loss and to guide conservation efforts, it is critical to understand which areas are priorities with respect to their biodiversity and ecosystem service values. This requires conservation planning that defines biodiversity and ecosystem services values generated by landscapes (that might or might not overlap spatially), and consideration of the effective conservation strategies given desired outcomes. There are many approaches to quantifying the importance of specific areas in terms of biodiversity. Two commonly used regional frameworks for large-scale areas are biodiversity hotspots and high-biodiversity wilderness areas (HBWAs). Finer-scale prioritizations include Key Biodiversity Areas (KBAs) and Alliance for Zero Extinction (AZE) sites. There are also legally designated areas for protection such as protected areas (PAs) and World Heritage Sites (WHS), each of which is described below.

Biodiversity hotspots are characterized as large regions with exceptional concentrations of endemic species—specific to an area and not found elsewhere—that are undergoing exceptional loss of habitat.[13] Currently, thirty-five biodiversity hotspots have been identified globally, most of them in tropical forests. Collectively, they represent just 2.3 percent of Earth's land surface, but contain around 50 percent of the world's endemic plant species and 42 percent of all terrestrial vertebrates. In addition to their exceptional biodiversity, these areas are also home to many vulnerable human populations that are directly dependent on nature to survive: by one estimate, hotspots account for 35 percent of the ecosystem services that vulnerable human populations depend on.[14]

Eight of the world's thirty-five biodiversity hotspots are in sub-Saharan Africa (see Figure 11.1):

1 Cape Floristic Province is located within South Africa; it is home to the greatest non-tropical concentration of higher plant species in the world.
2 The Coastal Forests of eastern Africa, though tiny and fragmented, stretch along the eastern edge of Africa, from small coastal patches of forest in southern Somalia, south through Kenya, into Tanzania, and along the coast of Mozambique.
3 The eastern Afromotane encompasses several widely scattered mountain ranges in eastern Africa and the Arabian Peninsula, from Saudi Arabia and Yemen in the north to Zimbabwe in the south.
4 The Guinean Forests of west Africa encompass all of the lowland forests of political west Africa, stretching from Guinea and Sierra Leone eastward to the Sanaga River in Cameroon.
5 The Horn of Africa covers most of Somalia; all of Djibouti; parts of Ethiopia, Eritrea, Kenya, Yemen, and Oman; and a small piece of far eastern Sudan.

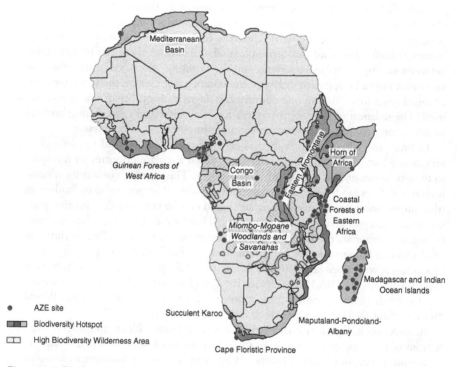

Figure 11.1 Biodiversity hotspots are in sub-Saharan Africa

6   Madagascar and the Indian Ocean islands.
7   Maputaland-Pondoland-Albany stretches along the east coast of southern Africa.
8   Succulent Karoo stretches along the Atlantic coast of Africa from southwestern South Africa north into southern Namibia.[15]

Like hotspots, HBWAs are also rich in endemic species. They are large regions (minimally 10,000 square kilometers), but have had relatively little human impact, and have lost 30 percent or less of their natural habitat.[16] Two of the five global HBWAs are found in Africa: (1) the Congo forests of Central Africa and (2) the Miombo-Mopane woodlands of southern Africa (including the Okavango Delta).

KBAs and AZE sites represent finer-scale priorities. KBAs are nationally defined, site-scale areas of global significance for biodiversity conservation based on the high irreplaceability and/or high vulnerability of the biodiversity they contain. Given their scale, (e.g., at the level of individual PAs) KBAs are often used as a starting point for landscape-level conservation planning.[17] The authors identified 881 KBAs in 49 countries in continental Africa and Madagascar, which excludes the island nations of Cabo Verde, Comoros, Mauritius, São Tomé and Príncipe, and the Seychelles. A subset of KBAs are AZE sites: considered the most critically important and threatened sites for biodiversity, AZE sites are characterized by the presence of 95 percent or greater of the known population of endangered or critically endangered species on the *International Union for Conservation of Nature* (IUCN) Red List of Threatened Species. If an AZE site were to disappear, it is almost guaranteed that a key species will become extinct in the wild.[18] In continental Africa and Madagascar, there are sixty-six AZE sites, with high concentrations of sites in Madagascar and eastern Africa (Figure 11.1).

All four categories—hotspots, HBWAs, KBAs, and AZE sites—provide methodologies for understanding where biodiversity is found and for prioritizing those areas for conservation or sustainable use. With such information on hand, governments are better able to make decisions about potentially competing land uses, including deciding where to prioritize conservation and where to situate development activity like mining.

The creation of legal PAs at a national or sub-national level is a common strategy for ensuring conservation of a nation's most important biodiversity areas. PAs are conservation management tools designed to safeguard important areas for nature and their associated ecosystem services and cultural values, through legal or other effective means.[19] Globally, there are more than 200,000 PAs, including national parks, nature reserves, and indigenous and communal areas, covering almost 15 percent of the world's land surface and almost 3 percent of its oceans.[20] Africa is home to 7,000 registered PAs. Despite this high level of commitment by African nations to the principles of conservation and sustainable development, in many cases where PAs are threatened by civil unrest, weak institutions, poorly trained staff, and limited budgets.[21] The PA systems in Africa have historically been based on a model whereby boundaries were established to keep species in and local communities out. Growing resistance to that model led to a rapid growth of community-based natural resource management and conservation activities related to PAs that strives to involve local communities more effectively, including initiatives to develop co-management regimes.[22]

Some national sites are also recognized under international agreements. WHS under the World Heritage Convention, for example, are places that have been designated as having outstanding universal value to humanity, and so have been inscribed on the World Heritage List by the United Nations Educational, Scientific and Cultural

Organization (UNESCO) to be protected for future generations. Globally, more than 750 natural, cultural, and mixed sites have been inscribed on the World Heritage List. Of the total number of WHS, 12 percent are located in Africa—93 sites that include places as diverse and unique as east Africa's Serengeti and the pyramids of Egypt.[23]

Unlike biodiversity, global- or national-scale priorities for ecosystem services have not been systematically defined. This is in large part because ecosystem services assessments require a good understanding of the socioeconomic context of a given area, and the identification of beneficiaries for individual services based on their specific needs; these data are not widely available at such fine scales. Additionally, ecosystem services assessments require careful consideration of the dynamics of ecosystem services flows in space and in time, as well as consideration of synergies and trade-offs, not only between different ecosystem services but also for different users of the same service.[24] Most common approaches for the assessment of ecosystem services rely on surveys and most frequently on biophysical modeling/mapping and valuation approaches. Ecosystem services mapping in Africa has focused most frequently on hydrological services (water and flow regulation), fodder provision, climate regulation, recreation, and tourism (see Box 11.2).[25]

Box 11.2 Climate change, mining risk, and shifting conservation priority areas

No continent will experience the impacts of climate change as harshly as Africa will. Climate change will dramatically amplify economic, social, and environmental vulnerabilities.[26] Projected impacts include temperature increases of 1 to 4.5°C, decreased rainfall, and intensified weather events such as heatwaves and drought, with gradients of severity across the continent. One of the most troubling impacts will be increased stress on water resources, amplifying existing stress on this essential resource from growing populations and heavy reliance on agriculture for both subsistence and livelihoods.[27] According to the UNEP, 75 million to 250 million people in Africa are projected to be exposed to increased water stress because of climate change.[28]

Flora and fauna can survive only within a specific range of climatic and environmental conditions, thus will be dramatically affected by climate change. Many regions throughout sub-Saharan Africa are already experiencing impacts from climate change, exposing the vulnerabilities of the continent's conservation priority areas. Studies have documented rapid shifts in geographic range, abundance, seasonal activities, and migration patterns of many terrestrial and freshwater species; projections indicate that these shifts will continue and accelerate in the coming years.[29] These shifts will expose the vulnerability of the millions of people that depend on the ecosystem services provided by conservation priority areas, including water, food, and livelihoods.

Mining is not immune to the impacts of climate change, with companies already reporting concerns about temperature increases that bump up cooling costs and endanger workers, extreme rainfall events that shut down mines for a week or more, storm surges that block ports, and persistent droughts that

> limit or restrict business operation.³⁰ As Africa's conservation priority areas and species shift due to the pressures of climate change, there will likely be new areas of intersection with mineral development, creating new opportunities for conflict. Declining ecosystem resilience will also make conservation priority areas more vulnerable to impacts from mining development. Where management of shared water resources might already be contentious, decreased water availability could further strain efforts to balance operational use and community needs.
>
> In order to maintain their social license to operate under a changing climate, companies will need to consider not only the impacts of their businesses, but also how their operations could compound climate-related impacts on nearby communities and ecosystems. Mining companies and national governments should work together to understand projected climate impacts and to integrate these considerations into land-use planning, risk management, and operational decision-making. Companies might consider climate modeling or analysis of existing data to better understand how climate change might impact the project site and surrounding area, in order to better anticipate potential conflicts.

Ecosystem services assessments generate data on the magnitude of ecosystem services (supplied or demanded), an understanding and estimate of potential bundles of ecosystem services, interactions and trade-offs between the services, and estimated market and non-market values of services, among others. These data in turn provide information and metrics that can be used toward other analytical approaches, including cost-effectiveness analysis of conservation interventions, environmental–economic accounting, and conservation decisions based on overlaps of biodiversity and human well-being.

Ecosystem services and biodiversity might coincide in space, and different conservation efforts might lead to different outcomes for biodiversity and services.³¹ Synergies and trade-offs are likely to occur, depending on priority. A careful exercise in conservation planning could help to identify biodiversity and ecosystem services values generated by landscapes, and could indicate the feasibility of potential conservation strategies. Indeed, one study argues that conservation planning could be used to identify the spatial alignment of benefits of biodiversity and ecosystem services—or lack thereof—and should include factors such as trade-offs and economic efficiency, including financial and opportunity costs. Conservation planning can inform complementary management options and solutions that are more effective than those focusing individually on societal objectives or nature conservation outcomes.³²

Trends in mining and the environment

The overlap of mining and African biodiversity

Over the past decades, population growth and rapid urbanization in China, India, and other emerging economies has driven unprecedented global demand for minerals and metals.³³ More recently, commodity prices have become more volatile and have trended downward, falling sharply during the global financial crisis in 2008, with a period of rebound before falling again in 2011 across many mineral and metal commodities. Despite the recent downward trend, many commodities remain at historically high prices (iron ore, gold, copper, nickel, zinc, platinum group metals, silver, and lead).[34]

Even with these price fluctuations, Africa is on the verge of an unprecedented mining boom.[35] According to a 2015 World Bank report, mining is one of sub-Saharan Africa's most important industries, hosting substantial portions of known global mineral reserves—22 percent of gold, 58 percent of cobalt, 7 percent of copper, 95 percent of platinum group metals, and 18 percent of uranium.[36] The continent's absolute spending on exploration rose more than 700 percent from 2002 to 2012, to about $3.1 billion in 2012.[37]

With demand for minerals on the rise, mining operations will likely expand into remote and undeveloped areas. To better understand how mining activity intersects with areas important for their biodiversity significance, in 2014 the authors conducted a spatial analysis overlaying mine projects and known conservation priority areas—namely biodiversity hotspots, PAs, WHS, HBWAs, KBAs, and AZE sites.[38]

From a total of 318 mine projects, the analysis revealed several instances of overlap between mine projects and conservation priority areas (see Table 11.1).[39] As expected, given their large scale, numerous overlaps occurred within hotspots and HBWAs, with forty-six mining operations coinciding with six of Africa's biodiversity hotspots, the majority falling within the Guinean Forests of west Africa biodiversity hotspot (see Figure 11.1). When examining finer-scale conservation priority areas, the analysis uncovered fourteen instances of mining exploration or exploitation within KBAs, with four active cases in South Africa and others in Botswana, Guinea, Liberia, Mauritania, Namibia, and Tanzania. Figure 11.2 illustrates 28 instances of direct overlap between mine projects and IUCN-category PAs. South Africa and Namibia contain the greatest number of coincidences, with seven overlap sites each. Mining activity was also found to coincide within one African WHS, which is also an AZE site—the Mount Nimba Strict Nature Reserve, shared by Guinea and Côte d'Ivoire—an overlap that led Guinea to degazette a portion of

Table 11.1 Results from analysis of mapping mine projects and conservation priority areas

Conservation Priority Areas	Coincidences with mine projects
CI hotspots	46
Protected areas	28
Key biodiversity areas	14
High biodiversity wilderness areas	11
Alliance for zero extinction sites	1
World heritage sites	1

Note: Total overlaps per conservation priority do not include closed or temporarily suspended mine projects, but do include those in the feasibility and advanced exploration stage.

Figure 11.2 Overlap between mine projects and IUCN-category protected areas

the WHS in 1992.[40] A separate study found at least twenty-one African WHS with mining or oil and gas operations within 50 kilometers of their borders.[41]

Mining's potential impact on priority conservation areas

While mining could have a positive impact on many African countries—as an economic catalyst, a significant revenue stream, and a creator of jobs and infrastructure—it also intersects with conservation priorities in important ways and could have negative impacts if not managed responsibly. The impacts typically associated with mining can vary significantly depending on the type of mining, the scale of mining activities, and how companies implement mitigation measures. Typical direct impacts include land clearance, water extraction, sedimentation, and pollution of terrestrial and aquatic ecosystems resulting in loss and/or degradation of habitat and associated species.[42] These impacts can be acute, in some cases irreversible, or chronic, lasting for decades.[43] Perhaps more challenging are the indirect, induced, and cumulative impacts that can occur beyond the boundaries of the project site and outside the project's direct control. Examples of induced impact include situations where a project has opened up a previously inaccessible natural frontier, resulting in habitat fragmentation and in-migration that can lead to increased collection of forest products and associated biodiversity loss; or bushmeat hunting, which has been linked to negative impacts on human health. Cumulative impacts could occur where mining projects are developed in environments that are influenced by other projects; examples include situations where multiple projects in the same water catchment collectively impact water quality or aquatic species health.[44]

The risk of not properly mitigating and managing for these impacts can be significant for companies, communities, and the environment. According to UNEP, at least 40 percent of all intrastate conflicts can be linked to natural resources.[45] Though environmental factors are not the sole reason for conflict, ongoing research suggests that environmental issues are among the most common issues precipitatingconflict, specifically pollution and access to or competition for environmental resources.[46]

Mitigating these impacts has become a central goal among civil society, environmental regulators, and leading extractive companies. Environmental and social impact assessments (ESIAs) are the prevailing tool used by governments and companies for evaluating projects, often driven by regulation in environment or mining codes. Impacts are identified, options to modify the project are assessed, and the project is either granted or denied a license. In many cases, projects' impacts on biodiversity and ecosystem services, and the consequences of these effects on human well-being, are not well understood; these critical interdependencies among the impact of mining, biodiversity, and affected stakeholders are often overlooked.[47] An ecosystem services approach to ESIAs advocates for a fuller understanding of both the socioeconomic dimensions of a project's environmental impacts and the implications of ecosystem change for affected stakeholders.[48] This approach provides a holistic, cross-sectoral, and integrated framework for assessing risks and impacts, as opposed to individually assessing each impact area separately.[49] Companies seeking to apply this approach are using tools such as "Weaving Ecosystem Services into Impact Assessment" and/or the "Natural Capital Protocol" to measure the impact of the business and its dependencies on natural capital, and to make decisions accordingly.[50]

Best practice also dictates that the mitigation hierarchy be applied for biodiversity management.[51] The mitigation hierarchy is an iterative approach to first avoiding significant impacts to biodiversity and ecosystem services, then minimizing and rehabilitating, and finally offsetting any remaining significant negative impacts.[52] Biodiversity offsets, the final step in the mitigation hierarchy, are conservation actions intended to compensate for the residual, unavoidable impacts of an activity on biodiversity values by generating an equivalent benefit elsewhere.[53] Many governments are increasingly requiring offsets for projects resulting in significant residual impacts to biodiversity: at least fifty-six countries have current laws or policies on biodiversity offsets or compensatory conservation.[54] A number of leading mining companies have also adopted voluntary commitments to biodiversity offsetting where there is no such regulatory requirements. An example is the Ambatovy mine in Madagascar, which is partnering with a number of conservation NGOs to implement a "multifaceted biodiversity offsets program [which] comprises a series of conservation areas, with a total area approximately fifteen times the size of the Mine Footprint."[55]

Multi-stakeholder initiatives such as the international collaboration Business and Biodiversity Offsets Programme have emerged to provide guidance on how to design and implement biodiversity offsets and to provide a platform for practitioners to share and exchange ideas on the subject. Through the full application of the mitigation hierarchy and best practice mitigation, a development activity could result in either a no-net-loss or a net-gain in biodiversity.[56] Despite the considerable discourse on biodiversity offsets, it's important for projects to apply a precautionary approach to the initial placement and environmental management of mining projects, emphasizing the avoidance step in the mitigation hierarchy.[57] This is particularly true in areas of high biodiversity and with endemic species, where the potential risks around irreversible impacts to critical environmental values are higher.

Beyond the no-go debate

Despite efforts to understand and mitigate environmental impacts from mining, extractive development remains contentious in areas where mineral and environmental values overlap. One topic that has received a lot of attention among mining companies, conservation groups, and governments is the issue of mining and PAs. As recognized by the Convention on Biological Diversity, first convened by the UNEP in 1988, PAs are intended to ensure the survival of a nation's most important natural wealth. PAs are legal mechanisms for conservation of specific areas and any development activity inside or adjacent to PAs should be strictly governed to minimize damage to the biodiversity values to be conserved. There are different categorizations of PAs, ranging from Strict Nature Reserves (Category IA) to Protected Area with sustainable use of natural resources (Category VI), which are typically more modified habitats and could allow for multiple uses (see Appendix I, this chapter, for details).[58] Approaches to mining vary depending on the categorization.

The pressures from mining and other major development activities on Africa's PAs has long been known to the international conservation community, which has increasingly called for policies that amount to making some PAs no-go areas for mining and other forms of development. The concept of no-go gained momentum in 2000 when the IUCN World Conservation Congress recommended that mining be banned from IUCN Category I–IV PAs (see Appendix I for more detail). To date, much of the focus on no-go areas has centered on natural or mixed WHS and IUCN Categories I–IV PAs. At the 2016 World Conservation Congress, a quadrennial event organized by the IUCN that brings together government officials, business leaders, academics, and civil society, a motion was passed that calls on governments to prohibit environmentally damaging industrial activities and infrastructure development in all IUCN PA categories; the motion also calls on the business community to respect all PA categories as no-go areas, withdrawing from current and avoiding future activities in those areas.[59] Though this motion is not legally binding, it sends a strong message to governments and business alike to closely manage and monitor industrial activities within or adjacent to all PAs (see Box 11.3).

> **Box 11.3 Protected Area downgrading, downsizing, and degazettement (PADDD)**
>
> It is also possible that the protection afforded to important biodiversity areas could be weakened to allow for extractive development. The phenomenon of PA downgrading, downsizing, and degazettement, known as PADDD, is a legal process by which national parks, nature reserves, and other PAs lose their protection due to conflicting land-use priorities including mining, agriculture, infrastructure, and oil and gas projects. PADDD can also be exacerbated by weak institutions, corruption, and the drive for economic growth even when development conflicts with environmental sustainability, all of which are relevant in many parts of Africa.
>
> A recent study on PADDD found that of the 342 events analyzed, 43 percent were driven by economic development, including forestry; industrial agriculture; and mining, oil, and gas exploration.[60] Mining is a particular threat to PAs in countries with high biodiversity and robust mining sectors.[61] This is the case in Central and west Africa, and is exemplified by the removal of protection from a portion of the Mount Nimba National Park in the face of prospective mining.[62]

Many international mining companies view environmental impact management through a lens of mitigating financial, legal, and reputational risk. The threats of lost profits, costly court cases, and damage to the social license to operate are real factors in determining corporate commitments to the environment. At the fifth World Parks Congress, in 2003, the International Council on Mining and Metals (ICMM), a consortium of twenty-three mining companies and more than thirty-four regional and national mining associations, adopted a resolution to avoid mining in WHS.[63] Uptake has not been uniform across the broader sector, nor have governments applied the same restrictions, however. At the 2014 World Parks Congress several conservation organizations called for all current and future mining activities to be barred from WHS, holding that the natural heritage of these areas is incompatible with mineral exploration and development.

Movements such as the push for no-go areas can be complex, however, when it comes to implementation, especially when considering issues such as PADDD (Box 11.3). Although barring future development from the most important biodiversity areas, which are not confined to PAs, seems reasonable, the fact that many mines already exist and operate within these areas necessitates more-nuanced solutions in some cases. In addition, comparatively junior operators tend to lag behind the major mining companies in terms of systematically applying best practice sustainability standards such as the mitigation hierarchy and avoiding important biodiversity areas. This situation represents a challenge because governments often fail to apply no-go policies in an even and consistent manner. In the absence of uniform application, larger companies with more-advanced environmental management systems might avoid operating in areas that are important for biodiversity and ecosystem services, leaving junior companies with less-rigorous standards to fill the gap, a scenario that could result in perversely worse environmental outcomes.[64] Governments might also forgo formal protection of important places to allow for the option to develop land in the future.

The public and private sectors, and civil society, need to collaborate to value natural wealth, evaluate the trade-offs between potential development pathways, plan for the sustainable use of natural resources, and create thoughtful policy to regulate the development agenda. This might necessitate expansion of government institutions' mandates, creating coordinating mechanisms among various government entities. It could also require creation of new entities that have mandates to and are capable of implementing policies that promote more-sustainable mineral development. Moreover, wherever extractive development is to occur, care must be taken to ensure that potential ESIAs are understood and all efforts are made to mitigate them. It is also important to move beyond the almost exclusive focus on business impacts on nature, and to address businesses' critical dependencies on nature as well. Failing to do so is likely to pose serious risks for businesses—and to their supply chain sustainability, while precluding opportunities for increased efficiency and innovative approaches.

Two case studies—one from Mozambique and one from Madagascar—illustrate the pressures, complexities, and responses in balancing mining development and conservation. The first case study takes a broad view of the current state in Mozambique, which is on the verge of a mineral boom, and the steps it is taking to address conflicting land uses. The second case study highlights the use of analytical and modeling approaches in a PA in Madagascar to understand the role of natural capital and its contribution to the provision of ecosystem services, and to analyze the role of ecosystem services as inputs to various economic sectors.

Case studies

Case study: conservation and mining development in Mozambique

Context

Mozambique has a population of 28 million people dispersed over approximately 800,000 square kilometers.[65] The country is richly endowed with a variety of mineral resources, especially gas, coal, oil, heavy-sand deposits, gold, and copper, among others. Mining is propelling current GDP growth, but there are major challenges related to poverty and inequitable income, environmental sustainability, and risk related to environmental disasters. Mozambique is in transition, with a president elected in October 2014 and the launch of expected natural gas projects that will likely alter the country's economic and social landscape. The government's National Strategy for Development (2015–2035) places particular emphasis on industrialization and key priority areas of agriculture, fisheries, industrial diversification, infrastructure, and extractive industries development.[66]

Mozambique is rich in biodiversity. The country is characterized by varied ecological systems: the north coast is marked by coral reefs, mangroves, and small islands, and the southern region is characterized by high dunes covered with dense forest. Three biodiversity hotspots can be found within the country: the Coastal Forests of eastern Africa, the Maputaland-Pondoland-Albany and the eastern Afromontane (see Figure 11.1). Mozambique is home to 5,500 species of flora and 4,271 species of terrestrial wildlife, several of which are endemic to Mozambique.[67] Conservation areas cover about 25 percent of Mozambique's surface area, and the country lays claim to the largest African coastal marine reserve, the Primeiras and Segundas Protected Areas.[68]

The extractive industry, namely mining and hydrocarbons, position Mozambique as an important player in the global mining sector.[69] In 2014 Mozambique's GDP grew by 7.6 percent, boosted by the construction, transportation, and communications sectors. This growth includes the establishment of large infrastructural development projects particularly related to mining in the center and north of the country.[70]

Rio Tinto, Vale, and Jindal have invested heavily in developing Mozambique's coal deposits, which are believed to be the fourth-largest untapped recoverable coal reserves in the world.[71] There have also been significant offshore gas finds in the north of Mozambique, where the total known reserves in the Rovuma Basin are 180 trillion cubic feet of gas, with production expected at 20 million tonnes per year. The 2014 Extractive Industries Transparency Initiative (EITI) report demonstrates that, in 2012, government revenues from the extractive industries reached approximately $100 million, increasing by almost 60 percent over the previous year.[72]

Conflict between extractive industries and the environmental sector

There has clearly been an influx of new wealth in Mozambique from oil, coal, and gas deposits, leading to a spike in on- and offshore exploration and mining in sensitive ecological areas. Mineral and hydrocarbon deposits frequently coincide in Mozambique with natural resources such as fisheries, mangroves, wetlands, estuaries, and coastal forests. Some examples of these competing or conflicting land uses include the following, among many others:

- The environmental considerations related to gas liquefaction in near proximity to the Quirimbas National Marine Park. This includes concerns over the development of onshore pipelines, dredging, and the development of port facilities in the Rovuma Basin.[73]
- Sasol's offshore gas exploration in Bazaruto Archipelago represents a conflict between gas exploration, tourism, sensitive biodiversity (dugong), and artisanal fishing communities.[74]
- Oil and gas prospecting and exploration in the Marromeu Complex (Ramsar Site) in the Zambezi Delta of central Mozambique. The Ramsar Mission's advisory report expressed concerns over drilling activities, including forest clearance for the construction of drill sites, as well as adjacent camps and roads, and recommended that the area be declared a priority zone and that a management plan be formulated to address these conflicting interests.[75]
- In southern Mozambique plans are proposed to develop a deepwater port in Techobanine. The site is in the Maputo Special Reserve and Ponta do Ouro Partial Marine Reserve, both popular tourist destinations that are rich in marine

life and endangered biodiversity. Botswana, Mozambique, and Zimbabwe signed a memorandum of understanding in 2012 on the construction of this port and a 1,100-kilometer railway line to link southern Mozambique to Botswana through Zimbabwe. Botswana specifically looks to the project to transport coal to Asian markets and to source fuel from Mozambique, reducing its reliance on South Africa.[76]

Responses to mineral development and environmental trade-offs

Mozambique has approached these conflicts and assessed trade-offs in a number of ways, including the use of tools and mechanisms to provide for integrated spatial planning tools; streamlined institutions and cross-sector platforms; and alignment of legal frameworks with best practice standards.

INTEGRATED SPATIAL PLANNING TOOLS

Spatial planning tools, such as strategic environmental assessments (SEA) and environmental management frameworks, are useful mechanisms to balance multiple objectives related to the management of natural resources.[77] Recent Mozambican reforms, such as the new mining law, include more-stringent environmental requirements, including requirements for differing levels of environmental impact assessments depending on the classification of the mine.[78] The government also intends the SEA to take account of environmental policy in Mozambique's broader planning guidelines (e.g., local spatial plans, municipal master plans, detailed plans, development plans) to avoid potential adverse environmental effects in the first case, and reinforced in the second. It appears that Mozambique as a nation is gradually moving toward more-rigorous environmental requirements.

This approach is evident, for example, in Mozambique's draft coastal SEA, which provides an overview of coastal zone dynamics, and contains important recommendations for harmonizing development and economic progress with the maintenance of natural coastal systems. The report considers divergent economic activities, including oil and gas exploration, mining and infrastructure development, fisheries and aquaculture development, tourism, biodiversity, and community livelihoods. It also identifies high-priority concerns to improve Mozambique's legislative framework for governing biodiversity in coastal areas.[79] The government has also initiated a localized SEA for the Bazaruto Archipelago National Park to assess the costs and benefits of conservation. The associated management strategy of that national park (2014–2018) includes tools to reduce the conflicts of interest and minimize environmental impacts.

STREAMLINED INSTITUTIONS AND CROSS-SECTOR PLATFORMS

In 2014 President Nyussi indicated that he wanted a leaner, more-technocratic public sector to better align national objectives, strategies, and mandates. As such there was a merger of the Ministry of Environmental Coordination and the Institutes for Land and Rural Development to form the Ministry of Land, Environment, and Rural Development, which could streamline future environmental planning.[80]

Centralized interdisciplinary forums of horizontal coordination are being developed within and across departments in the national government to facilitate coordination and to examine overlap between economic development, energy, water, and biodiversity. Mozambique has established a National Council for Sustainable Development (CONDES), with a mandate to promote and coordinate the sustainable use of natural resources. This council includes representatives of all ministries involved in environmental affairs and other relevant departments, including the ministries of Energy and Mineral Resources. It has also established environmental units and focal points for the environment in the ministries of Agriculture, Energy, Health, Mineral Resources, and Public Works.

Equally important is a devolved, vertical system of governance among spheres of government that includes partnerships and co-management structures between municipalities, local governments, and communities. Communities play an increasingly prominent role in the management of resources through co-management institutions such as the community fisheries councils that are on the ground and can disseminate key lessons to coastal communities, clarify park boundaries, and enforce fishing regulations.[81]

ALIGNMENT OF LEGAL FRAMEWORKS WITH BEST PRACTICE STANDARDS

In 2014 Mozambique revised its fiscal and legal framework for the mining and hydrocarbons sector to align with international standards. The country introduced requirements for local participation in foreign direct investment–funded projects that will increase revenues from the sector. Over the past years legislation and policy has also evolved to improve the contribution of mining and hydrocarbons to good governance, socioeconomic uplift, and environmental sustainability.[82]

Mozambique's Conservation Areas Act of 2014 both safeguards critical areas for conservation from development and requires best practice mitigation measures such as biodiversity offsets. The Act divides conservation areas into areas of total conservation—that cover nature reserves and national parks, where no hunting, agriculture, logging, or mining are permitted, and areas of sustainable use that include special reserves, environmental protection areas, official hunting areas, community conservation areas, wildlife sanctuaries, and private wildlife farms. This Act also stipulates that any exploitation of natural resources within conservation areas or its buffers must compensate for impacts and ensure no net loss of biodiversity.[83]

Case study: valuing ecosystem services in Madagascar

Context

Madagascar is an island nation of 538,295 square kilometers off the eastern coast of Africa, with a population of 23.5 million people in 2014.[84] Known for its high degree of endemic biodiversity, it is home to nearly 5 percent of the world's biodiversity, despite only constituting an area of less than 0.5 percent of the Earth's total landmass. Madagascar is home to 212 KBAs and 21 AZE sites, as well as several ecosystem services important to the nation's population. These services, including water-related services, fuel wood, wildlife, and fish for food provision and carbon sequestration, are also of international importance.[85]

Madagascar's main exports are agricultural and mineral, with overall GDP in 2014 at $10.6 billion and mining as 17.8 percent of total export contributions.[86] The first-ever development license in the country for on- and off-shore oil and gas exploration and development was issued in April 2015 by Madagascar Oil.[87] In spite of this resource richness, the United Nations Development Program's (UNDP's) Human Development Index—which measures relative life expectancy, education, and income per capita using a range of indicators—ranked the country low globally at 158 out of 188 in 2015.[88] Political unrest in the first part of the twenty-first century, marked by a coup in 2009, has compounded these development challenges. A democratic presidential election in 2013 was seen by many as a step toward government stability, but lingering uncertainty has contributed to reduced foreign direct investment in the mining, tourism, and textile sectors.[89]

With the exceptional natural resources of Madagascar, it is no surprise that conservation and extractive interests overlap with some frequency, creating potential for conflict.

Assessment of ecosystem services values in the CAZ

In 2011, as part of the Wealth Accounting and the Valuation of Ecosystem Services (or WAVES) program of work, the World Bank commissioned a preliminary rapid assessment of ecosystem services in Madagascar's Ankeniheny-Zahamena Forestry Corridor (CAZ).[90] The assessment demonstrated the economic importance of selected ecosystem services including water supply, sediment retention, and carbon storage and sequestration. Using both biophysical modeling and economic analyses, the complex nature of ecosystem service provision and the role of ecosystem services as inputs to sectors such as agriculture, mining, tourism, and hydroelectricity were analyzed.[91] Unlike many other ecosystem services assessment studies, the methodology for this biophysical analysis emphasized the spatial dynamics of ecosystem services flows and the use of those services by beneficiary groups.[92]

The CAZ is a PA located in the eastern part of Madagascar. It harbors the largest remaining contiguous patch of humid forest in the country, and is surrounded by multiple land uses such as agriculture, mining, villages, national parks, and reserves. With a surface area of 381,000 hectares, the CAZ's forests, wetlands, and rivers are home to more than two thousand species of plants, mammals, amphibians, and birds, many of which are endemic to the region. Located atop an escarpment, the CAZ provides substantial services in the provision of water and control of sediment flow to the surrounding areas. Approximately 350,000 people live in the CAZ, primarily members of rural communities who practice both subsistence agriculture and cash crop production. This PA is in close proximity to the Ambatovy nickel and cobalt mining project (roughly 70 kilometers to the north-east). This mining project is Madagascar's largest foreign investment project and one of the largest for all of sub-Saharan Africa.[93]

Conservation International used biophysical modeling, combining both physical data (such as precipitation, slope, and soil type) and social data (such as the level of need and the location of beneficiary groups) to connect the landscape and social data with ecosystem service physical outputs.[94] Researchers used economic analysis, focusing on climate regulation and water supply, to complete the link between the landscape and the economy in a scaled-up context.[95]

The study findings showed the following: (1) Based on the biophysical analysis of water and sediment regulation, the CAZ site demonstrated the potential to sustain much greater water demand compared to sites outside the corridor, where there are critical demands for water. (2) Water quality, measured as reduced sediment load, was estimated to be significantly better inside the CAZ area than outside, highlighting the role the PA plays in preventing sediment contamination of the water supply. (3) The economic value of water varies by sector. Water use efficiency was found to be greater in the region's agricultural and tourism sectors. The marginal economic value of water (an estimate of the economic value per unit of output of the production sector in question) was greatest in the mining sector. (4) Carbon sequestration values were found to be very high in the CAZ, suggesting the area has high value as a carbon pool and sink.

Implications

The results from this study are relevant given the need to balance competing land use demands in Madagascar, as well as in other countries. While rapid and preliminary in scope, it nevertheless provides insights for climate mitigation strategy and PA planning—revealing the high economic value of carbon sequestration that CAZ offers, and highlighting the potential benefits of engaging in carbon markets to generate revenues for PA management and improved livelihoods. The study demonstrates the ability to quantify the contribution of natural capital to a regional economy, and eventually a national accounting framework that appreciates the economic benefits of PAs. The assessment conducted is also informative for regional and national integrated water resources management planning and policy, whereby policymakers need to address the individual and competing needs of different sectors. From an economic standpoint, because the marginal economic value of water is the greatest for the mining sector, the relative importance of water for mining is clear. At the same time, the case study of Madagascar demonstrates the importance of measures that would ensure the sustainability and quality in water supply, a role clearly played by forest ecosystems in the region.

Ultimately, this analysis demonstrates that conservation of CAZ provides important benefits to mining (as well as other sectors, including agriculture, tourism, and energy production) and hints at the need for careful consideration of individual and competing needs of different sectors. Though beyond the scope of this study, further research that also assesses equity in the distribution of water resources and potential negative impacts of mining on water resources could provide a more holistic view of these complex values, user needs, and potential risks.

RECOMMENDATIONS FOR HOW TO ADDRESS TRADE-OFFS

As extractive development continues to play a significant role in the economies of many African nations, the continent's conservation priority areas will continue to be at odds with this and other competing and potentially threatening land uses.

The importance of biodiversity-rich areas and the ecosystem services they provide cannot be overemphasized—from their intrinsic value, to the multitude of benefits they provide to people and economies, to the role they play in underpinning stable global systems. The question, then, is: How can we both develop minerals for economic growth and support healthy natural systems, for contemporary and future generations? We offer some recommendations to address potential trade-offs, improve decision-making and prevent conflicts:

- Understand and value natural capital. Before engaging in a discussion about competing land uses and trade-offs, it is important to understand the economic, environmental, social, and cultural values and assets found within a landscape or at a national level. Too often, natural capital (the stock of natural assets and the flows of ecosystem services they provide) is not well understood, and in many cases goes unvalued.[96] For governments, at a national level, this can mean engaging in natural capital accounting to appreciate and account for the full contribution that nature provides in a country's wealth. The Madagascar case study demonstrates the utility of such assessments in identifying nature benefits with tangible economic value to different sectors and stakeholders. Initiatives like the Wealth Accounting and the Valuation of Ecosystem Services (WAVES) partnership led by the World Bank and the "Gaborone Declaration for Sustainability in Africa" are engaging many African governments to advance and demonstrate this concept in practice.[97]
- Conduct integrated spatial planning. At local and regional levels, a better understanding of complex natural capital values, along with other assets, is critical to inform effective integrated spatial planning processes (including SEAs, land use, or development plans). As highlighted in the Mozambique case study, governments could consider leading integrated planning processes, which aim to maximize not only traditional economic objectives but also seek to integrate biological and sociocultural, ones as well.
- Promote cross-sectoral linkages. Given the multiple dimensions inherent in land use and development planning processes, governments should consider establishing formal mechanisms that facilitate discourse, collaboration, coordination, and planning across different constituencies. Mozambique's National Council for Sustainable Development, with a mandate to promote and coordinate the sustainable use of natural resources is an example of one such mechanism. Similarly, Madagascar set up an interministerial committee to address some of the overlaps between mineral exploitation concessions and PAs in an effort to address these potentially competing interests.[98]
- Create a tiered approach to PAs and development. Governments could consider establishing a tiered approach to PAs and development whereby they give certain areas strict protection and apply no-go rules for the most critical areas (e.g., PAs I–IV, WHS, and AZE sites), while deeming other areas as more flexible, allowing extractive activity to overlap with environmental values; in those areas governments would mandate that good management practices be applied. Around the world, tiered approaches to PA systems are common. In Liberia, for example, the Protected Forest Area Network Act of 2003 commits the country to establishing a network of biologically representative PAs covering 30 percent of forest area. The Act specifically precludes

mining from several categories of PAs—national park, nature reserve, communal forest, cultural site, and game reserve—that align with higher-level IUCN PA categories.
- Require responsible practices for extractive development. Where extractive development does occur, governments could consider mandating that companies implement leading environmental and social management practices and companies could voluntarily apply these best practices. The 2012 International Finance Corporation's performance standards provide a good benchmark for how governments can frame expectations for companies and guidelines that companies should strive toward.[99] Themes such as comprehensive ESIAs that also consider health and human rights, and application of the mitigation hierarchy including biodiversity offsets, are particularly relevant. Comprehensive ESIAs that systematically include biodiversity and ecosystem services could provide regulators and other concerned stakeholders with an opportunity to review predicted impacts and mitigation measures for a development proposal before it is approved, thereby avoiding potential conflicts. A precautionary approach to ESIAs whereby important areas receive de facto protection while proposed activities are adequately assessed can also be an effective approach. Namibia, among other countries, modeled this when it placed an eighteen-month moratorium on planned marine phosphate mining off its coast in 2013 until an environmental impact assessment could be performed that demonstrated that mining would not harm the country's fishing industry.[100] The International Finance Corporation's Performance Standard 6 specifically sets a target for companies to demonstrate no net loss to biodiversity for natural habitat and net gain for critical habitat by applying the mitigation hierarchy, and, where applicable, implementing biodiversity offsets.[101] Performance Standard 1 also requires companies to consider a project's vulnerability to climate change in addition to the potential to increase the vulnerability of ecosystems and communities as part of the risks and impacts of the identification process.[102] Companies should also consider mainstreaming natural capital into business decisions. The Natural Capital Coalition has recently launched the Natural Capital Protocol—a standardized approach for business assessment of their impacts and dependencies on nature, led by the Natural Capital Coalition—with an aim to support better management and decision-making, and to facilitate identification of risks that might constrain business development or provide business opportunities for innovation, new markets, and better financing. Piloting the Protocol for a chosen application, such as assessment of risks and opportunities, comparison of options, assessment of impact on stakeholders, among other uses, has the great potential to support companies' enhanced understanding of biodiversity and ecosystem service values—both to business and to society within their projects' areas of influence.
- Utilize sound data and effective stakeholder engagement. For any of these efforts to be successful, it is critical to have sound data and thorough and inclusive stakeholder engagement. Best available scientific information on biodiversity

and ecosystem services, based on biological and sociocultural data, is needed to make informed land use decisions and to underpin robust ESIAs, yet there is often a shortage of baseline data, particularly in developing countries.[103] Equally important is stakeholder engagement. Individuals and communities are often important sources of information on biodiversity and ecosystem services, as well as potential partners in land use and management plans. To be meaningful, engagement needs to be transparent and to begin early in the process. Civil society groups have a role to play, engaging with communities, governments, and the private sector, to support collection of scientific information and to facilitate effective stakeholder engagement processes.

Conclusion

In conclusion, as the analysis and case studies in this chapter illustrate, conservation and development planning are complex processes that involve addressing the competing—and sometime seemingly incompatible—interests of different natural resource stakeholders. Making decisions about the environment and its management typically involves difficult trade-offs resulting in both winners and losers. Taken together, our recommendations call for more-transparent and more-inclusive dialogues among government, private sector companies, and communities; and for rigorous assessments on the values of biodiversity and ecosystem services, to support better-informed, value-based policies and practices as the first step to prevent or mitigate conflict among mining and conservation priorities in Africa.

Appendix I

IUCN protected area management categories[104]	Description
IA—Strict nature reserve	"Strictly protected areas set aside to protect biodiversity and also possibly geological/geomorphological features, where human visitation, use and impacts are strictly controlled and limited to ensure protection of the conservation values."
IB—Wilderness area	"Large unmodified or slightly modified areas, retaining their natural character and influence, without permanent or significant human habitation, which are protected and managed so as to preserve their natural condition."
II—National park	"Large natural or near natural areas set aside to protect large-scale ecological processes, along with the complement of species and ecosystems characteristic of the area, which also provide a foundation for environmentally and culturally compatible spiritual, scientific, educational, recreational and visitor opportunities."

(continued)

Appendix I (continued)

IUCN protected area management categories[104]	Description
III—Natural monument or feature	"Areas set aside to protect a specific natural monument, which can be a landform, sea mount, submarine cavern, geological feature such as a cave or even a living feature such as an ancient grove They are generally quite small protected areas and often have high visitor value."
IV—Habitat/ species management area	"Protected areas aiming to protect particular species or habitats and management reflect this priority. Many category IV protected areas will need regular, active interventions to address the requirements of particular species or to maintain habitats, but this is not a requirement of the category."
V—Protected landscape/ seascape	"Areas where the interaction of people and nature over time has produced an area of distinct character with significant ecological, biological, cultural and scenic value and where safeguarding the integrity of this interaction is vital to protecting and sustaining the area and its associated nature conservation and other values."
VI—Protected area with sustainable use of natural resources	"Generally large, with most of the area in a natural condition, where a proportion is under sustainable natural resource management and where low-level non-industrial use of natural resources compatible with nature conservation is seen as one of the main aims of the area."

Source: UNEP-WCMC.

Notes

1 African Development Bank Group (ADBG), "African Natural Resources Center Strategy – Revised Version" (African Development Bank Group, June 2015), www.afdb.org/fileadmin/uploads/afdb/Documents/Generic-Documents/African_Natural_Resources_Center%E2%80%99s_Strategy_for_2015-2020.pdf.
2 United Nations Environment Programme (UNEP), "Toolkit and Guidance for Preventing and Managing Land and Natural Resource Conflict" (Nairobi: UN Framework Team for Preventive Action, 2012); Daniel Franks, *Mountain Movers: Mining, Sustainability and the Agents of Change* (London: Routledge, 2015); UNEP, "From Conflict to Peacebuilding the Role of Natural Resources and the Environment" (UNEP, 2009), www.unep.org/pdf/pcdmb_policy_01.pdf.
3 UNEP, "Biodiversity in Africa," *Africa Environment Outlook 2: Our Environment, Our Wealth* (Malta: Progress Press Inc., 2011). www.eoearth.org/view/article/150570/.
4 UNEP, *Africa: Atlas of Our Changing Environment* (Malta: Progress Press Inc., 2008).
5 UNEP, "Biodiversity in Africa."
6 N. Benis Egoh, J. Patrick O'Farrell, Aymen Charef, Leigh Josephine Gurney, Thomas Koellner, Henry Nibam Abi, Mody Egoh, and Louise Willemen. "An African Account of Ecosystem Service Provision: Use, Threats and Policy Options for Sustainable Livelihoods," *Ecosystem Services* 2 (2012): 71–81

7 Department of Environmental Affairs (DEA), Department of Mineral Resources, Chamber of Mines, South African Mining and Biodiversity Forum, and South African National Biodiversity Institute, "Mining and Biodiversity Guideline: Mainstreaming biodiversity Into the Mining" (Pretoria: Department of Environmental Affairs, 2013), www.environment.gov.za/sites/default/files/legislations/miningbiodiversity_guidelines 2013.pdf.
8 Convention on Biological Diversity (CBD), "Convention on Biological Diversity" (Montreal, Canada: Secretariat of the Convention on Biological Diversity, 1992), 3.
9 R. Hassan, R. Scholes, and N. Ash, *Millennium Ecosystem Assessment: Ecosystems and Human Wellbeing, Volume 1, Current State and Trends* (Washington, DC: Island Press, 2005), 1.
10 R. Hassan, R. Scholes and N. Ash (eds.), "Ecosystems and Human Well-being: Synthesis" (Washington DC: Millennium Ecosystem Assessment Series, Island Press, 2005).
11 "fully exploited or overexploited": FAO, "Review of the state of world marine fishery resources." *FAO Fisheries and Aquaculture Technical Paper No. 569* (Rome: FAO, 2011).
12 Millennium Ecosystem Assessment (MA). R. Hassan R. Scholes N. Ash (eds.), "Ecosystems and Human Well-being: Synthesis."
13 Norman Myers, Russell A. Mittermeier, Cristina G. Mittermeier, Gustavo A. B. da Fonseca, and Jennifer Kent. "Biodiversity Hotspots for Conservation Priorities," *Nature* 403 (2000): 853–858.
14 Conservation International, "Hotspots." www.conservation.org/How/Pages/Hotspots.aspx (accessed November 12, 2015).
15 CEPF, "Biodiversity Hotspots in Africa," www.cepf.net/resources/hotspots/africa/Pages/default.aspx (accessed November 9, 2015).
16 R. A. Mittermeier, C. G. Mittermeier, T. M. Brooks, J. D. Pilgrim, W. R. Konstant, G. A. B. da Fonseca, and C. Kormos, "Wilderness and Biodiversity Conservation," *Proceedings of the National Academy of Sciences* 100:18 (2003): 10309–10313.
17 Langhammer, P. F., M. I. Bakarr, L. A. Bennun, T. M. Brooks, R. P. Clay, W. Darwall, N. De Silva, G. J. Edgar, G. Eken, L. D. C. Fishpool, G. A. B. da Fonseca, M. N. Foster, D. H. Knox, P. Matiku, E. A. Radford, A. S. L. Rodrigues, P. Salaman, W. Sechrest, and A. W. Tordoff, "Identification and gap Analysis of Key Biodiversity Areas: Targets for Comprehensive Protected Area Systems," *Best Practice Protected Area Guidelines Series* 15 (Gland: IUCN, 2007).
18 AZE, "Home," (AZE, 2013), www.zeroextinction.org/index.html.
19 UNEP-WCMC, "Protected Areas," *Biodiversity A-Z*, (November 20, 2014). http://biodiversitya-z.org/content/protected-areas.
20 IUCN, "What are Protected Areas?" (IUCN, 2014), http://worldparkscongress.org/about/what_are_protected_areas.html.
21 IUCN, "IUCN World Commission on Protected Areas – West and Central Africa Region" (IUCN: December 3, 2015), www.iucn.org/about/work/programmes/gpap_home/gpap_wcpa/gpap_wcparegion/gpap_westandcentralafrica/.
22 Ibid.; J. Hanks, and C. A. M. Atwell, "Financing Africa's Protected Areas. Introduction to the Sustainable Finance Stream and the Policy Context for Protected Area Financing" (Durban: Prepared for Vth World Parks Congress, September 2003).
23 UNESCO World Heritage List, "Search by Region – Africa", http://whc.unesco.org/en/list/CID=31&SEARCH=&SEARCH_BY_COUNTRY=&REGION=5&ORDER=&TYPE=&&order=year (accessed August 30, 2017).
24 Ferdinando Villa, "Assessing Biophysical and Economic Dimensions of Societal Value: An Example for Water Ecosystem Services in Madagascar," in *Water and Ecosystem Services*, Julian Martin-Ortega, Robert C. Ferrier, Iain J. Gordon, and Shahbaz Khan (eds.) (Cambridge, UK: Cambridge University Press, 2015).
25 Egoh et al., "An African Account," 71–81.
26 AMCEN Secretariat, "Fact Sheet Climate Change in Africa – What is at Stake?" (Excerpts from IPCC reports, the Convention, & BAP Compiled by AMCEN Secretariat).

27 Climate & Development Knowledge Network, "The IPCC's Fifth Assessment Report: What's in it for Africa?" (Climate & Development Knowledge Network, 2014).
28 Kamal Baher, "Climate: Africa's Human Existence is at Severe Risk." *Inter Press Service News Agency*, April 21, 2016.
29 I. Niang, O. C. Ruppel, M. A. Abdrabo, A. Essel, C. Lennard, J. Padgham, and P. Urquhart, "Africa," in *Climate Change 2014: Impacts, Adaptation, and Vulnerability. Part B: Regional Aspects* (Contribution of Working Group II to the Fifth Assessment Report of the Intergovernmental Panel on Climate Change: Barros, V. R., C. B. Field, D. J. Dokken, M. D. Mastrandrea, K. J. Mach, T. E. Bilir, M. Chatterjee, K. L. Ebi, Y. O. Estrada, R. C. Genova, B. Girma, E. S. Kissel, A. N. Levy, S. MacCracken, P. R. Mastrandrea, and L. L. White (eds.), Cambridge, UK; New York, USA: Cambridge University Press, 2014), 1199–1265; Thulani Dube, Philani Moyo, Moreblessings Ncube, and Douglas Nyathi. "The Impact of Climate Change on Agro-Ecological Based Livelihoods in Africa: A Review," *Journal of Sustainable Development* 9:1 (2016): 256–267.
30 ICMM, "Adapting to a Changing Climate: Implications for the Mining and Metals Industry" (ICMM, March, 2013); Julia Nelson and Ryan Schuchard, "Adapting to Climate Change: A Guide for the Mining Industry" (Business for Social Responsibility); Tristan Pearce, James D. Ford, Jason Prno, and Frank Duerden, "Climate Change and Canadian Mining: Opportunities for Adaptation. Summary for Decision-Makers" (David Suzuki Foundation, August 2009).
31 Jonah Busch and Hedley Grantham, "Parks Versus Payments: Reconciling Divergent Policy Responses to Biodiversity Loss and Climate Change from Tropical Deforestation," *Environmental Research Letters* 8:3 (2013): 1–15.
32 H. S. Grantham, R. Portela, M. Alm, D. Juhn, and L. Connell, "Maximizing Biodiversity and Ecosystem Service Benefits in Conservation Decision-Making" in *Routledge Handbook of Ecosystem Services*, M. Potschin, R. Haines-Young, R. Fish and K. Turner (eds.) (London: Routledge, 2016).
33 ICMM, "Trends in the Mining and Metals Industry" (Mining's Contribution to Sustainable Development, October 2012), www.icmm.com/document/4441.
34 ICMM, "The Role of Mining in National Economies" (Mining's Contribution to Sustainable Development, October 2014), www.icmm.com/document/8264.
35 D. P. Edwards, S. Sloan, L. Weng, P. Dirks, J. Sayer, and W. F. Laurance, "Mining and the African Environment," *Conservation Letters* 7:3 (June 2014): 302–311.
36 Sudeshna Ghosh Banerjee, Zayra Romo, Gary McMahon, Perrine Toledano, Peter Robinson, and Inés Pérez Arroyo, "The Power of the Mine: A Transformative Opportunity for Sub-Saharan Africa" (Washington, DC: World Bank Group, Directions in Development, Energy and Mining, 2015), https://openknowledge.worldbank.org/bitstream/handle/10986/21402/9781464802928.pdf?sequence=3.
37 Ibid.
38 This analysis focused on commercial-scale mining versus artisanal and small-scale mining. Source for data on mine projects was the World Bank's Africa Power-Mining Projects Database. World Bank, "The Africa Power – Mining Database" (World Bank Group, 2014), https://databox.worldbank.org/Extractives/Africa_Map_Mines_Location/yd4g-ie4d. Biodiversity hotspots as of 2011, Protected Areas as of 2015, World Heritage Sites inscribed as of August 2012, High Biodiversity Wilderness Areas as of 2003, Key Biodiversity Areas as of 2015, and Alliance of Zero Extinction sites as of 2010 were included. Conservation International (CI), Conservation Synthesis, Center for Applied Biodiversity Science at Conservation International, "Biodiversity Hotspots Revisited," (CI, 2011), http://databasin.org/datasets/23fb5da158614110 9fa6f8d45de0a260; IUCN and UNEP-WCMC, "The World Database on Protected Areas (WDPA)" (Cambridge, UK: UNEP-WCMC, 2015), www.protectedplanet.net; IUCN and UNEP-WCMC, "The World Database on Protected Areas (WDPA): World Heritage Sites" (Cambridge, UK: UNEP-WCMC, 2012), www.protectedplanet.net; R. A. Mittermeier, C. G. Mittermeier, T. M. Brooks, J. D. Pilgrim, W. R. Konstant, G. A.

B. da Fonseca, and C. Kormos. "Wilderness and Biodiversity Conservation," *Proceedings of the National Academy of Sciences* 100:18 (2003): 10309–10313; BirdLife International, Conservation International, and partners, "Global Key Biodiversity Areas" (Cambridge, UK; Arlington, VA: BirdLife International and Conservation International, 2015). These data represent the combination of global Important Bird Areas developed and maintained by the BirdLife partnership and Key Biodiversity Areas developed and maintained by Conservation international and partners. For a full list of collaborators and supports please contact data@conservation.org; Alliance for Zero Extinction (AZE). "2010 AZE Update," 2010, www.zeroextinction.org.

39 Three hundred and eighteen mine projects from the World Bank's Africa Power-Mining Projects Database, excluding past production and temporarily suspended projects, as well as mine projects with no coordinates.

40 The word *degazette* means to remove an official status from something by publishing the fact in an official gazette. The word *degazettement* refers to a loss of legal protection for a PA (biodiversitya-z.org/content/degazettement). David P. Mallon, Michael Hoffmann, Matthew J. Grainger, Fabrice Hibert, Nathalie van Vilet, and Philip J. K. McGowan, "An IUCN Situation Analysis of Terrestrial and Freshwater Fauna in West and Central Africa," *Occasional Paper of the IUCN Species Survival Commission* 54 (Gland; Cambridge, UK: IUCN, 2015).

41 UNEP-WCMC. "Identifying Potential Overlap Between Extractive Industries (Mining, Oil and Gas) and Natural World Heritage Sites," (UNEP-WCMC, 2013), www.icmm.com/document/6950.

42 DEA, "Mining and Biodiversity Guideline: Mainstreaming Biodiversity into the Mining."

43 America P. Durán, Jason Rauch, and Kevin J. Gaston, "Global Spatial Coincidence Between Protected Areas and Metal Mining Activities," *Biological Conservation* 160 (2013): 272–278.

44 IFC, "Cumulative Impact Assessment and Management: Guidance for the Private Sector in Emerging Markets," (IFC, 2013).

45 UNEP, "Toolkit and Guidance for Preventing and Managing Land and Natural Resource Conflict" (Nairobi: UN Framework Team for Preventive Action, 2012); Franks, Daniel. *Mountain Movers: Mining, Sustainability and the Agents of Change.*

46 Franks, *Mountain Movers: Mining, Sustainability and the Agents of Change.*

47 Florence Landsberg, Mercedes Stickler, Norbert Henninger, Jo Treweek, and Orlando Venn, "Weaving Ecosystem Services into Impact Assessment" (World Resources Institute, 2013), www.wri.org/sites/default/files/weaving_ecosystem_services_into_impact_assessment.pdf.

48 Ibid.; J. C. S. Rosa, and L. E. Sanchez, "Revisiting the EIS of a Mining Project Using Ecosystem Services" (Calgary, Canada: IAIA13 Conference Proceedings, Annual Meeting of the International Association for Impact Assessment, 2013), www.iaia.org/conferences/iaia13/proceedings/Final%20papers%20review%20process%2013/Revisiting%20the%20EIS%20of%20a%20mining%20project%20using%20ecosystem%20services.pdf?AspxAutoDetectCookieSupport=1.

49 Landsberg et al., "Weaving Ecosystem Services"; Guy Henley, "The Transition of Ecosystem Services into Impact Assessment: Perceptions from The Oil and Gas Sector" (PhD dissertation, University of East Anglia, 2013).

50 Landsberg et al., "Weaving Ecosystem Services"; Natural Capital Coalition, "The Natural Capital Protocol" (Natural Capital Coalition, 2016), http://naturalcapitalcoalition.org/protocol/.

51 ICMM, "Good Practice Guidance for Mining and Biodiversity," 2006.

52 CSBI, "A Cross-Sector Guide for Implementing the Mitigation Hierarchy" (White paper prepared by The Biodiversity Consultancy, 2015).

53 Business and Biodiversity Offsets Programme (BBOP), *Biodiversity Offset Design Handbook-Updated* (Washington, DC: BBOP, 2012).

54 Becca Madsen, Nathaniel Carroll, Daniel Kandy, and Genevieve Bennett, "State of Biodiversity Markets: Offset and Compensation Programs Worldwide" (Washington, DC: Forest Trends, 2011), www.thegef.org/gef/sites/thegef.org/files/publication/Bio-Markets-2011.pdf; Kerry T. Kate and M. L. A. Crowe, "Biodiversity Offsets: Policy Options for Governments. An Input Paper for the IUCN Technical Study Group on Biodiversity Offsets" (Gland, Switzerland: IUCN, 2014).
55 Ambatovy, "Overview" (Ambatovy, 2015), www.ambatovy.com/docs/?p=373.
56 Business and Biodiversity Offsets Programme (BBOP), "Resource Paper: No Net Loss and Loss-Gain Calculations in Biodiversity Offsets" (BBOP, 2012).
57 IUCN, "IUCN Policy on Biodiversity Offsets" (IUCN, 2016).
58 IUCN, "Protected Areas Category VI" (IUCN, 2012), www.iucn.org/about/work/programmes/gpap_home/gpap_quality/gpap_pacategories/gpap_category6/.
59 IUCN, "Motion 026 – Protected areas and Other Areas Important for Biodiversity in Relation to Environmentally Damaging Industrial Activities and Infrastructure Development," https://portals.iucn.org/congress/motion/026.
60 William S. Symes, Madhu Rao, Michael B. Mascia, and L. Roman Carrasco, "Why Do We Lose Protected Areas? Factors Influencing Protected Area Downgrading, Downsizing and Degazettement in the Tropics and Subtropics," *Global Change Biology* (2015): 1–10.
61 America P. Durán, Jason Rauch, and Kevin J. Gaston, "Global Spatial Coincidence Between Protected Areas and Metal Mining Activities."
62 PADDD Tracker, "Mount Nimba National Park, Guinea," www.padddtracker.org/view-paddd/paddd-events/b196.
63 United Nations Educational, Scientific and Cultural Organization (UNESCO), "UNESCO Welcomes 'No-Go' Pledge from Leading Mining Companies," (UNESCO, August 22, 2003), http://portal.unesco.org/en/ev.php-URL_ID=14151&URL_DO=DO_TOPIC&URL_SECTION=201.html.
64 UNEP-WCMC, "Protected Areas and The Extractive Industry: Challenges and Opportunities" (Cambridge, UK: UNEP-WCMC, 2014).
65 WB, "Mozambique Data," (WB, 2016), http://data.worldbank.org/country/mozambique.
66 André Almeida Santos, Luca Monge Roffarello, Manuel Filipe, "Mozambique" (African Economic Outlook, 2015), www.africaneconomicoutlook.org/fileadmin/uploads/aeo/2015/CN_data/CN_Long_EN/Mozambique_GB_2015.pdf.
67 Convention on Biological Diversity (CBD), "Convention on Biological Diversity" (Montreal, Canada: Secretariat of the Convention on Biological Diversity, 1992), www.cbd.int/convention/articles/default.shtml?a=cbd-02.
68 Peace Parks Foundation, "Mozambique?" www.peaceparks.co.za/story.php?pid=1318&mid=1332 (accessed December 29, 2015); Monica Echeverria, "Mozambique Creates Africa's Largest Coastal Marine Reserve" (World Wildlife Fund, November 6, 2012), www.worldwildlife.org/press-releases/mozambique-creates-africa-s-largest-coastal-marine-reserve. Mozambique has seven national parks, eight national reserves, thirteen forest reserves, and ten hunting areas.
69 There also are significant investments under way in agribusiness (biofuels production) and non-hydrocarbon minerals and fisheries.
70 Almeida Santos et al., "Mozambique."
71 Pascal Fletcher, "Mozambique Turns to Coal Exports Despite Competition," *Business Day Live*, July 29, 2014, www.bdlive.co.za/africa/africannews/2014/07/29/mozambique-turns-to-coal-exports-despite-competition.
72 Manuel Mucari, "Mozambique Sees $30 bln Investment for 2018 LNG Exports Startup," *Reuters*, August 21, 2014, www.reuters.com/article/mozambique-gas-idUSL-5N0QR49C20140821.
73 Environmental Resources Management (ERM) and Impacto, "Environmental Impact Assessment (EIA) Report for the Liquefied Natural Gas Project in Cabo Delgado,"

Final EIA Report, (ERM, February 1, 2014), www.erm.com/contentassets/9f1c634 c714f419384baea6dcdb492bd/volume-1/vol-i--front-pages-nts_lng-final-eia_sept-2014_eng.pdf.
74 Stuart Heather-Clark, Albert de Jong, and Marta Henriques. "Sasol's Offshore Gas Exploration Project, Bazaruto Mozambique" (Southern African Institute for Environmental Assessment, 2009), www.saiea.com/case_studies09/09%20sasoloffshore gas.pdf.
75 David Pritchard, "Ramsar Advisory Mission No 62: Marromeu Complex, Ramsar Site Mozambique," *August Mission Report* (2009), www.ramsar.org/sites/default/files/documents/library/ram62e_tanzania_2009.pdf.
76 Keikantse Lesemela, "Mozambique Withdraws Maputo Port Offer," *Mmegi Online*, February 3, 2012, www.mmegi.bw/index.php?sid=4&aid=295&dir=2012/february/friday3; TradeMark Southern Africa (TMSA), "Botswana, Mozambique Agree On Construction Of Techobanine Deepwater Port" (TMSA, 2011), www.trademarksa.org/news/botswana-mozambique-agree-construction-techobanine-deepwater-port.
77 Strategic environmental assessments refer to the "analytical and participatory approach that aims to integrate environmental considerations into policies, plans, and programs and evaluate the inter linkages with economic and social considerations." Organization for Economic Co-operation and Development (OECD), *Applying Strategic Environmental Assessment: Good Practice Guidance for Development Co-operation* (Paris: DAC Guidelines and Reference Series, OECD Publishing, 2006).
78 Couto, Graçia, and Associados, "Mozambique's New Mining Law and The Key Changes it Introduces," (Couto, Graçia, and Associados, 2015), www.cga.co.mz/wp-content/uploads/2015/03/CGA-Mozambiques-new-Mining-Law.pdf; Shearman and Sterling, "Mozambique's New Mining Law: A Re-Balancing Act," (Sherman and Sterling, October 27, 2014), www.shearman.com/~/media/Files/NewsInsights/Publications/2014/10/Mozambiques-New-Mining-Law--A-ReBalancing-Act-PDF-102714.pdf.
79 Ministry of Coordination of Environmental Affairs (MICOA), Mozambique, "Strategic Environment Assessment of the Coastal Zone of Mozambique," *Impacto*, September 2012.
80 Joseph Nhachote, "Reshaped Government: Who Is Who & Details of Changes," *Mozambique's Political Process Bulletin 57* (Maputo: Centro de Integridade Pública and the European Parliamentarians with Africa, 2015).
81 Alex Benkenstein, "Small Scale Fisheries in a Modernizing Economy: Opportunities and Challenges in Mozambique" *Research Report 13* (South African Institute of International Affairs, Governance of Africa's Resources Programme, August 2013).
82 Almeida Santos et al., "Mozambique." This includes the development of specific environmental laws and instruments including: Petroleum Law, Law no. 3/2001; Regulations on Petroleum Operations, Decree no. 24/2004; Environmental Regulations for Petroleum Operations, Decree no. 56/2010; Strategy for the Concession of Areas for Petroleum Operations, resolution no. 27/2009; Mining Law, Law no. 14/2002; Regulations of the Mining Law, Decree no. 62/2006; Decree no. 61/2006; Environmental Regulations for Mining, Decree no. 26/2004; the Law on Mining Taxes, Law no. 11/2007; Regime on Fiscal Incentives for the Mining and Petroleum areas, Law no. 13/2007; Ministerial Diploma no. 189/2006, on Basic norms of Environmental Management.
83 Government of Mozambique, "Law of Conservation Areas No.16/2014" (Government of Mozambique, 2014), http://faolex.fao.org/docs/pdf/moz134834.pdf.
84 WB, "World Development Indicators: Madagascar," (WB, 2014), http://databank.worldbank.org/data/reports.aspx?source=2&country=MDG&series=&period=.
85 "212 KBAs and 21 AZE sites" and "a number of ecosystem services": Critical Ecosystem Partnership Fund (CEPF), "Ecosystem Profile: Madagascar and Indian Ocean Islands" (CEPF, December, 2014), www.cepf.net/SiteCollectionDocuments/madagascar/

EcosystemProfile_Madagascar_EN.pdf; AZE, "Madagascar," 2013, www.zeroextinction.org/search_results_country.cfm.

86 Trading Economics, "Madagascar Exports," www.tradingeconomics.com/madagascar/exports (accessed December 2, 2015); WB, "World Development Indicators: Madagascar"; ICMM, "The Role Of Mining in National Economies" (Mining's Contribution to Sustainable Development, 2nd edition, October, 2014), www.icmm.com/document/8264.

87 Madagascar Oil, "Tsimiroro Block 3104," (Madagascar Oil, 2015), www.madagascaroil.com/operations/tsimiroro-block/.

88 United Nations Development Program (UNDP), "Human Development Report 2016: Human Development for Everyone" (UNDP, 2016), http://hdr.undp.org/sites/default/files/2016_human_development_report.pdf.

89 Drazen Jorgic and Lovasoa Rabary, "Madagascar Faces Uphill Struggle to Revive Bruised Mining Sector," *Reuters*, August 25, 2015, www.reuters.com/article/2015/08/25/madagascar-mining-idUSL3N10W42720150825#LIHWGvQQQOrTifLT.99.

90 Wealth Accounting and the Valuation of Ecosystem Services (WAVES) is a World Bank-led initiative that aims to integrate natural capital values into national account systems, and thereby encourage better, more efficient decision-making and planning.

91 Rosimeiry Portela, Paulo A. L. D. Nunes, Laura Onofri, Ferdinando Villa, Anderson Shepard, and Glenn-Marie Lange, "A Demonstration Case Study for the Wealth Accounting and the Valuation of Ecosystem Services (WAVES) Global Partnership" (Conservation International, June 2012).

92 Ferdinando Villa, "Assessing Biophysical and Economic Dimensions of Societal Value."

93 Zachary Portela, "A Demonstration Case Study."

94 The ARIES (Artificial Intelligence for Ecosystem Services) modeling tool was utilized to conduct the biophysical analysis.

95 The preliminary nature of the biophysical study meant only a subset of possible ecosystem services were investigated, and the rapid nature of the assessment did not allow for an in-depth industrial economic analysis of each economic sector.

96 Katherine Bolt, Gemma Cranston, Thomas Maddox, Donal McCarthy, James Vause, and Bhaskar Vira, "Biodiversity at The Heart of Accounting for Natural Capital: The Key to Credibility" (Cambridge Conservation Initiative, July 20, 2016).

97 "Gaborone Declaration for Sustainability in Africa," www.gaboronedeclaration.com/about/ (accessed December 29, 2015).

98 CEPF, "Ecosystem Profile: Madagascar and Indian Ocean Islands."

99 International Finance Corporation (IFC), "IFC Performance Standards on Environmental and Social Sustainability" (IFC, January 1, 2012).

100 D. R. Wilburn, and K. A. Stanley, "Exploration Review," *Mining Engineering* 66:5 (May 2014): 18–118.

101 IFC, "IFC Performance Standards on Environmental and Social Sustainability."

102 International Finance Corporation (IFC), "International Finance Corporation's Guidance Notes: Performance Standards on Environmental and Social Sustainability" (IFC, January 1, 2012).

103 Nicolas King, Asha Rajvanshi, Selwyn Willoughby, Ruben Roberts, Vinod B. Mathur, Mandy Cadman, and Vishwas Chavan, "Improving Access to Biodiversity Data for, and from, EIAS – A Data Publishing Framework Built to Global Standards," *Impact Assessment and Project Appraisal* 3:3 (2012): 148–156.

104 Quotes are from the following: (IA) UNEP-WCMC, "IUCN Category IA – Strict Nature Reserve," *Biodiversity A-Z*, November 20, 2014, http://biodiversitya-z.org/content/iucn-category-ia-strict-nature-reserve. (IB) UNEP-WCMC, "IUCN Category IB – Wilderness Area," *Biodiversity A-Z*, November 20, 2014, http://biodiversitya-z.org/content/iucn-category-ib-wilderness-area. (II) UNEP-WCMC, "IUCN Category

II – National Park," *Biodiversity A-Z*, November 20, 2014, http://biodiversitya-z.org/content/iucn-category-ii-national-park. (III) UNEP-WCMC, "IUCN Category III – Natural Monument or Feature," *Biodiversity A-Z*, December 9, 2014, http://biodiversitya-z.org/content/iucn-category-iii-natural-monument-or-feature. (IV) UNEP-WCMC, "IUCN Category IV – Habitat/Species Management Area," *Biodiversity A-Z*, November 20, 2014, http://biodiversitya-z.org/content/iucn-category-iv-habitat-species-management-area. (V) UNEP-WCMC, "IUCN Category V – Protected Landscape/Seascape," *Biodiversity A-Z*, November 20, 2014, http://biodiversitya-z.org/content/iucn-category-v-protected-landscape-seascape. (VI) UNEP-WCMC, "IUCN Category VI – Protected Area with Sustainable Use of Natural Resources," *Biodiversity A-Z*, November 20, 2014, http://biodiversitya-z.org/content/iucn-category-vi-protected-area-with-sustainable-use-of-natural-resources.

12 Ebola and other emerging infectious diseases
Managing risks to the mining industry

Osman A. Dar, Francesca Viliani, Hisham Tariq, Emmeline Buckley, Abbas Omaar, Eloghene Otobo, and David L. Heymann

In brief

- The outbreak of Ebola virus in west Africa in 2014–2016 demonstrates the continued risk of emerging infectious diseases (EIDs) in Africa. It also illustrates the vulnerability of mining companies to the economic impacts of EIDs: that the outbreak forced several companies to scale back and/or terminate projects.
- Through its impacts on land use, human population dynamics, and biodiversity alterations, among many other behavioral and ecological changes, the mining industry creates risks for the spread of EIDs. The industry can mitigate these risks through cross-sectoral partnerships involving other private sector companies, NGOs, and local and international health authorities to help prevent, prepare for, and respond to EID outbreaks. The success of such cross-sectoral partnerships depends in part on a good understanding of past effective collaborations.
- Policymakers and companies should focus more on risks of EIDs during routine impact assessment of mining projects, and should better identify how the mining industry can support countries' adherence to international frameworks and guidelines for preventing and responding to EID outbreaks.

Introduction

The response to the Ebola crisis of 2014–2016 in west Africa serves as a reminder of the importance of public and private sector collaboration to manage risks, ensure business continuity, and foster sustainable development. That outbreak infected more than 28,000 people across Guinea, Liberia, and Sierra Leone, killing more than 11,300 of those infected. It had a significant impact on already fragile local and national health systems. To tackle Ebola, these systems diverted resources away from maternal and child health, HIV/AIDS, and tuberculosis programs, among others, thus compromising other health care services.[1] In addition to the health sector, the outbreak had a substantial impact on socioeconomic development in the affected countries: the World Bank estimates that the Ebola-attributable GDP loss in 2014–2015 for the three countries was approximately $2.8 billion.[2] The World Health Organization Ebola Response Team (WHOER) estimates individual country losses as $240 million for Liberia, $535 million for Guinea, and $1.4 billion for Sierra Leone.[3] For the extractive industry and the mineral sector, which together are a major source of tax revenues for governments in the region, the epidemic has had significant consequences, with many companies forced to suspend activities, absorb significant losses, or, in some instances, cease trading altogether.[4]

Ebola and other emerging infectious diseases

The 2014–2016 Ebola outbreak demonstrated a clear need for improved engagement between two entities with seemingly disparate interests, if there is to be business continuity: first, business-focused extractive companies and second, government authorities, including public health agencies. Ebola is just one of many emerging infectious diseases (EIDs) that can cause widespread disruption to trade, travel, and economic development across the tropics, including tropical Africa. EIDs can be known diseases that are becoming more common or are spreading into new areas as a result of changes in the disease-causing microorganisms, changes in land use (including use for farming, lumbering, and mining), and a changing climate. Other EIDs have existed in wild or domesticated animals and are newly identified in humans; an example is HIV, identified in the late twentieth century, which is thought to have emerged in non-human primates before infecting humans.

A review of EIDs in the past 60 years suggests that over 60 percent of these diseases were transmitted from animals; of those animal-transmitted diseases, just over 70 percent came from wild animals and the rest from domesticated animals.[5] An example of an EID is Zika fever, a mosquito-borne viral disease. Zika was first identified in a non-human primate in the Zika Forest of Uganda in 1947, and gradually moved east across the Pacific before appearing in Brazil in 2014. Since that year it has spread rapidly across communities in south and central America and the Caribbean. Zika is associated with microcephaly and Guillain-Barré syndrome, among other neurologic syndromes, and has had a detrimental effect on trade and commerce in the affected areas.[6]

Other important examples of EIDs and reemerging infectious diseases reported since the late twentieth century include the Middle-East respiratory syndrome coronavirus, West Nile fever, dengue fever, chikungunya, severe acute respiratory syndrome, and novel influenza virus strains such as H1N1 (also known as swine flu) and H5N1 (also known as highly pathogenic avian influenza). Advances in medical sciences and technology have improved our ability to identify and detect these pathogens, and monitor their spread through the development of increasingly sophisticated disease surveillance systems. Globalization and increasing international travel have resulted in new convergences of people, wild and domestic animals, and the environment, altering ecosystems and providing some microbes with greater opportunities to breach species barriers (Figure 12.1).[7] Humans and their domesticated animals have infringed on wild animal habitats, increasing contacts among them. At the same time, international trade has lengthened supply chains and increased the speed with which infectious diseases can spread and even transition into pandemics, with potentially deadly consequences.

The Ebola and Zika outbreaks demonstrate that EIDs are not restricted to rare health events affecting a small subset of the population. Since 2000 EID events have increased in frequency, creating high-impact, high-risk emergencies that can have global consequences: the World Health Organization (WHO) estimates that of 2,797 international health hazards recorded worldwide between January 2001 and September 2013, 84 percent were EID outbreaks.[8] In 2016 the World Economic Forum identified the spread of infectious diseases as one of the greatest potential threats to global business activities and economic growth.[9] The economic burden of controlling outbreaks of six major EIDs between 1997 and 2009 was at least $80 billion, according to an extensive study by the World Bank. Those six outbreaks were Nipah virus, West Nile fever, severe acute respiratory syndrome, highly pathogenic avian influenza, bovine spongiform encephalopathy, and Rift Valley fever. The study highlighted the importance of proactive early control of EID outbreaks (ideally in animal populations before the outbreaks extend into human communities) to limit both the costs and the impacts of diseases (Figure 12.2). These controls are key considerations for mining companies seeking to better manage business continuity risks in the face of public health emergencies such as EIDs.[10] Indeed, any emergency response system developed to deal with EIDs would also be relevant and helpful in case of endemic, more-frequent outbreaks such as cholera or measles, which also have serious cost and health implications.

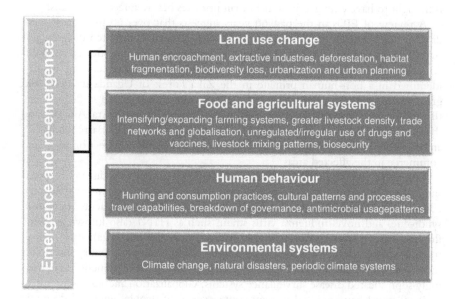

Figure 12.1 EIDs and reemerging infectious diseases: human–animal–environment pathways
Source: adapted from Matthew A. Dixon et al., *Veterinary Record* 174 (2014): 546–551

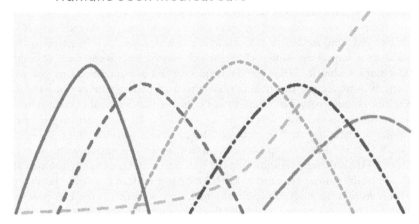

Figure 12.2 Rationale for early control of EID outbreaks

Source: adapted from David L Heymann, A. Dar Osman, "Prevention is Better Than Cure for Emerging Infectious Diseases," *BMJ,* 2014 348: g1499

Managing risk for the extractives industry: prevention, preparedness, and response

The mining industry causes changes in the social and natural environments that can lead to more-frequent EID outbreaks.[11] Extractive projects fundamentally deal with land-use change, human population dynamics, and biodiversity alterations, among many other behavioral and ecological changes, all of which are key drivers in disease emergence and outbreaks. For example, extractive projects can result in greater contact with wildlife through encroachment into previously uninhabited areas, thereby altering the distribution and abundance of wildlife and their associated pathogens and increasing the movement of and interactions between wild and domesticated animals.[12] Many of the mineral deposits in Africa are located in areas that are hotspots for EIDs, as well as for reemerging infectious diseases. Furthermore, several countries on the continent continue to struggle with developing fully integrated health-care systems. All stakeholders—including mining companies and governmental authorities—should be aware that outbreaks can lead to suspensions or shutdowns in extractive projects, significantly disrupting business continuity and fiscal revenues. In turn, these disruptions can adversely affect the profitability and even the survivability of firms in the industry as well as the affected countries' socioeconomic development.[13]

Currently many companies in the mining sector do partial assessments of the risks that diseases could pose to their operations at the start of their projects, adopt evidence-based recommendations to manage risks, and invest in health programs in the countries where they operate.[14] All these exercises are a fundamental component of a preventive approach that ensures that risks are identified early and subsequently managed. For example, the International Council on Mining and Metals (ICMM) guidance on HIV/AIDS, tuberculosis, and malaria provides guidance on development of a management plan process for EID control.

The main process mandated by national governments worldwide to identify, assess, and mitigate health risks associated with project developments is the environmental impact assessment (EIA), which incorporates some limited health considerations within it. When governments conduct assessments, many companies actively mitigate the potential adverse effects of their operations on wildlife and promote biodiversity, but often do not consider the potential transmission of zoonotic pathogens and the capacity of the public health and veterinary systems to respond to these threats.[15] In fact, the EIA process is generally poor at covering health at large unless companies commission a comprehensive health impact assessment (HIA), thereby transforming the EIA into an integrated environmental, social, and health impact assessment (ESHIA).[16] Most African countries, however, have no requirements or guidance for the inclusion of health in the EIA process, and very few African countries even mention community health issues within their mining codes.[17] Therefore the majority of public health risks are not correctly and promptly identified and managed, and, as such, little is done to address these risks in the wider area of industry operations.

The mining sector's approach to infectious disease control has until recently been primarily based on industrial practices and programs focused on "inside the fence" measures.[18] Mining companies might not, therefore, take into consideration the status of WHO International Health Regulations (IHR) core competencies and capacities required by national health systems to deal with outbreaks, unless they do so during the ESHIA. Nevertheless, the industry increasingly recognizes that to improve both risk assessment and management around EIDs and health emergencies, national health, animal, environmental, and emergency response authorities need to engage more widely.

A narrative literature review carried out in 2016 assessed the extent of collaboration between mining companies and government authorities in terms of strengthening outbreak prevention and management processes and systems.[19] The review found anecdotal evidence of collaboration in the 106 sustainability reports of ICMM members reviewed, and demonstrated that there is very little evidence published in peer-reviewed literature to date regarding effective collaboration for outbreak management: only thirty-four gray literature articles satisfied all the inclusion criteria used in the study. These documents described either general infectious diseases control outside the fence (fifteen articles), or outbreak management in collaboration with other stakeholders (nineteen documents). Only two of those documents described collaboration not related to the 2014–2016 Ebola outbreak in west Africa.

According to the review, influenza pandemic risk generated the most collaborative preparedness efforts. Furthermore, while most mining companies mentioned in their annual sustainability reports that their projects had developed emergency response plans, these plans primarily focused on accidents and natural disasters. Although they included some preparedness for disease outbreak, the extent to which they considered outbreaks was often not clear from the annual sustainability reports. The review also found the main drivers for company policy changes were business continuity of their operations, duty of care toward their workers, and the social responsibility agenda.

The 2014–2016 Ebola outbreak in west Africa was a game changer for the role played by the private sector, including mining companies, in the management of outbreaks and potentially other health emergencies.[20]

Mining companies directly affected by the Ebola outbreak in west Africa are now learning the important lessons around improving EID control and outbreak management. International mining companies generally responded by either progressively phasing down their operations in the region and not getting involved in response efforts, or by remaining as fully operational as possible during the outbreak.[21] The latter group classed the outbreak as a serious operational risk and collaborated with multiple stakeholders to help deal with it.[22]

The primary focus of mining companies that remained operational during the outbreak was the protection of their investment, project sites, and their workers and their dependents. Collaborative activities between the public sector and these mining companies varied during the outbreak and can be broadly framed under five categories: (1) immediate relief assistance, either directly to affected people or through clinics or hospitals with medical equipment, cash donations, or transportation assistance; (2) support to community training and capacity-building activities, including public awareness campaigns, with a focus on improving hygiene and fighting misperceptions about the disease; (3) monitoring to feed into daily situation reports and improve risk management; (4) financial contributions to national governments to support the response; and (5) international advocacy to elicit a stronger response from the international community. Collaborative approaches became an essential aspect of the business resilience to the risks posed by the outbreak because corporations realized that working in silos did not protect people or investments.[23] Collaborations were often established at the project, country, and global levels.

Some non-extractive companies that went through the experience of the outbreak now play an active role in contributing ideas and recommendations to post-Ebola recovery efforts in the region.[24] For instance, the International Air Transport Association has recommended introducing greater flexibility into the airline industry's framework for responding to future public health emergencies in order to protect the continuation of services in west Africa and to support recovery.[25] Furthermore, the International Finance Corporation post-Ebola plan includes steps to preserve the private sector in Ebola-affected countries by investing $250 million for the Ebola Crisis Response Facility to support small and medium enterprises.[26]

Importantly, the 2014–2016 Ebola crisis coincided with unfavorable changes in market conditions such as the decrease in the global price of iron ore. Iron ore prices plummeted to five-year lows in September 2014, with analysts expecting prices to continue into decline in the medium term.[27] These unfavorable commodity prices posed an even bigger challenge to mining operations in the region that were fighting Ebola in the short run and struggling to continue mining operations in the medium and long terms. In Liberia, for example, ArcelorMittal Liberia, the largest mining company in that country, delayed its planned investment to expand production capacity from 5.2 million tonnes of iron ore to 15 million tonnes. China Union, the other iron ore mining firm operating in Liberia, shut down its operations in August 2014.[28] As a result, the country's mining sector was expected to contract over the course of the outbreak compared with an initial projection for growth above 4 percent.[29]

To prevent interruptions such as these in business continuity, the extractives sector at large must develop an improved understanding at the senior management level of how EIDs and health emergencies can impact activities. Once they have this understanding, decision makers will be able to use accurate cost estimates to factor into any cost benefit analyses for building operational resilience.[30]

Box 12.1   Case study: EID control in Katanga Province, Democratic Republic of the Congo

The Chatham House Centre on Global Health Security in partnership with the University of Lubumbashi conducted a qualitative study in the mining industry in Katanga Province, Democratic Republic of the Congo (DRC) in 2014 to assess the acceptability and perceived usefulness of introducing specific EID vulnerability assessment tools and mitigation management practices to mining companies.[31] The study participants were a variety of senior DRC-based managers from four multinational mining companies operating in Katanga: Freeport McMoRan (TenkeFungurume Mining), Mawson West (Dikulushi & Pweto—two different projects), Tiger (Kipoi), and MMG (Kinsevere). All four companies operate through local subsidiaries and collectively account for 32 percent of the total cobalt and copper production in Katanga Province. These companies varied in terms of their operational size, levels of capital investments management structure, and geographical remoteness; the projects were at different stages of the project cycle, thus allowing for the study to explore a range of experiences. Researchers interviewed respondents on a range of topics, including their views on the main health risks and vulnerability affecting the mining industry, their experiences of disease outbreaks, current measures implemented to prevent and control infectious diseases, and barriers to and facilitators of introducing further infection prevention and control measures.

A key finding to emerge from the study was the fact that, despite generally having extensive infection prevention and control measures in place inside the fence, respondents acknowledged the vulnerability of mining operations to disease threats occurring outside the fence, in the local community. They attributed this vulnerability to weak public health and social infrastructure, underdeveloped surveillance systems, and low levels of education and awareness. All the companies in the study were engaged in trying to address this problem through community health improvement and disease prevention programs. The study revealed tension surrounding the roles and responsibilities of the state and private sector in terms of health service provision, however. Respondents indicated a sense of futility in the face of outbreaks: the health needs of the local population far outstripped the services provided by the mining companies.

Nevertheless, respondents also recognized the importance of increasing resilience inside and outside the fence, both in terms of reducing mortality and morbidity and in economic terms. The study concluded that there is significant potential for the extractive industry to strengthen local state-run health services and outbreak response mechanisms as a way of avoiding dependency, and as a means of promoting long-term community health. Investment could include providing resources to local communities in preparation for outbreaks (e.g., infrastructure, medical equipment, and drugs), training local health workers and other stakeholders on the management of EIDs, and sharing information and guidelines. In the event of disease outbreaks, the study respondents suggested that mining companies could play a bigger role, using their access to public health expertise and their reach into local communities as a means of rapidly identifying cases, tracing contacts, and managing outbreaks, particularly in remote settings. Moreover, mining companies could offer more support to health authorities by conducting health promotion campaigns in the community to encourage early referrals to care, adopting preventive strategies, and addressing cultural practices that increase the risk of transmission.

To ensure better sustainability, individual companies could plan and implement these activities in partnership with other mining companies and local authorities to reduce the impact on any single organization, and to ensure consistency. For example, mining companies operating in an area could pool their financial and human resources to improve laboratory capacity and efficient sample processing that would allow early diagnosis of infectious diseases. The study suggested the establishment of a network that would promote greater integration and coordination of all relevant stakeholders, including mining companies, the Ministry of Health, veterinary services, NGOs, the WHO, research institutions, and United Nations bodies.[32]

The way forward—multisectoral collaboration in a globalized system for EID control

EIDs, as discussed, often have an animal reservoir; for a variety of reasons they can sometimes breach the species barrier and enter human populations. A subset of EIDs, such as HIV, Ebola, and Zika, have the potential to become major outbreaks or even pandemics. There has been a growing global focus, therefore, in developing disease control strategies that focus on an integrated manner at the interface between animals, humans, and the ecosystems in which they live while being sensitive to the demands of international trade and commerce. This is the One Health approach.[33] The One Health approach is defined as a collaborative effort of multiple disciplines to attain optimal health for people, animals, and the environment. A key plank of this approach at global and national levels is to encourage the building of robust and well-governed human and animal public health systems compliant with the World Organisation for Animal Health (OIE) and the IHR standards.

IHR is the international legal framework created to help the international community prevent and respond to acute, global public health risks. In 2005 WHO revised IHR to reflect the changes brought about by globalization and the increased level of knowledge on infectious disease control.[34] The IHR helps WHO member states better prepare and respond to outbreaks; WHO is required to collaborate with and support countries in implementing the IHR and coordinating all competent intergovernmental organizations or international bodies in support of this role. The focus of the IHR is on the identification and containment of health emergencies and disease threats at the source. Achievement of these goals requires WHO member states to develop core national public health capacities. The eight core capacities covered by the IHR are (1) legislation, policy, and financing; (2) coordination; (3) surveillance; (4) response; (5) preparedness; (6) risk communications; (7) human resources; and (8) laboratories.

Many countries with mining operations in areas that are at an increased risk of EID outbreaks have not fully met these core capacities, particularly in relation to animal health. Many, therefore, are unable to quickly and efficiently detect and respond to potential disease threats. Mining companies should take these weaknesses into consideration in any business risk mitigation strategy. Alongside national commitments to the implementation of the IHR, there are global agreements around emergency and disaster response capability through the Sendai Framework on Disaster Risk Reduction.[35] United Nations member states are committed to developing robust systems for disaster prevention, management, and response under the Sendai Framework. In addition, the WHO secretariat has set a course that weaves together disaster risk reduction and health emergency response, including infectious disease outbreaks. Commitments include the health sector providing greater input and participation into disaster risk reduction fora at the national, regional, and global levels; WHO promoting an all-hazards approach; and WHO promoting an integrated multisectoral response to emergencies.[36] The relationship of the health sector to broader disaster response mechanisms at national and subnational levels has gained increasing prominence as a result of the 2014–2016 Ebola outbreak in west Africa; the health sector's role is being more clearly conceptualized and defined in the context of all-hazards approaches to risk management (Figure 12.3).[37]

Importantly, the United Nations' Sustainable Development Goals (SDGs) Agenda 2030, which came into effect at the start of 2016, provides the framework for potentially integrating the private sector and mining companies into the global agreements around IHR and disaster risk reduction.[38] These goals provide an overarching set of development objectives for all countries around the world including for those related to improved health and disease control. An extensive list of targets support countries monitor the progress being made toward each SDG. SDG targets 17.16 and 17.17 in particular provide the relevant guidance for integrating

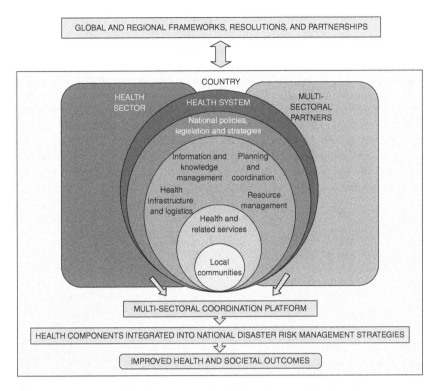

Figure 12.3 A Framework for integrating health into disaster risk reduction strategies
Source: adapted from Dar et al., *Am J Public Health*. 104:10 (2014): 1811–1816
Note: A concept diagram for managing health risks in a multisectoral environment.

the private sector into national systems for disaster response and risk mitigation in line with the principles outlined by a multidisciplinary and multi-stakeholder One Health approach. Target 17.16 recognizes "multi-stakeholder partnerships as important vehicles for mobilizing and sharing knowledge, expertise, technology, and financial resources to support the achievement of the SDGs." Target 17.17 calls for the "encouragement and promotion of effective public, public–private and civil society partnerships, building on the experience and resourcing strategies of previous initiatives."[39] The extractives industry could, therefore, use the SDG framework to support national governments' efforts to develop sustainable systems for disease control while also improving its own ability to assess and manage health emergency and broader disaster risks.

The African mining sector operates in a context where the capacity to detect and respond to disease threats is still low; it is therefore in the interest of companies, for business continuity purposes, to collaborate with the relevant responsible organizations to mitigate those threats.

In summary, establishing integrated One Health public–private partnerships under the appropriate global frameworks outlined is likely to be the most acceptable and operationally cost-effective approach to EID control and emergency response for all relevant stakeholders; both within the private sector (i.e., the mining industry) and the countries concerned. Success, however, is not guaranteed, and partnerships will have to adapt to local conditions, across the spectrum of disease control from prevention to recovery. This process should involve identifying past instances of successful collaboration around EID control between the mining and public sectors, and ensuring that planned impact assessments of future mining projects sufficiently consider health and EID potential within their remit. It should also include conduction of locally relevant health systems and epidemiological research to examine the mining sector's potential role in supporting IHR implementation, piloting collaborations under an accountable and transparent governance framework, and having robust metrics and validated tools available for evaluations of partnerships.

Notes

1 World Health Organization (WHO), "Ebola Data and Statistics" (WHO, 2016).
2 World Bank Group, "2014–2015 West Africa Ebola Crisis: Impact Update," http://pubdocs.worldbank.org/en/297531463677588074/Ebola-Economic-Impact-and-Lessons-Paper-short-version.pdf.
3 World Health Organization Ebola Response Team (WHOER), "Ebola Virus Disease in West Africa — the First 9 Months of the Epidemic and Forward Projections," *New England Journal of Medicine* 371:16 (2014).
4 "Governments in the region": Mark Roland Thomas, Gregory Smith, Francisco H. G. Ferreira, David Evans, Maryla Maliszewska, Marcio Cruz, Kristen Himelein, Mead Over, "The Economic Impact of Ebola on Sub-Saharan Africa: Updated Estimates for 2015," *Working Paper 93721* (Washington, DC: World Bank Group, 2015), http://documents.worldbank.org/curated/en/2015/01/23831803/economic-impact-ebola-sub-saharan-africa-updated-estimates-2015#; "cease trading": J Allouche, "Ebola and the Extractive Industry," *Practice Paper in Brief 21* (Institute od Development Studies, 2015).
5 Kate E. Jones et al., "Global Trends in Emerging Infectious Diseases," *Nature* 451:7181 (2008).
6 European Centre for Disease Prevention and Control (ECDC), "Zika Virus Infection" (ECDC, 2017), http://ecdc.europa.eu/en/healthtopics/zika_virus_infection/pages/index.aspx.
7 Matthew A. Dixon, Osman A. Dar, and David L. Heymann, "Emerging Infectious Diseases: Opportunities at the Human-Animal-Environment Interface," *Veterinary Record* 174:22 (2014); David L. Heymann and Osman A. Dar, "Prevention is Better Than Cure for Emerging Infectious Diseases," *BMJ* 348 (2014).
8 Christopher Dye, "After 2015: Infectious Diseases in a New Era of Health and Development," *Philosophical Transactions of the Royal Society B: Biological Sciences* 369 (2014): 1645.
9 The Global Competitiveness and Risks Team, "The Global Risks Report 2016 11th Edition," (The Global Competitiveness and Risks Team, 2016); World Economic Forum (WEF), "The Global Risks 2015 10th Edition" (The Global Competitiveness and Benchmarking Network, 2015).
10 The World Bank, "People, Pathogens and Our Planet: The Economics of One Health," in *The Economics of One Health* (World Bank, 2012).
11 Daniel G. Bausch and Lara Schwarz, "Outbreak of Ebola Virus Disease in Guinea: Where Ecology Meets Economy," *PLoS Negl Trop Dis* 8:7 (2014).
12 M. Hamburg, M. S. Smolinski, and J. Lederberg, *Microbial Threats to Health: Emergence, Detection, and Response* (Washington, DC: The National Academies Press, 2003); Stephen

S. Morse, "Factors in the Emergence of Infectious Diseases," *Emerging Infectious Disease Journal* 1:1 (1995); J. A. Daszak, P. Patz, G. M. Tabor, A. A. Aguirre, M. Pearl, J. Epstein, N. D. Wolfe, A. M. Foufopoulos, J. Kilpatrick, D. Molyneux, and D. J. Bradley, "Unhealthy Landscapes: Policy Recommendations on Land Use Change and Infectious Disease Emergence," *Environ Health Perspect* 112:10 (2004).

13 Omayra Bermúdez-Lugo and William D. Menzie, "The Ebola Virus Disease Outbreak and the Mineral Sectors of Guinea, Liberia, and Sierra Leone," in *Fact Sheet* (Reston, VA: USGS, 2015).

14 "Pose to their operations": International Council on Mining & Metals (ICMM), "Good Practice Guidance on Occupational Health Risk Assessment" (ICMM, 2009); ICMM, "Good Practice Guidance on Health Impact Assessment" (ICMM, 2010). "Manage risks": ICMM, "Good Practice Guidance on Hiv/Aids, Tuberculosis and Malaria," (ICMM, 2008); IOGP; IPIECA, "Vector-Borne Disease Management Programmes" (IPIECA, 2012). "Health programs": ICMM, "Community Health Programs in the Mining and Metals Industry" (ICMM, 2013); WEF, "Global Health Initiative: Catalyzing Partnerships to Tackle Hiv/Aids, Tuberculosis & Malaria," http://web.worldbank.org/archive/website00818/WEB/OTHER/GLOBAL_H.HTM.

15 Francesca Viliani, M. Edelstein, E. Buckley, A. Llamas, and O. Dar, "Mining and Emerging Infectious Diseases: Reflections from the Infectious Disease Risk Assessment and Management (IDRAM) Initiative Pilot" (Special Issue of *Extractive Industry and Society Journal*, forthcoming).

16 Patrick Harris, Francesca Viliani, and Jeff Spickett, "Assessing Health Impacts within Environmental Impact Assessments: An Opportunity for Public Health Globally Which Must Not Remain Missed," *International Journal of Environmental Research and Public Health* 12:1 (2015).

17 Mirko Winkler, Gary Krieger, Mark Divall, Guéladio Cissé, Mark Wielga, Burton Singer, Marcel Tanner, and Jürg Utzinger, "Untapped Potential of Health Impact Assessment," *Bulletin of the World Health Organization* 91:4 (2013).

18 "Inside the fence" is used in the document as any activity targeting the workforce within the mining site. "Outside the fence" is used for any activities carried out outside of the mining site and targeting the people living in the area (they could be workers, dependants or general community). The ICMM health and safety performance indicators exclusively cover workers and contractors, an environment that is easier to control and modify; Alan C. Emery, "Good Practice in Emergency Preparedness and Response," (UNEP, ICMM, 2005); ICMM, "Health and Safety Performance Indicators," *Health and Safety* (2014).

19 Carried out as part of the Infectious Disease Risk Assessment and Management (IDRAM), an initiative led by the Chatham House Centre on Global Health Security (in press).

20 Boston Consulting Group (BCG) World Economic Forum, "Managing the Risk and Impact of Future Epidemics: Options for Public-Private Cooperation" (BCG, 2015); UN Global Impact, "Un Mission for Ebola Emergency Response. Meeting Report: Un-Business Collaboration for Global Ebola Response" (UN Global Impact, 2014).

21 Philippe Calain, "What Is the Relationship of Medical Humanitarian Organisations with Mining and Other Extractive Industries?" *PLoS Med* 9:8 (2012).

22 D. Nabarro, "Office of the United Nations Secretary-General's Special Envoy on Ebola" (United Nations, 2015).

23 "business resilience to the risk": Ebola Private Sector Mobilisation Group, "Being Clever by Being Simple: The Ebola Private Sector Mobilisation Group Story," www.epsmg.com/media/6220/epsmg-being-clever-by-being-simple-final-june-2015.pdf; Andre Willemse, "How the Petroleum Industry Can Learn from the Ebola Crisis of 2014" (Society of Petroleum Engineers). "people or investments": C. Esbenshade, "Business Interests Are Human Interests," (Ebola Private Sector Mobilisation Group, 2015).

24 S. Fischer and E. Brandt, "The United Nations and the Global Ebola Response. The Un Emergency Health Mission 'Unmeer': Lessons Learned and Recommendations for Similar Future Health Emergencies" (Master's thesis, Verlag, 2015).

25 International Air Transport Association (IATA), "Aviation's United Reponse to Ebola Epidemic," *Analysis* (IATA, 2014), http://airlines.iata.org/analysis/aviation%E2%80%99s-united-response-to-ebola-epidemic.
26 International Finance Corporation (IFC), "Responding to the Unexpected: IFC's Support to Ebola-Affected Countries," in *Rebuilding for Tomorrow: Private Sector Development in Fragile and Conflict-Affected Situations in Africa* (Washington, DC: IFC, 2015), www.ifc.org/wps/wcm/connect/7440fe80485e6a868cacfd299ede9589/IFC_CASA_SmartLessons_Booklet.pdf?MOD=AJPERES.
27 Omayra Bermúdez-Lugo and William D. Menzie, "The Ebola Virus Disease Outbreak and the Mineral Sectors of Guinea, Liberia, and Sierra Leone."
28 J. Allouche, "Ebola and the Extractive Industry."
29 Mark Roland Thomas, "The Economic Impact of Ebola."
30 Bingunath Amaratunga Ingirige, Dilanthi Kumaraswamy, Mohan Liyanage, Aslam Perwaiz, Peeranan Towashiraporn, and Gayan Wedawatta, "Prepared for the 2015 Global Assessment Report on Disaster Risk Reduction," (Global Assessment Report on Disaster Risk Reduction, UNISDR, 2014).
31 Francesca Viliani, Michael Edelstein, Emmeline Buckley, Ana Llamas, and Osman Dar, "Mining and Emerging Infectious Diseases: Results of the Infectious Disease Risk Assessment and Management (IDRAM) Initiative Pilot" *The Extractive Industries and Society Volume* 4:2 (April 2017): 251–259, www.sciencedirect.com/science/article/pii/S2214790X16301277.
32 More information can be found in "The Infectious Disease Risk Assessment and Management (IDRAM) Initiative: Reflections from DRC Pilot."
33 One Health Global Network, 2016.
34 WHO, *International Health Regulations*, 2nd edition (WHO, 2005).
35 UNISDR, "The Sendai Framework for Disaster Risk Reduction 2015–2030," www.unisdr.org/we/coordinate/sendai-framework.
36 WHO, "WHO: Statement Made at the Global Platform for Disaster Risk Reduction" (news release, 2011), http://preventionweb.net/go/21928.
37 O. Dar et al., "Integrating Health into Disaster Risk Reduction Strategies: Key Considerations for Success," *Am J Public Health* 104:10 (2014).
38 UN, "Partnership for Sustainable Development Goals: A Legacy Review Towards Realizing the 2030 Agenda," (UN, 2015).
39 Sustainable Development Goals, "SDG Indicators," 17.16 and 17.17, http://indicators.report/targets/17-16/ and http://indicators.report/targets/17-17/

13 Mineral investment decision-making in Africa

A real options approach in integrating price and environmental risks

Kwasi Ampofo and Alidu Babatu Adam

In brief

- Mining companies face numerous challenges in managing commodity price and environmental uncertainties.
- We develop a framework to incorporate environmental, social, and fiscal uncertainties into the mineral investment decision-making process.
- We develop new valuation techniques that promote sustainable and responsible mining with little or no impact to the environment.
- We apply the real options analytical method to the valuation of three hypothetical mine projects under scenarios in Ghana, Tanzania, and South Africa.
- The valuation model we have developed enhances the strategic planning and investment decision-making processes of mining companies in Africa.

Introduction

The United Nations Sustainable Development Goals (SDGs) encourage responsible mining operations that adequately support the local communities they operate in and that take responsibility for environmental protection. The two primary goals that underline these principles are SDGs 8 and 15.[1] SDG 8 calls for actions that "Promote sustained, inclusive and sustainable economic growth, full and productive employment and decent work for all." SDG 15 addresses the urgent need to "Protect, restore and promote sustainable use of terrestrial ecosystems, sustainably manage forests, combat desertification, and halt and reserve land degradation and halt biodiversity loss."

To achieve these goals, mining companies must ensure their operations are economically sustainable in order to provide development opportunities for their host countries and to support environmentally sustainable activities that protect and restore the environment. From 2003 to 2013 two main risks that hampered the realization of these SDGs in Africa were commodity price and environmental uncertainties.[2]

Most African countries are overdependent on metals and minerals for public revenue, making them vulnerable to the cyclic nature of the commodity markets. This dependence, over the years, has resulted in boom-and-bust commodity-driven economic development, whereby high growth performances are realized with high commodity prices, followed by economic vulnerabilities and fragility when commodity prices fall. Compounding the problem are weak environmental frameworks in most African countries, enabling mining companies to avoid their primary mandate of protecting and restoring the environment. Acquisition of environmental permits has become a politically motivated process in some countries, which hampers the development of new mining projects.[3] This chapter applies the *real options technique,* an economic term introduced by Stewart Myers in 1977. The real options technique uses additional information to introduce flexibility in the valuation of mineral projects.[4] It has significantly improved how firms introduce flexibility into investment decision-making and strategic planning to avert risk and manage uncertainties. We use that method to develop an integrated model that includes commodity price risk and environmental risk in determining the value of mineral projects in Ghana, Tanzania, and South Africa. Cumulatively, these three countries account for over 80 percent of gold produced in Africa and fairly represent the 3 main economic regions of sub-Saharan Africa.[5] Lessons drawn from these specific country-context scenarios could potentially be extended into a framework for project valuation in the wider African continent.

Project valuation using real options

The mining industry has been working toward a consistent approach to determine the effect of project risk on value and operating policy, using a variety of valuation techniques. Different attempts span from the use of multiples as a valuation technique in the 1960s, to the discounted cash-flow approach in the 1970s, to the Monte Carlo simulations and decision trees in the late 1990s.[6] Project analysts are attempting to build mine valuation models that successfully integrate market information about risk with a detailed description of project structure.[7] Mining companies are increasingly adopting project valuations techniques from the financial industry so that mine valuation models can successfully integrate with market information about risk with a detailed description of project structure.[8]

Risk is the ability to assign numerical probabilities to random economic events, whereas uncertainty results from random events to which agents cannot assign probabilities.[9] Or, put simply, "The future is subject to risk; the future is uncertain."[10] Every process in the minerals industry has an element of risk. Mining is thus correctly referred to as a high-risk venture, that calls for highly effective minerals project valuation techniques.[11] There is an increasing realization among industry patrons that the capital-intensive nature of mining projects, several years of preproduction investment and uncertainties, and relatively long project lifespans make it very difficult to accurately evaluate projects using traditional techniques.[12]

Although it is easy to apply traditional approaches to estimating project value, they are mostly built on faulty assumptions.[13] Increasingly, researchers call for improved and more-effective valuation techniques because current investment projects are becoming riskier, with more cases of declining ore grades and higher volatility in commodity prices.[14] The failure of traditional valuation methods is attributed to those methods' inability to allow for a flexible (but agile) strategic decision-making process.[15] Faulty investment decisions can lead to unsustainable operational costs or bankruptcy. Therefore, good financial management combined with good capital investment decisions are critical to the survival of mining projects.[16]

It is crucial for valuation techniques to evolve from traditional to sophisticated cutting-edge techniques because of prevailing challenges in tracking resource valuation, mine and process scheduling, marketing, and financial modeling into one integrated mineral project valuation process.[17] Additionally, the failure of traditional mineral project valuation methods to capture the true value of mining investments could lead to the rejection of potentially good investment opportunities or allocation of capital to marginal projects.[18] The application of risk interpretation in conventional valuation processes is based more on subjective judgment than on a quantitative approach that eliminates prejudice in decision making. The subjective approach to project valuation could result in biased outcomes and the inability of mineral practitioners to fully capture project risks.[19] This challenge in project valuation has called for improved valuation techniques leading to a more efficient system for mine planning in the face of multiple uncertainties that minimize subjective judgments.[20]

There is a degree of controversy today as to which is the most reliable real options valuation model.[21] The option to change operating scale (to expand, contract, shut down, or restart) provides decision-makers with the ability to expand scale of production or to accelerate resource extraction during favorable market conditions.[22]

The real options method has proven to be a reliable approach to integrating other elements of uncertainty and risk in the evaluation of minerals projects. One unique feature of the real options approach is its ability to adjust for risk within cash-flow components. The discounted cash-flow method, on the other hand, discounts for risk in the aggregate net cash flow.[23] This means that project analyst adjusts for risk through probabilities rather than through discounting at a blanket risk premium.[24] We cannot eliminate risk in the mining industry, but we can minimize it.[25]

One key success of real options valuation is its ability to include the value of decisions associated with uncertainties into its initial model. This increases its reliability and helps managers to consider future events so that change finds them prepared and ready to act accordingly when certain red-light indicators are triggered.

Mining in Ghana

Ghana is the second-largest producer of gold in Africa, with about fifteen medium- to large-scale operations mining in commercial quantities, in addition to others in the exploration and reconnaissance stages.[26] In 2014 the total minerals export was estimated at $4.516 billion, with the mining and quarry sector contributing 8 percent of total GDP. The sector contributed over 16 percent of fiscal receipts to the Ghana Revenue Authority, the mandated revenue collector for the Ghanaian government.[27] The mining industry in Ghana is regulated and managed by twelve key laws and regulations, including these six:

- Constitution of the Republic of Ghana, 1992;
- Minerals and Mining (Amendment) Act of 2010 (Act 794);
- Minerals and Mining (Health, Safety, and Technical) Regulations of 2012;
- Minerals and Mining (Compensation and Resettlement) Regulations of 2012;
- The Ghana Revenue Act of 2009 (Act 791); and
- The Environmental Protection Agency Act of 1994 (Act 490).

The Ghanaian government receives mining-related income through taxes and levies, including the Mineral Right Licence, Property Rate, Ground Rent, Mineral Royalty, Corporate Tax, and Dividends.

Mining in Tanzania

Tanzania possesses significant deposits of gold, silver, and diamonds. It is the fourth-largest producer of gold after South Africa, Ghana, and Mali. After nationalization of most of the mines in 1972, the mining sector was liberalized with the introduction of the Mineral Policy of 1997. This liberalization saw the influx of foreign investment in the mining sector with gold representing 90 percent of mineral production in the country as of this writing.[28] The mineral industry in Tanzania is primarily governed by the Mining Act No. 15 of 2010, with eleven subsequent regulations to guide operations, management, and regulation. Notable among these regulations are

- Mining (Mineral Rights) Regulations of 2010;
- Mining (Safety, Occupational Health And Environmental Protection) Regulations of 2010; and
- Mining (Mineral Trading) Regulations of 2010.

In addition to these, the Environmental Management Act of 2004, the Income Tax Act of 2004, and the Tanzania Investment Act of 1997 provide key guidelines in the administration of the industry. In 2014 the mining sector contributed 3.3 percent of the total GDP of the Tanzanian economy and a direct contribution of 7,000 skilled and semi-skilled jobs.

The Tanzanian government receives mineral revenue through taxes and levies, including the corporate tax, royalty, and annual rent fee.

Mining in South Africa

South Africa has recorded significant decreases in its mining sector from 2003 to 2013; yet mining remains a key pillar in the economy. In 2013 mining accounted for 26 percent of South Africa's total merchandise exports. On the Johannesburg Stock Exchange mining stocks contributed about 18.7 percent of the total market capitalization at the end of 2014. Also, the sector contributed a total of 7.6 percent of the GDP of the South African Economy.[29]

Table 13.1 Major sources of mineral taxes and levies in Ghana, Tanzania, and South Africa

Tax/Levy	Ghana	Tanzania	South Africa
Corporate tax	35%[a]	30%	28%[b]
Royalty	5%[c]	4%	0.5% to 5%
Dividend	10%[d]	Vary	0
VAT	0	18%	15%
Withholding tax (dividend)	8%	10%	15%
Withholding tax (services)	20%	5%	15%
Withholding tax (royalty)	15%	10%	12%
Withholding tax (interest)	8%	10%	15%
Capital gains tax	0	30%	Vary
Depreciation allowance	20%[e]	100%[f]	100%[g]

Note: a = net profit; b = varies; c = gross revenue; d = carried interest in company shares; e = straight line for five years; f = indefinitely for exploration and development costs; g = immediate expensing.

The South African government generates income from the mining sector through the collection of corporate income tax, royalty, license fees, and withholding tax.

Table 13.1 summarizes the major sources of taxes and levies retrieved by governments in the mining sector for Ghana, Tanzania, and South Africa. In cases where a sliding scale rate is used, the highest value is selected.

Modeling gold price uncertainty

In making investment decisions, mining companies apply the gold price model universally to all projects irrespective of the host country. The model used in this chapter simulates gold prices for a twenty-five year period using historical data from 1990 to 2015. The simulation is based on the theory that gold prices follow a Geometric Brownian Motion. This model is used to simulate the gold prices along twenty-five independent paths. In Figure 13.1 the first section of the graph (single

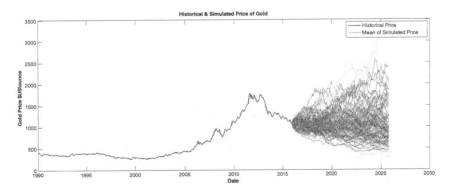

Figure 13.1 Price simulation of gold for ten years based on a twenty-five year historical price path

blue line) represents the historical gold prices, followed by the expanding fan of mean of simulated prices with independent price paths.

Modeling environmental uncertainty

Environmental risks and uncertainty have become a big challenge in the mineral industry since the beginning of the twenty-first century for two main reasons:

1. The increase in awareness of the damaging effects of human activities on the environment has led to the strengthening of regulatory processes to curb the harm caused to the environment. In the mining industry, researchers have made groundbreaking discoveries in the areas of toxic chemicals and acid mine drainage, to name only two. These discoveries have contributed significantly to the improvement of various environmental laws across several jurisdictions, leading governments to issue more mining permits.
2. Second, the exponential increase in the scale of extraction processes during the mining super cycle has presented entirely new environmental challenges. Mega-projects require huge equipment to haul materials, which in turn increases emissions and waste, and widens the emission footprints of mining operations. The dam failure at Brazil's Samarco in late 2015 proves that in spite of all the advancements made in safety, large corporations today can still cause accidents that harm the environment.[30] These environmental casualties are becoming eminent due to the continuous increase in commodity extraction capacity, demand, and supply, and have highlighted the environment as a major source of uncertainty in the mineral industry.

Unlike other market uncertainties that are universal, such as commodity prices and foreign exchange, environmental uncertainties are region and country specific. Even within some countries, environmental challenges and regulations vary from state to state or province to province. This chapter addresses the challenge of building the environmental uncertainty and risk model posed by environmental issues by developing a standardized environmental model that can be extended to evaluate gold mining projects across Africa.

The chapter develops an environmental model that embodies seven main criteria common to most African mining regions and that are specifically present in Ghana, Tanzania, and South Africa (Table 13.2). These criteria fall under Epstein and Rejc's broad framework of strategic risks, operational risks, reporting, and compliance risks.[31] We weigh these criteria on their anticipated likelihood and consequence and report them in a risk map.

- Environmental and Safety Management: This uncertainty criterion reviews the environmental and safety management practices of randomly selected mines in the host country, including the expertise of the mines' management with regard to ensuring good environmental practices. In addition, this criterion reviewed and assessed corporate governance structure and its component addressing environmental challenges.

Africa's mineral project valuation 201

Table 13.2 Criteria used to build environmental uncertainty and risk model

Risk/Uncertainty	Criteria	Code
Operational risk/uncertainty	Environmental and safety management	A
	Air and water quality	B
	Toxic and waste disposal	C
Strategic risk/uncertainty	Valuation and monitoring	D
	Corporate social responsibility	E
Reporting & compliance risk/ uncertainty	Prevention and control	F
	Laws and regulations	G

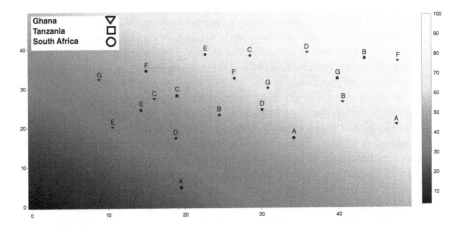

Figure 13.2 Coordinates for each of the criteria in the study country

- Air and Water Quality: This criterion benchmarks existing standards in the country review against other peer countries to determine the quality of the country's processes: How are mining companies mandated to manage water bodies on their concessions? The criterion reviews existing blasting standards of the country and air quality benchmarks: Are the activities of mining companies in tandem with globally accepted standards?
- Toxic and Waste Disposal: Waste management is one of the mineral industry's greatest challenges. For each ounce of mineral produced, several tonnes of waste are discharged into the environment. Recent waste management disasters have confirmed that even within the same companies, corporations discriminate in their waste management practices depending on the host country's oversight and focus on toxic and waste disposal. As an example, although Shell is noted for encouraging standard practices in its operations globally, it has poorly managed its waste and toxic disposal strategy in Nigeria, creating one of the worst environmental hazards in the region. Amnesty International reports, "In any other country, this would be a national emergency. In Nigeria,

it appears to be standard operating procedure for the oil industry. The human cost is horrific—people living with pollution every day of their lives."[32]

- Valuation and Monitoring: This criterion reviews the countrywide mechanism for the valuation and monitoring of environmental practices and their impacts: How effective are the structures that the country's regulator has put in place to handle issues pertaining to the environment? Are they being implemented and monitored effectively? Are the feedback mechanisms, if any, efficient? These are the basic areas this criterion addresses to arrive at a peer-to-peer benchmark to review how these countries effectively and efficiently monitor and evaluate the setting in which mining companies operate.
- Corporate Social Responsibility: This criterion assesses the culture of social contributions of corporations to the communities in which they operate. This area has attracted significant debate from researchers and practitioners. Although countries such as Australia require mining companies to formalize interventions to communities, others such as Canada consider such interventions voluntary and therefore non-binding. This criterion assesses the legal requirement of the host country and the existing culture of companies toward mining catchment areas.
- Prevention and Control: This criterion reviews operational procedures for activities such as handling explosives, and assesses the regulatory frameworks to manage and protect the environment as well as the compliance history of various companies. The criterion's review of mining accidents and disasters contributed to analysis of the compliance rate and prevention measures in place. To manage incidents and accidents, the criterion also includes response framework to control disasters to individually determine the capability of host countries prevention and control processes in managing the environment.
- Laws and Regulations: This final criterion assesses in a matrix all the laws and regulations that guide the effects of mining activities with regard to the environment. This assessment leads to the estimation of the impact and effectiveness of initiatives implemented to protect the environment from mineral operations. This criterion also considers and factors compliance to these laws and regulations.

The model

We estimate the value each criterion based on the associated benefits and costs using the framework developed by Epstein Marc and Adriana Rejc Buhovac.[33] The framework estimates the value of risk and uncertainty by quantifying the probability of each risk and uncertainty criterion as a product of its expected probability of occurrence. The framework uses the loss distribution analysis to estimate the cash flow at risk from environmental uncertainty. The aggregate value of the environmental risk and uncertainty is incorporated in the real options value of the project.

With the help of Matlab tools, these analyses collectively provide a robust output tool to determine the value of the project. The methodology employed for this valuation comprises

- a risk-adjusted distribution of gold prices for each year of production obtained through simulations;
- an environmental framework with eight major uncertainties;
- an environmental model based on the loss distribution analysis;
- a distribution of risk-adjusted project cash flows across the life of the mine;
- a net present value distribution analysis of the project; and
- a risk assessment based on distributions of the economic outcomes.

Hypothetical case study: a gold mining project

We use a hypothetical gold mine to analyze the project value for each country scenario, holding constant the same production and cost assumptions for each country scenario. The project is brownfield with a ten-year life of the mine. Its average output is 550,000 ounces per annum, and the mine managers anticipate an all-in sustaining cost of less than $1,000. We use the in-country tax requirements in the valuations, and assume market risks such as foreign exchange to be negligible.

Discussions

Based on the models developed, we estimate the mean value of each project and the cash flow at risk due to environmental risks and uncertainties. Our results

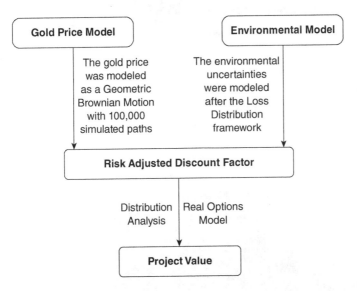

Figure 13.3 Framework for estimating value of projects using real options and distribution analysis

show that the country with the highest project value is Tanzania with a value of $336 million, albeit with a 5 percent probability of losing $55 million through environmental uncertainties and risks. Ghana has the next-highest project value with a $298 million mean project value with a 5 percent probability of losing $60 million of the project value through environmental risks and uncertainty. South Africa has the highest environmental risk probability of $62 million and returned the least project value, at $267 million. The results are illustrated in Figures 13.4, 13.5, and 13.6.

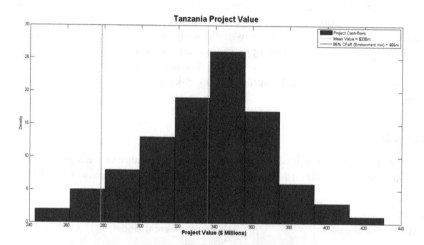

Figure 13.4 Tanzania project returned mean value of $336 million

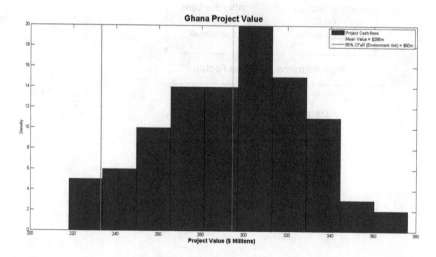

Figure 13.5 Ghana Project returned mean value of $298 million

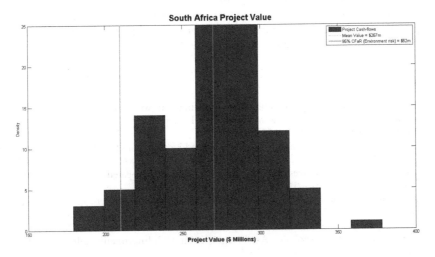

Figure 13.6 South Africa project returned mean value of $267 million

Conclusion

While recognizing the differentials in the risk landscape as well as the regulatory and institutional structures of the various countries, our analysis and valuation of potential mine projects, using the real option model, allow for composite understanding and integration of financial and environmental risk in corporate decision making with regard to whether to invest in a project. In all cases, however, the potential for mining and multinational corporations to contribute to advancing the development objectives of their host countries in emerging resource-dependent economies depends on their ability to guarantee sustainable operations in the midst of price volatility and changing social and environmental uncertainties.

We note that the real option model sidesteps complex sociocultural and political factors that affect corporate decisions and on-the-ground mining operations in commodity states. Many mining companies in Africa are challenged in their ability to gain access to land and other natural resources that they share with local communities. Nevertheless, the ability of multinational mining corporations to navigate these complexities depends on deploying techniques that allow them to comprehensively account for major fiscal and environmental risks in the corporate stage-gate decision-making process.

The chapter has focused on integrating the financial and environmental uncertainties faced by gold mining projects into a valuation model using the real options approach, Monte Carlo simulations, and the statistical distribution analysis. This has enabled us to determine a critical value for investment for the three of the major producers of gold in Africa: Ghana, Tanzania, and South Africa. In addition, the potential environmental risk exposure of a portfolio is singled out to determine its economic impact on project cash flows. This will enable companies to identify their value metrics, minimize risk, and hence contribute immensely to sustainable and responsible mining. In addition, incorporating environmental uncertainties in the investment decision-making process will encourage an inclusive growth in mining communities as well as protect and promote sustainable use of our resources, a major ambition set out under SDGs 8 and 15.

Notes

1 United Nations General Assembly, "Transforming our world: the 2030 Agenda for Sustainable Development," www.un.org/ga/search/view_doc.asp?symbol=A/RES/70/1 &Lang=E (accessed November 11, 2015).
2 "commodity price": Jan Dehn, "Commodity Price Uncertainty in Developing Countries" (World Bank). "environmental uncertainties": Julie Wroblewski and Hendrik Wolff, "Risks to Agribusiness Investment in Sub-Saharan Africa" (Prepared for the Agricultural Policy and Statistics Team, Bill & Melinda Gates Foundation).
3 Julie Wrobleski, and Hendrik Wolff, "Risks to Agribusiness Investment in Sub-Saharan Africa," *Working Papers No. UWEC-2011–02* (Washington, DC: University of Washington, Department of Economics, 2010), http://faculty.washington.edu/hgwolff/Evans%20UW_Request%2057_Agribusiness%20and%20Investment%20Risks_02-12-2010.pdf (accessed November 11, 2015).
4 Adam Borison, "Real Options Analysis: Where are the Emperor's Clothes?" (Washington, DC: Paper presented at Real Options Conference, July 2003). An option is a security that provides the right but not the obligation to buy or sell an asset, subject to certain conditions, within a specified period of time. See Fischer Black and Myron Scholes, "The Pricing of Options and Corporate Liabilities," *Journal of Political Economy* 81:3 (May–June, 1973): 637–654.
5 Dick Warren and Prinesha Naidoo, "Update: World Top 10 Gold Producers – Countries and Miners," *Mine Web*, March 29, 2016, www.moneyweb.co.za/news/industry/world-top-10-gold-producers-countries-and-miners/ (accessed September 4, 2017).
6 Soussan Faiz, "Real Options Application: From Successes in Asset Valuation to Challenges for an Enterprise-Wide Approach" (Proceedings of the SPE Annual Technical Conference and Exhibition, 2000), 243–250.
7 V. Blais, R. Poulin and M. R. Samis, "Using Real Options to Incorporate Price Risk into the Valuation of a Multi-Mineral Mine" (The Australasian Institute of Mining and Metallurgy, 2007), 21–27.
8 Soussan Faiz, "Real Options Application" ; "project structure": V. Blais, R. Poulin, and M. R. Samis, "Using Real Options to Incorporate Price Risk."
9 Knight F. H., *Risk, Uncertainty and Profit*, (United States: Courier Corporation, 1921).
10 Stanley Block. "Are Real Options Actually Used in The Real World?" *Engineering Economist* 52:3 (2007): 255–267.
11 T.Y. Dube, "How the Use of Market Based Risk Metrics Can Undervalue Good Mining Projects and Overvalue Poor Ones" (Melbourne: Conference proceedings of Project Evaluation 2012, 24–25 May 2012).
12 Hesam Dehghani and Majid Ataee-Pour, "Determination of the Effect of Economic Uncertainties on Mining Project Evaluation Using Real Option Valuation," *International Journal of Mining and Mineral Engineering* 4:4 (2013): 265–277.
13 Avinash K. Dixit and Robert S. Pindyck, "The Options Approach to Capital Investment" (Cambridge, UK; London: MIT Press, 2001).
14 "valuation techniques": Bartolomeu Fernandes, Jorge Cunha, and Paula Ferreira, "The Use of Real Options Approach in Energy Sector Investments," *Renewable and Sustainable Energy Reviews* 15:9 (2011): 4491–4497. "becoming riskier": J. A. Botin, M. F. Del Castillo, R. R. Guzmn, and M. L. Smith, "Real Options: A Tool for Managing Technical Risk in a Mine Plan" (SME Annual Meeting and Exhibit 2012, Meeting Preprints, 2012), 572–578.

15 R. R. Bhappu and J. Guzman, "Mineral Investment Decision-Making - A Study of Mining Company Practices," *E&MJ-Engineering and Mining Journal* 196:7 (1995): 36–38.
16 Bartolomeu Fernandes, Jorge Cunha, and Paula Ferreira, "The Use of Real Options Approach in Energy Sector Investments," *Renewable and Sustainable Energy Reviews* 15:9 (2001): 4491–4497.
17 Stanley Block. "Are Real Options Actually Used in the Real World?"
18 Graham A. Davis, "Option Premiums in Mineral Asset Pricing: Are They Important?" *Land economics* 72:2 (1996): 167–186.
19 M. A. Haque, E. Topal, and E. Lilford, "A Numerical Study for a Mining Project Using Real Options Valuation Under Commodity Price Uncertainty," *Resources Policy* 39:1 (2014): 115–123.
20 S. A. Abdel Sabour, R. G. Dimitrakopoulos, and M. Kumral, "Mine Design Selection Under Uncertainty," *Transactions of the Institutions of Mining and Metallurgy, Section A: Mining Technology* 117:2 (2008): 53–64.
21 P. Guj and A. Chandra, "Real Option Valuation of Mineral Exploration/Mining Project Using Decision Trees – Differentiating Market Risk from Private Risk" (Melbourne: Project Evaluation 2012, 24–25 May 2012).
22 Avinash K. Dixit and Robert S. Pindyck, "The Options Approach to Capital Investment."
23 Eduardo S. Schwartz and Gonzalo Cortazar, "Monte Carlo Evaluation Model of an Undeveloped Oil Field," *Journal of Energy Finance & Development* 3:1 (1998): 73–84.
24 B. Groeneveld and E. Topal, "Flexible Open-Pit Mine Design Under Uncertainty," *Journal of Mining Science* 47:2 (2011): 212–226.
25 Ibid.
26 Projects in the exploration and reconnaissance stages include those in western, Ashanti, central and Brong Ahafo regions. R. Bloch, & G. Owusu, "Linkages in Ghana's Gold Mining Industry: Challenging the Enclave Thesis," *Resources Policy* 37:4 (2012): 434–442, http://dx.doi.org/10.1016/j.resourpol.2012.06.004.
27 "$4.516 billion" and "16 percent of fiscal receipts": Ghana Extractive Industries Transparency Initiative (GHEITI), "Mining Reconciliation Report 2014, Ministry of Finance" (GHEITI, December 2015).
28 Tanzania Extractive Transparency Initiative, "Sixth Report of the Tanzania Extractive Industries Transparency Initiative" (Prepared by BDO East Africa, November 2015).
29 Percentages given in this paragraph: Chamber of Mines of South Africa, "Annual Report 2013/2014: Adding Value in the Right Places Contributing Positively" (Chamber of Mines of South Africa, 2014).
30 "Samarco's Bill for Brazil Dam Failure Could Grow," *The Wall Street Journal*, December 8, 2015, www.wsj.com/articles/samarcos-bill-for-brazil-dam-failure-could-grow-14496 18310 (accessed on February 3, 2016).
31 Marc J. Epstein, and Adriana Rejc Buhovac. "Identifying, Measuring, and Managing Organizational Risks for Improved Performance: Management Accounting Guidelines" (Society of Management Accountants of Canada, 2005).
32 Nigeria: Hundreds of Oil Spills Continue to Blight Niger Delta, www.amnesty.org/en/latest/news/2015/03/hundreds-of-oil-spills-continue-to-blight-niger-delta/ (accessed on 7 February, 2016).
33 Marc J. Epstein and Adriana Rejc Buhovac. *Making Sustainability Work: Best Practices in Managing and Measuring Corporate Social, Environmental, and Economic Impacts* (Oakland, CA: Berrett-Koehler Publishers, 2014).

14 The potential of Zambian copper-cobalt metallophytes for phytoremediation of minerals wastes

Antony van der Ent, Peter Erskine, Royd Vinya, Jolanta Mesjasz-Przybyłowicz, and François Malaisse

In brief

- The Copper-Cobalt Belt of the Democratic Republic of the Congo and Zambia is one of the most important metallogenic regions in the world. In addition, it hosts the world's richest metallophyte flora.
- There are more than 600 metallophytes in the Copper-Cobalt Belt, including many species unique to the area.
- The phenomenon of abnormal copper-cobalt accumulation in certain plants has been observed since the 1950s in the Copper-Cobalt Belt, with more than thirty hyperaccumulator plants identified. Hyperaccumulator plants might be useful for important phytotechnologies.
- There is also a wide range of Excluder-type metallophytes in the Copper-Cobalt Belt, especially grasses and sedges, that might be suitable for phytostabilization of minerals wastes.

Metallophytes and hyperaccumulator plants

Plants require micronutrients such as copper and cobalt for optimal growth and cell metabolism. In high concentrations, however, both these elements can be toxic.[1] A few plant species can grow on metal-enriched, or metalliferous, soils; we refer to these plants as metallophytes.[2] Facultative metallophytes are often ecotypes of weedy plant species that have evolved metal tolerance, and obligate metallophytes are restricted to metalliferous soils. In response to excessive concentrations of trace elements in the soil (e.g., copper, cobalt, nickel, or zinc) plants might respond by excluding, marginally accumulating, or even hyperaccumulating these elements in their living tissues.[3] The rarest category of metallophytes are hyperaccumulator plants, which accumulate exceptional concentrations of trace elements, such as copper and cobalt, in their shoots.[4] Approximately 500 plant species have been identified to hyperaccumulate arsenic, cadmium, cobalt, copper, manganese, zinc, nickel, lead, thallium, or selenium.[5] Compared to the total number of known plant species in the world, these hyperaccumulators are rare. The threshold level for assigning hyperaccumulator status is a concentration value in the dried leaves defined as at least an order of magnitude higher than normal plants, thus the threshold level differs per element.[6] For example, the hyperaccumulation criterion for nickel is 1,000 ppm (parts per million), for zinc it is 3,000 ppm (0.3 Wt percent [percentage by weight]), and for manganese it is 10,000 ppm (1 Wt percent).[7]

Hyperaccumulation thresholds for copper and cobalt were initially defined at 1,000 ppm, but that level was revised downward to 300 ppm to better correspond to the low concentrations found in most plants.[8] More than 95 percent of the known copper and cobalt hyperaccumulator species are found in the Copper-Cobalt Belt of the Democratic Republic of the Congo (DRC) and Zambia, which hosts the famous copper flora.[9] Extensive research has so far identified thirty-two copper-cobalt hyperaccumulators from that region.[10]

Compared to the number of nickel hyperaccumulator plants (accounting for at least 350 of 500 known hyperaccumulator plant species), copper-cobalt hyperaccumulators are scarce. This might be because of the lack of a suitable field spot test for screening copper and cobalt in plant materials similar to the spot test that exist for nickel in the form of dimethylglyoxime-impregnated paper. As a result, systematic knowledge of the copper and cobalt concentrations in plant species is limited, which hinders a comprehensive understanding of the occurrence of the hyperaccumulation phenomenon of these two elements.

Mining industry and environmental legacy in Zambia

The Central African Copper-Cobalt Belt is the largest and highest-grade sediment-hosted stratiform copper province in the world; north Zambia accounts for approximately 46 percent of the region's production and reserves.[11] The mining industry in this area has been operating for more than a hundred years and has experienced significant growth over that time, growth that has come at significant environmental cost.[12] Intense base metal mining and ore processing have resulted in enormous pollution, much of it related to the disposal sites for mining and smelting waste that includes mine tailings and slags.[13] Heavy metal pollution presents a substantial environmental threat in Zambia, with health implications for humans and livestock.[14] Not surprisingly, Zambia's long history as a mining nation has resulted in the establishment of numerous historical mining legacy sites that cause environmental problems. More than 12,000 hectares of land in Zambia contain mineral waste from past mining processes.[15] Of those, more than 400 hectares represent more than 25 waste rock dumps, 300 hectares are covered with 10 slag dumps, hundreds of abandoned mines (Figure 14.1), and 10,000 hectares are covered with 50 tailing dams (Figure 14.2). The land where historic mine sites are found is devoid of vegetation cover, and is not available for other important uses such as housing, agriculture, or forestry. A lack of vegetation cover on mineral wastes also makes these sites prone to erosion and associated environmental problems due to dust and run-off contamination. Therefore, establishing a sustainable plant cover is an important component of impact minimization.

Figure 14.1 Abandoned mines (such as this water filled open pit) number in the hundreds in Zambia

Figure 14.2 Expansive tailings facilities cover thousands of hectares in northern Zambia

In Zambia, with a distinct short wet season and a long dry season, the wind-blown contamination of dust from unrehabilitated tailings storage facilities is a major problem (Figure 14.2). Moreover, many (historical) tailings storage facilities with sparse vegetation cover are now located in urban and residential areas. The tailings materials are, however, not extremely toxic in most cases in Zambia, due to the high limestone/dolomite content and hence alkaline pH diminish prevailing metal toxicities (i.e., from high total substrate copper/cobalt concentrations). In order to establish plants, we need to be able to address the very low nutrient status, high salinity, and other physical attributes on these sites (Figure 14.2).

Research on the metallophytes of the Copper-Cobalt Belt in Zambia

Research on copper-cobalt metallophytes from the Copper-Cobalt Belt has almost exclusively focused on the DRC. The bulk of that research has been by the academics of the Université Libre de Bruxelles and the Université de Liège–Gembloux Agro-Bio Tech. The focus of the research by these teams has been mainly on plant taxonomy, vegetation, and conservation (http://copperflora.org/). Subjects of research include phytostabilization of copper-contaminated soil and metallophyte niches.[16]

Plant collection on Zambian copper sites began in 1911, with the fieldwork of R. E. Fries, who collected a new subspecies of *Lobelia* (*L. trullifolia* subsp. *rhodesica*) on the Bwana Mkubwa mine. Publications about Zambian metallophytes then started to appear from 1969 onward with studies on *Ocimum centrali-africanum* (formerly called *B. homblei*), a species that was used successfully for geobotanical prospections for copper in Zambia in the 1950s and 1960s.[17] No one has yet compiled an exhaustive list of voucher specimens collected on Zambian copper sites; reports of concentrations of copper and cobalt in Zambian metallophytes exist, but are rare.[18]

Current state of metallophytes in Zambia

At least sixty-seven copper metalliferous sites have been identified in Zambia, but numerous other geological anomalies exist that are not reported in the literature.[19] In recent fieldwork, we assessed a number of sites (Figure 14.1) covering a large portion of the Copper-Cobalt Belt region in northern Zambia. We conducted fieldwork on Zambian-type anomalies including Bwana Mkubwa and Roan Antelope, minerals waste areas near Kitwe and Ndola, and Katangan type outcrops of Kansanshi near the town of Solwezi. We collected and identified representative plant specimens from the flora at these sites (Figure 14.3 and Appendix Table 14.1).

Recent publications have emphasized that many of the unique metallophytes of the Copper-Cobalt Belt are under significant threat as a result of widespread mining activities and habitat destruction.[20] Action toward conserving the metallophyte resource of the Copper-Cobalt Belt is, therefore, imperative. These metallophytes are an important asset to the mining industry in the DRC and Zambia. The original flora on natural copper-cobalt mineralized sites has been destroyed on all former mining sites we studied in Zambia (Figure 14.4), including Katangan-type

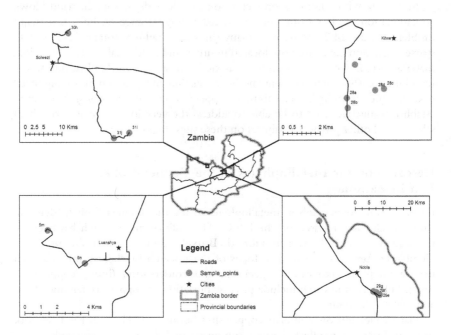

Figure 14.3 Map showing the localities and sites visited during 2014 fieldwork in Zambia

outcrops (e.g., Kansanshi) and Zambian-type anomalies (e.g., Bwana Mkubwa, Roan Antelope, Kitwe, and many other sites). It appears that untouched Zambian-type anomalies still exist in some places, according to local information. It is therefore urgent to both identify these sites and collect plant material from them before they are mined, and these plants destroyed. A second Katangan-type outcrop exists in Zambia at Luamata, west of the Kabompo dome, but no information is available about the plant species that occur on this site.[21]

The vegetation on copper-enriched sites in Zambia

In Zambia the copper ore occurs underground, although it is often shallow buried, with a halo of copper near the surface where copper clearings occur in the miombo open forest.[22] Soils with more than 1,000 ppm copper permit the development of shrubby vegetation dominated by perennial species such as *Olaxobtusifolia* and *Brachystegia taxifolia*. Although this vegetation superficially looks like the copper vegetation from the DRC, it is not the same from an ecological point of view. Many copper metallophyte species that are characteristic in the DRC are lacking in Zambia. In contrast, the vegetation in Zambia is often marked by toxicity symptoms: for example, shrubs like *Cryptosepalum* have yellow patches that are indicative of chlorosis due to copper toxicity.[23]

Ocimum centraliafricanum, the true copper flower of Zambia, has been used widely by geologists for prospecting purposes. We found this species occurring in

Figure 14.4 Artisanal mining activities are common on minerals wastes throughout northern Zambia

all but two sites, in thirty copper clearings, over an area from Lusaka to the border with the DRC.[24] In our field survey, we observed this plant in a miombo open forest on a site with 800 ppm of copper in the surface soil. This flower is highly copper tolerant and can grow on soils with more than 70,000 ppm copper (7 Wt percent), but it displays a typical Excluder-type behavior by accumulating only 324 ppm copper in the leaves.[25]

Phytostabilization of minerals wastes using metallophytes

The ability of metallophytes to cope with potentially toxic metal concentrations suggests they could be used to rehabilitate minerals wastes.[26] If so, the minerals industry could introduce plant species that occur naturally on metalliferous soils to rehabilitate sites economically. The pioneering work of Anthony Bradshaw on the revegetation of mineral wastes with metal-tolerant grasses has demonstrated that metallophytes can be used to revegetate large areas of mineral wastes, without the need to intensely ameliorate the substrate conditions.[27] When metallophytes establish, they can stabilize mineral wastes and thereby provide improved conditions for natural succession by other plant species. In seasonally dry environments like Zambia, selected plant species need to be not only metal tolerant but also drought resistant. The addition of some soil amendments, such as organic matter (compost) and nutrients (nitrogen, phosphorus, and potassium) often substantially enhances the establishment and growth of plants on minerals wastes such as tailings.[28] Several researchers have studied this issue: Leteinturier, Baker, and Malaisse have described a typology of the mining workings and metalliferous pollutions during precolonial, early colonial, and recent periods.[29] Short and Milton, and Leteinturier discuss aspects of phytostabilization of copper wastes in Zambia in an audit and a thesis, respectively.[30]

Potential for phytomining on minerals wastes using hyperaccumulator plants

Hyperaccumulator species can be used to extract metallic elements from minerals wastes or sub-economic ore bodies.[31] As such, growing hyperaccumulator plants on an agricultural scale with subsequent harvesting and biomass incineration generates a metal-rich product termed bio-ore.[32] Extensive trials have been undertaken for nickel phytomining around the world, which demonstrated yields up to 200 kilograms of nickel metal per hectare per year.[33] The technology for nickel is now reasonably well established and understood, but the application to other metals remains largely untested. Phytomining, in principle, could be used to produce metals such as copper, cobalt, manganese, nickel, and zinc, because hyperaccumulator plants are known for all of these elements.[34] Economic feasibility ultimately depends on the element market price, the annual yield per unit area of biomass and the target element, and the availability of surface areas enriched in this element.[35] The metal value of elements such as copper, manganese, and zinc are low, and hence phytomining is unlikely to be economical. However, phytomining, or rather phytoextraction, could be undertaken as part of a remediation exercise on tailings in which the value does not lie in the metal obtained, but in the soil and minerals wastes cleaned up. The metal value of cobalt is, however, sufficient to warrant further work on the viability of cobalt phytomining.

During our fieldwork we confirmed other researchers' findings that many tailings in north Zambia, especially those of older mining activities, have relatively high cobalt concentrations (500–1,000 ppm).[36] Many oxide-type surficial deposits that are below cut-off grade for conventional mining could also be high in cobalt. Local communities might undertake phytomining on such substrates to create an additional income stream farming metals.[37] The plant species with the highest potential to be used as the cobalt-metal crop is *Haumaniastrum katangense* (Lamiaceae), which is widespread in the Copper-Cobalt Belt, both in the DRC and in Zambia. It can accumulate up to 10,200 ppm (1.2 Wt percent) cobalt in its leaves in the field, although it accumulates lower concentrations in lab cultures for reasons not completely understood.[38] This annual plant produces copious amounts of seeds and is well adapted to the local climatic conditions. It probably could be grown on cobalt-rich minerals waste to extract residual cobalt. The phytomining process would entail starting seeds in a nursery, transplanting seedlings to a tailings facility, harvesting biomass at the end of the growth season, and drying and incinerating the biomass to produce an ash bio-ore. This process is theoretical: demonstrations are needed to provide real-life evidence of the process before the mining industry undertakes it.

Advanced scientific methods to understand hyperaccumulator plants

Almost all reports of abnormal accumulation of copper-cobalt in certain plants from the Copper-Cobalt Belt since the 1950s were from bulk analysis techniques such as AA (atomic absorption) or ICP (inductively coupled plasma). A study in 2007 on plant material from the DRC has shown that earlier plant analyses were done on contaminated material and that the number of copper-cobalt hyperaccumulators might be more limited than initially reported.[39] Preliminary identification of copper-cobalt hyperaccumulating plants during field studies is not as easy as it is for nickel, where dimethylglyoxime-impregnated paper allows fast screening in the field. Field analysis using a portable XRF instrument is very useful, although such results have to be treated as qualitative rather than quantitative due to technical restrictions in instrument calibration and geometry of analysis (Appendix Table 14.2).

To clearly define whether a particular plant meets hyperaccumulator criteria requires advanced laboratory methods to provide unambiguous quantitative information on the concentration of copper and cobalt in bulk volume of particular plant parts, and elemental distribution in tissues and cells. Such data allow us to explain how hyperaccumulating plants uptake, transport, and store potentially toxic levels of copper and cobalt. Relatively popular chemical methods such as AA, ICP-AES (inductively coupled plasma–atomic emission spectroscopy), or ICP-MS (inductively coupled plasma–mass spectrometry) are currently the methods of choice for determining bulk values. Further studies require one or more of the following microanalytical methods: histochemical techniques; autoradiography; LA-ICP-MS (laser ablation–inductively coupled plasma–mass spectrometry); SIMS (secondary ion mass spectrometry); SEM (scanning electron microscopy) and TEM (transmission electron microscopy) combined with EDS (and energy dispersive x-ray spectroscopy); PIXE (particle induced x-ray emission); and various synchrotron techniques such as X-ray fluorescence spectroscopy, differential absorption tomography, fluorescence tomography or XAS (X-ray absorption spectroscopy).[40]

It is far more difficult to collect and prepare specimens for microanalytical studies than to evaluate whole plant tissue concentration levels, however. Proper specimen sampling and subsequent procedures and preparations are crucial parts of microanalysis; only cryotechniques are recommended. The aim of specimen preparation in these analyses is the preservation of elemental distribution as close as possible to its native (in vivo) state. This requirement applies to all microanalytical techniques.

When plants are growing in a laboratory, proper fast cryofixation does not present a problem, but preservation of living plant status when collecting plant material in the field is a challenge. There are two options that have been applied successfully: (1) collecting whole plants with soil in pots from the metalliferous outcrops and rapid transport to a properly equipped cryopreparation laboratory; or (2) freezing specimens in the field using metal-mirror cryofixation method and transport to laboratory in a cryoshipper, at liquid nitrogen temperature.[41]

The first elemental microanalysis done on a copper and cobalt hyperaccumulating plant from the Copper-Cobalt Belt—*Haumaniastrum robertii*—was performed with the use of the proton microprobe using the PIXE method.[42] There have been few reports on copper and/or cobalt elemental distribution and/or speciation in hyperaccumulating since that initial report.[43]

Conservation of metallophytes in Zambia

The restriction of metallophytes to localized copper-cobalt mineralized outcrops, which are mining targets, means that they are acutely threatened.[44] Natural copper outcrops, or copper clearings, once existed in Zambia, for example at Roan Antelope (Figure 14.5), but are now destroyed due to mining activities. New copper-cobalt deposits are most likely to be found in the more remote areas of the north-west province of Zambia near the border with the DRC. The International Council on Mining and Metals (ICMM) Sustainable Development Framework specifically acknowledges:

> Local extinctions can be caused by any sectoral activity, but there is one group of plants that is likely to go extinct as a result of mining activity alone—metallophytes. Mining activities can easily obliterate them, resulting in the loss of a potentially valuable resource.[45]

Recent fieldwork revealed that the high rates of human-induced disturbances due to industrial mining has led to major losses in the metallophytes and hyperaccumulator plant diversity in Zambia. For example, *Cheilanthes perlanata* var. *perlanata* observed in 2001 at Kansanshi has since disappeared.[46] This species was

Figure 14.5 Copper clearing at Roan Antelope before mining activities commenced
Source: Image reproduced from Duvigneaud and Denayer-De Smet (1963)

formerly recorded on ten sites, and the two Katangan sites are now destroyed. The subsequent artisanal mining activities (Figure 14.4) have without doubt resulted in further loss of the plant populations that once thrived under the local ecological settings.#

Outlook and future opportunities

The concomitant occurrence of copper-cobalt hyperaccumulator plants and large areas with metal-enriched and metal-contaminated soils makes a compelling case for developing phytotechnologies in Zambia. In addition, the minerals industry could manage copper-cobalt metallophytes originating from naturally mineralized areas to actively colonize industrial minerals wastes and mined-out areas, thereby establishing vegetative cover as part of their rehabilitation strategy. It is essential that systematic screening and cataloguing of metallophytes take place in remaining habitats, especially in the remote north-west province. Local knowledge is indispensable for the identification of metallophytes, and is vital to assess priorities for those most under threat.[47]

Acknowledgments

The authors would like to acknowledge the financial support from the International Mining for Development Centre (IM4DC) that made this project possible.

Appendix

Appendix Table 14.1 Preliminary identifications of collected herbarium specimens from the sites

Voucher	Refer.	Site	Family	Name
1	28A	Ushi dam, west Aka 16ft bridge	Orobanchaceae	*Alectra sessiliflora* (Vahl) Kuntze
2	28A	Ushi dam, west Aka 16ft bridge	Campanulaceae	*Lobelia erinus* L.
3	28A	Ushi dam, west Aka 16ft bridge	Polygonaceae	*Persicaria capitata* (Buch.-Ham. ex D. Don) H.Gross
4	28A	Ushi dam, west Aka 16ft bridge	Asteraceae	*Anisoppapus chinensis* (L.) Hook & Arn. subsp. *chinensis*
5	28A	Ushi dam, west Aka 16ft bridge	Convolvulaceae	*Ipomoea petitiana* Lejoly&Lisowski
7	28B	Ushi dam, west Aka, stream	Pteridaceae	*Pteris vittata* L.
8	28B	Ushi dam, west Aka, stream	Asteraceae	*Conyza pyrhopappa* A.Rich.
9	28B	Ushi dam, west Aka, stream	Portulacaceae	*Portulaca oleracea* L.

(continued)

Appendix Table 14.1 *(continued)*

Voucher	Refer.	Site	Family	Name
10	28B	Ushi dam, west Aka, stream	Cyperaceae	*Cyperus dives* Delile
11	28B	Ushi dam, west Aka, stream	Amaranthaceae	*Celosia trigyna*L.
12	28C	Uchi east, tailing	Polygonaceae	*Oxygonum sinuatum* (Hochst. &Steud. ex Meissn.) Dammer
13	28C	Uchi east, tailing	Poaceae	*Setaria* sp.
14	28C	Uchi east, tailing	Poaceae	*Loudetia*s sp.1
15	28C	Uchi east, tailing	Poaceae	*Loudetia*s sp.2
18	28C	Uchi east, tailing	Poaceae	*Cymbopogon densiflorus* (Steud.) Stapf
19	28C	Uchi east, tailing	Poaceae	*Hyparrhenia*s sp.
20	28C	Uchi east, tailing	Cyperaceae	*Bulbostylis*s sp.
22	28C	Uchi east, tailing	Amaranthaceae	*Celosia trigyna* L.
25	29E	Bwana Mkubwa, Rompod mine	Cyperaceae	*Bulbostylis pseudoperennis* Goetgh.
26	29E	Bwana Mkubwa, Rompod mine	Convolvulaceae	*Ipomoea cairica* (L.) Sweet var. *cairica*
27	29E	Bwana Mkubwa, Rompod mine	Convolvulaceae	*Ipomoea obscura* (L.) Ker-Gawl.
28	29E	Bwana Mkubwa, Rompod mine	Convolvulaceae	*Ipomoea* sp.1
29	29E	Bwana Mkubwa, Rompod mine	Asteraceae	*Tridaxprocumbens* L.
30	29E	Bwana Mkubwa, Rompod mine	Asteraceae	*Pseudognaphalium luteo-album* (L.) Hilliard & Burtt
31	29E	Bwana Mkubwa, Rompod mine	Ochnaceae	*Brackenridgea arenaria* (De Wild. & T. Durand) N. Robson
32	29E	Bwana Mkubwa, Rompod mine	Fabaceae	*Tephrosia bracteolata* Guill. & Pers. var. *strigulosa* Brummitt
33	29E	Bwana Mkubwa, Rompod mine	Boraginaceae	*Trichodesma zeylanicum* (Burm.) R. Br.
34	29E	Bwana Mkubwa, Rompod mine	Poaceae	*Arthraxon micans* (Nees) Hochst.
35	29E	Bwana Mkubwa, Rompod mine	Tiliaceae	*Triumfetta dekindtiana* Engl.
38	29F	Old concentrator, facing station	Pteridaceae	*Pityrogramma calomelanos* (Sw.) Link var. *aureoflava* (Hook.) Weath. ex Bailey
39	29F	Old concentrator, facing station	Phormidiaceae	*Porphyrosiphon notarsii* Kütz. ex Gomont

41	29F	Old concentrator, facing station	Apiaceae	*Diplolophium* sp.
42	29G	Mine pit summit	Asteraceae	*Vernonia* sp.
43	29G	Mine pit summit	Fabaceae	*Rhynchosia hirta* (Andr.) Meikle & Verdc.
44	29G	Mine pit summit	Anacardiaceae	*Rhus* sp.
46	29G	Mine pit summit	Asteraceae	*Vernonia* sp. 2
47	29G	Mine pit summit	Rubiaceae	cf. *Spermacoce*
49	30H	South-east dome	Lamiaceae	*Ocimum centrali-africanum* R. E. Fries
51	30H	South-east dome	Orobanchaceae	*Buchnera henriquesii* Engl.
52	30H	South-east dome	Rubiaceae	*Fadogia cienkowski* Schweinf.
53	30H	South-east dome	Tiliaceae	*Triumfetta digitata* (Oliv.) Sprague & Hutch.
54	30H	South-east dome	Annonaceae	*Annonastenophylla* Engl. & Dielssub sp. *nana* (Exell) N.Robson
55	30H	South-east dome	Fabaceae	*Erisosema engleriana* Harms
56	30H	South-east dome	Commelinaceae	*Cyanotis* sp.
57	31I	Miombo to pit	Euphorbiaceae	*Acalypha* sp. 1
58	31I	Miombo to pit	Lamiaceae	*Becium* sp.
60	31I	Miombo to pit	Cucurbitaceae	*Trochomeria macrocarpa* (Sond.) Hook.
61	31I	Miombo to pit	Aristolochiaceae	*Aristolochia heppii* Merxm.
80	31I	Miombo to pit	Malavaceae	*Hibiscus rhodanthus* Gürke
86	3L	Mwekera	Passifloraceae	*Adenialobata* (Jacq.) subsp. *rumicifolia* (Engl. & Harms) Lye
87	3L	Mwekera	Asparagaceae	*Asparagus africanus* Lam. var. *africanus*
88	3L	Mwekera	Vitaceae	*Cyphostemma* sp.
89	3L	Mwekera	Rubiaceae	*Fadogia cienkowski* Schweinf.
90	3L	Mwekera	Fabaceae	*Albizia adianthifolia* (Schumach.) W. Wight
91	3L	Mwekera	Smilacaceae	*Smilax anceps* Willd.
92	3L	Mwekera	Campanulaceae	*Wahlenbergia* sp.
93	3L	Mwekera	Passifloraceae	*Adenia gummifera* (Harv.) Harms var. *gummifera*
94	3L	Mwekera	Commelinaceae	*Commelina* sp.
95	3L	Mwekera	Convolvulaceae	*Ipomoea* sp.
96	3L	Mwekera	Cucurbitaceae	*Trochomeria macrocarpa* (Sond.) Hook.
97	3L	Mwekera	Loganiaceae	*Strychnos spinosa* Lam.
99	3L	Mwekera	Acanthaceae	*Thunbergia* sp.
101	3L	Mwekera	Commelinaceae	*Cyanotis* sp.

(continued)

Appendix Table 14.1 (continued)

Voucher	Refer.	Site	Family	Name
102	3L	Mwekera	Fabaceae	*Tephrosia* sp.
137	4A	Bwana Mkubwa	Lauraceae	*Cassytha filiformis* L.
140	4A	Bwana Mkubwa	Vitaceae	*Cyphostemmas* p. 2
141	4A	Bwana Mkubwa	Vitaceae	*Cyphostemmas* p. 3
142	4A	Bwana Mkubwa	Amaranthaceae	*Celosia trigyna* L.
143	4A	Bwana Mkubwa	Convolvulaceae	*Ipomoea cairica* (L.) Sweet var. *cairica*
144	4A	Bwana Mkubwa	Asteraceae	*Ageratum conyzoides* L.
147	4A	Bwana Mkubwa	Adiantaceae	*Pityrogramma calomelanos* (Sw.) Link. var. *aureoflava* (Hook.) Weath. ex Bailey
149	4B	Bwana Mkubwa	Typhaceae	*Typha domingensis* Pers.
150	4B	Bwana Mkubwa	Poaceae	*Phragmites mauritianus* Kunth
153	4B	Bwana Mkubwa	Polygonaceae	*Persicaria capitata* (Buch.-Ham. ex D. Don) H.Gross
154	4B	Bwana Mkubwa	Asteraceae	*Pseudognaphalium luteo-album* (L.) Hilliard &Burtt
107	4B	Bwana Mkubwa	Vitaceae	*Cyphostemma* sp. 1
108	4B	Bwana Mkubwa	Convolvulaceae	*Ipomoea* sp. 5
109	4B	Bwana Mkubwa	Solanaceae	*Nicandra physaloides* (L.) Gaertn.
110	4B	Bwana Mkubwa	Poaceae	*Digitaria* sp.
112	4B	Bwana Mkubwa	Adiantaceae	*Pityrogramma calomelanos* (Sw.) Link. var. *aureoflava* (Hook.) Weath. ex Bailey
114	4B	Bwana Mkubwa	Anacardiaceae	*Rhus* sp.
124	4L	Bwana Mkubwa	Convolvulaceae	*Ipomoea obscura* (L.) Ker.-Gawl.
126	4L	Bwana Mkubwa	Tiliaceae	*Grewia* sp. 1
133	4L	Bwana Mkubwa	Myrtaceae	*Psidium guajava* L.
158	5M	Roan Antelope shaft 14 (a)	Cyperaceae	*Bulbostylis pseudoperennis* Goetgh.
159	5M	Roan Antelope shaft 14 (a)	Phormidiaceae	*Porphyrosiphon notarsii* Kütz ex Gomont
160	5M	Roan Antelope shaft 14 (b)	Lauraceae	*Cassytha filiformis* L.
164	5M	Roan Antelope shaft 14 (c)	Asteraceae	*Tridax procumbens* L.
165	5M	Roan Antelope shaft 14 (c)	Fabaceae	*Crotalaria* sp. 4
166	5M	Roan Antelope shaft 14 (c)	Convolvulaceae	*Ipomoea cairica* (L.) Sweet var. *cairica*
167	5M	Roan Antelope shaft 14 (c)	Phyllanthaceae	*Phyllanthus* sp.

168	5M	Roan Antelope shaft 14 (c)	Amaranthaceae	*Celosia trigyna* L.
169	5M	Roan Antelope shaft 14 (c)	Asteraceae	*Pseudognaphalium luteo-album* (L.) Hilliard &Burtt
173	5M	Roan Antelope shaft 14 (d)	Asteraceae	Genus 1
174	5N	Roan Antelope wet site	Typhaceae	*Typha australis* K.Schum. & Thon.
175	5N	Roan Antelope wet site	Onagraceae	*Ludwigia leptocarpa* (Nutt.) H. Hara
176	5N	Roan Antelope wet site	Polygonaceae	*Polygonum* sp. 1
177	5N	Roan Antelope wet site	Lauraceae	*Cassytha filiformis* L.
180	5N	Roan Antelope wet site	Poaceae	*Typha australis* K.Schum. & Thon.
181	5N	Roan Antelope wet site	Orobanchaceae	*Alectra sessiliflora* (Vahl) Kuntze
182	5N	Roan Antelope wet site	Polygonaceae	*Polygonum* sp. 2
183	5N	Roan Antelope wet site	Cyperaceae	*Cyperus dives* Delile

Appendix Table 14.2 Preliminary plant material analytical results from XRF measurements

Refer.	Family	Name	As	Co	Cu	Se	Zn
28A	Orobanchaceae	*Alectra sessiliflora*	<LOD	<LOD	++	<LOD	++
28A	Asteraceae	*Anisoppapus chinensis* sub sp. *chinensis*	<LOD	++	++	<LOD	++
28A	Convolvulaceae	*Ipomoea* sp.	<LOD	<LOD	++	<LOD	++
28A	Polygonaceae	*Persicaria capitata*	<LOD	<LOD	+++	<LOD	++
28B	Amaranthaceae	*Celosia trigyna*	<LOD	+++	++	<LOD	++
28B	Asteraceae	*Conyza pyrhopappa*	<LOD	++	+++	<LOD	++
28B	Cyperaceae	*Cyperus dives*	<LOD	<LOD	++	++	++
28B	Portulacaceae	*Portulacca* sp.	<LOD	<LOD	+	<LOD	+
28B	Pteridaceae	*Pteris vittata*	++	<LOD	++	<LOD	+
28C	Amaranthaceae	*Celosia* sp.	<LOD	<LOD	++	<LOD	<LOD
28C	Poaceae	*Loudetia* sp.2	<LOD	<LOD	++	+	<LOD
28C	Polygonaceae	*Persicaria* sp.	<LOD	++	++	+	+
28C	Asteraceae	*Pseudognaphalium luteo-album*	<LOD	<LOD	+++	<LOD	++
29E	Convolvulaceae	*Ipomoea cairica* var. *cairica*	<LOD	<LOD	++	<LOD	+
29E	Convolvulaceae	*Ipomoea* sp.1	<LOD	<LOD	++	<LOD	+
29E	Convolvulaceae	*Ipomoea* sp.1	<LOD	<LOD	+++	<LOD	+

Refer.	Family	Name	As	Co	Cu	Se	Zn
29E	Ochnaceae	Ochna sp.	<LOD	<LOD	++	<LOD	+
29E	Asteraceae	Pseudognaphalium luteo-album	<LOD	<LOD	++	<LOD	++
29E	Fabaceae	Tephrosia sp.	<LOD	<LOD	++	<LOD	++
29E	Boraginaceae	Trichodesma sp.	<LOD	<LOD	++	<LOD	+
29E	Asteraceae	Tridax procumbens	<LOD	<LOD	++	<LOD	++
29F	Apiceae	cf. Diplolophium sp.	<LOD	<LOD	+++	<LOD	++
29F	Pteridaceae	Pityrogramma calomelanos var. aureoflava	+	<LOD	++	<LOD	++
29G	Fabaceae	Abrus sp.	<LOD	+	++	<LOD	++
29G	Rubiaceae	cf. Spermacoce	<LOD	<LOD	++	<LOD	++
29G	Anacardiaceae	Rhus sp.	<LOD	<LOD	++	<LOD	+
29G	Asteraceae	Vernonia sp. 1	<LOD	<LOD	++	+	++
29G	Asteraceae	Vernonia sp. 2	<LOD	<LOD	++	<LOD	++
30H	Annonaceae	Annona stenophylla sub sp. nana	<LOD	<LOD	++	<LOD	+
30H	Lamiaceae	Ocimum centrali-africanum	<LOD	<LOD	++	<LOD	+
30H	Orobanchaceae	Buchnera henriquesii	<LOD	<LOD	++	<LOD	+
30H	Commelinaceae	Cyanotis sp.	<LOD	<LOD	++	<LOD	+
30H	Fabaceae	Erisosema engleriana	<LOD	<LOD	++	<LOD	++
30H	Rubiaceae	Fadogia cienkowski	<LOD	<LOD	++	<LOD	+
30H	Tiliaceae	Triumfetta digitata	<LOD	<LOD	++	<LOD	+
31I	Euphorbiaceae	Acalypha sp. 1	<LOD	<LOD	++	<LOD	++
31I	Euphorbiaceae	Acalypha sp. 1	<LOD	<LOD	++	<LOD	+
31I	Euphorbiaceae	Acalypha sp. 1	<LOD	<LOD	++	<LOD	+
31I	Aristolochiaceae	Aristolochia heppii	<LOD	<LOD	++	<LOD	++
31I	Lamiaceae	Becium sp.	<LOD	<LOD	++	<LOD	++
31I	Fabaceae	Crotalaria sp.	<LOD	<LOD	++	<LOD	+
31I	Malavaceae	Hibiscus rhodanthus	<LOD	<LOD	++	<LOD	++
31I	Fabaceae	Indigofera sp.	<LOD	<LOD	++	<LOD	+
31I	Cucurbitaceae	Trochomeria macrocarpa	<LOD	<LOD	++	<LOD	+

Note: As = arsenic; Co = cobalt; Cu = copper; Se = selenium; Zn = zinc

Preliminary plant material analytical results from XRF measurements (semi-quantitative results as either +, ++ or +++ to denote relative concentrations). None of the plants recorded are hyperaccumulators of either As, Co, Cu, Se, or Zn. <LOD is below instrument Limit of Detection which corresponds to approximately 50 ppm for most transition elements, such as copper (Cu) and cobalt (Co)

Notes

1 F. J. Stevenson, *Cycles of Soil-Carbon, Nitrogen, Phosphorus, Sulfur, Micronutrients* (New York: John Wiley and Sons, 1986); H. Marschner, *Mineral Nutrition of Higher Plants* (London: Academic Press, 1995).
2 A. J. M. Baker and S. N. Whiting, "Metallophytes—A Unique Biodiversity and Biotechnological Resource in the Care of the Minerals Industry," A. Fourie, M. Tibbett,

I. Weiersbye and P. Dye (eds.) (Johannesburg: Proceedings of the Third International Seminar on Mine Closure, 14–17 October 2008, Nedlands: Australian Centre for Geomechanics, 2008), 13–20; A. J. M. Baker, W. H. O. Ernst, A. van der Ent, F. Malaisse, and R. Ginocchio, "Metallophytes: The Unique Biological Resource, its Ecology and Conservational Status in Europe, Central Africa and Latin America," in *Ecology of Industrial Pollution*. L. C. Batty and K. B.Hallberg (eds.) (Cambridge, UK: Cambridge University Press, 2010), 7–40.

3 A. J. M. Baker, "Accumulation and Excluders—Strategies in the Response of Plants to Heavy Metals," *Journal of Plant Nutrition* 3 (1981): 643–654; A. J. M. Baker, "Metal Tolerance," *New Phytologist* 106 (1987): 93–111.

4 R. D. Reeves, "Tropical Hyperaccumulators of Metals and Their Potential for Phytoextraction," *Plant and Soil* 249 (2003): 57–65.

5 A. Van der Ent, A. J. M. Baker, R. D. Reeves, A. J. Pollard, and H. Schat, "Hyperaccumulators of Metal and Metalloid Trace Elements: Facts and Fiction," *Plant and Soil* 362 (2013): 319–334, doi: 10.1007/s11104-012-1287-3.

6 R. D. Reeves, "Tropical Hyperaccumulators of Metals."

7 A. Van der Ent et al. "Hyperaccumulators of metal."

8 "defined at 1,000 ppm": F. Malaisse, J. Grégoire, R. R. Brooks, R. S. Morrison, and R. D. Reeves, "*Aeolanthus biformifolius* De Wild.: A Hyperaccumulator of Copper from Zaire," *Science* 199 (1978): 887–888. "in most plants": A. Van der Ent et al., "Hyperaccumulators of metal," doi: 10.1007/s11104-012-1287-3; U. Krämer, "Metal Hyperaccumulation in Plants." *Plant Biology* 61 (2010): 517–534, doi: 10.1146/annurev-arplant-042809-112156.

9 See also P. Duvigneaud, "La végétation du Katanga et de ses sols métallifères," *Bulletin de la Société Royale de Botanique de Belgique* 90 (1958): 127–186; P. Duvigneaud and S. Denaeyer-De Smet, "Cuivre et végétation au Katanga," *Bulletin de la Société Royale de Botanique de Belgique* 96:2 (1963): 93–231; R. R. Brooks, R. D. Reeves, R. S. Morrison and F. Malaisse, "Hyperaccumulation of Copper and Cobalt—A Review," *Bulletin de la Société Royale de Botanique de Belgique* 113 :2 (1980): 166–172, www.jstor.org/stable/20793840; M.-P. Faucon, A. Meersseman, M. N. Shutcha, G. Mahy, M. N. Luhembwe, F. Malaisse, and P. Meerts, "Copper Endemism in the Congolese Flora: A Database of Copper Affinity And Conservational Value of Cuprophytes." *Plant Ecology and Evolution* 143:1 (2010): 5–18.

10 R. R. Brooks, R. D. Reeves, R. S. Morrison and F. Malaisse, "Hyperaccumulation of Copper and Cobalt—A Review"; R. R. Brooks, "Copper and Cobalt Uptake by *Haumaniastrum* Species," *Plant and Soil* 48 (1977): 541–544; R. S. Morrison, R. R. Brooks, R. D. Reeves and F. Malaisse, "Copper and Cobalt Uptake by Metallophytes from Zaire," *Plant and Soil* 53 (1979): 535–539; R. S. Morrison, "Aspects of the Accumulation of Cobalt, Copper and Nickel By Plants" (PhD thesis, 1980), 287; R. R. Brooks, F. Malaisse and A. Empain, "The Heavy Metal Tolerant Flora of Southcentral Africa" (Rotterdam; Boston: A.A. Balkema, 1985), i–x, 199; R. R. Brooks, S. M. Naidu, F. Malaisse and J. Lee, "The Elemental Content of Metallophytes from the Copper/Cobalt Deposits of Central Africa," *Bulletin de la Société Royale de Botanique de Belgique* 119:2 (1987): 179–191, www.jstor.org/stable/20794087; R. D. Reeves, "Hyperaccumulation of Trace Elements By Plants," in *Phytoremediation of Metal-Contaminated Soils. NATO Science Series (IV): Earth and Environmental Sciences,* 68, J.L. Morel, G. Echevarria, N. Goncharova (eds.) (Dordrecht: Springer, 2006), 25–52.

11 D. Selley, D. Broughton, R. Scott, M. Hitzman S. Bull, R. Large, P. McGoldrick, M. Croaker, N. Pollington, and F. Barra, "A New Look at the Geology of the Zambian Copperbelt" in *Society of Economic Geologists. 100th Anniversary Volume*, J. W. Hedenquis, J. F. H. Thompson, R. J. Goldfarb and J. R. Richards (eds.) (Littleton, CO: SEG Publications, 2005), 965–1000.

12 N. Simutanyi, "Copper Mining in Zambia: The Developmental Legacy of Privatization," *Institute of Security Studies (ISS) July. Paper 165* (2008).

13 B. D. Tembo, K. Sichilongo, and J. Cernak, "Distribution of Copper, Lead, Cadmium and Zinc Concentrations in Soils Around Kabwe Town in Zambia," *Chemosphere* 63:3 (2006): 497–501, doi: 10.1016/j.chemosphere.2005.08.002.
14 J. Yabe, S. M. M. Nakayama, Y. Ikenaka, K. Muzandu, M. Ishizuka, and T. Umemura, "Uptake of Lead, Cadmium, and Other Metals in the Liver and Kidneys of Cattle Near a Lead-Zinc Mine in Kabwe, Zambia," *Environmental Toxicology and Chemistry* 30 (2011): 1892–1897.
15 Environmental Council of Zambia, *Zambia Environment Outlook Report 3* (Lusaka: Government of the Republic of Zambia, 2008); J. Lindahl, "Environmental Impacts of Mining in Zambia: Towards Better Environmental Management and Sustainable Exploitation of Mineral Resources," *Geological Survey of Sweden* (2014).
16 "phytostabilization of copper-contaminated soil": M. N. Shutcha, M. N. Mubemba, M.-P. Faucon, M. N. Luhembwe, M. Visser, G. Collinet, and P. Meerts, "Phytostabilisation of Copper-Contaminated Soil in Katanga: An Experiment with Three Native Grasses and Two Amendments," *International Journal of Phytoremediation* 12:6 (2010): 616–632, doi: 10.1080/15226510903390411. "metallophyte niches": L. Saad, I. Parmentier, G. Colinet, F. Malaisse, M.-P. Faucon, P. Meerts, and G. Mahy, "Investigating the Vegetation-Soil Relationships on the Copper-Cobalt Rock Outcrops of Katanga (D.R. Congo), An Essential Step in a Biodiversity Conservation Plan," *Restoration Ecology* 20 (2011): 405–415, doi: 10.1111/j.1526–100X.2011.00786.x; E. Ilunga, M. Séleck, G. Colinet, M.-P. Faucon, P. Meerts, and G. Mahy, "Small-Scale Diversity of Plant Communities and Distribution of Species Niches on a Copper Rock Outcrop in Upper Katanga, D.R. Congo," *Plant Ecology and Evolution* 146 (2013): 173–182, doi: http://dx.doi.org/10.5091/plecevo.2013.816; M. Séleck, J.-P. Bizoux , G. Colinet, M.-P. Faucon, A. Guillaume, P. Meerts, J. Piqueray, and G. Mahy, "Chemical Soil Factors Influencing Plant Assemblages Along Copper-Cobalt Gradients: Implications for Conservation and Restoration," *Plant and Soil* 373:1 (2013): 455–469, doi: 10.1007/s11104-013-1819-5; M.-P. Faucon, E. Ilunga, B. Lange, S. Boisson, M. Séleck, S. Le Stradic, M. Shutcha, C. Grison, O. Pourret, P. Meerts, and G. Mahy, "Implications des relations sol-plante en ingénierie écologique des habitats et sols métallifères dégradés, le cas des habitats riches en cuivre du Katanga (République Démocratique du Congo)," *Bulletin Séances* ARSOM 61 (2015); D. K. Kaya Muyumba, A. Liénard, G. Mahy, M. N. Luhembwe, and G. Colinet, "Characterization of Soil-Plant Systems in the Hills of the Copper Belt in Katanga. A Review," *Biotechnologie, Agronomie, Société et Environnement* 19 (2015): 204–214.
17 C. Reilly, "The Uptake and Accumulation of Copper by *Becium homblei* (De Wild.) Duvign. &Plancke," *New Phytologist* 68 (1969): 1081–1087, doi:10.1111/j.1469–8137.1969.tb06509.x; C. Howard-Williams, "The Ecology of *Becium homblei* in Central Africa with Special Reference to Metalliferous Soils," *Journal of Ecology* 58:3 (1970): 745–763, doi: http://doi.org/10.2307/2258533; C. Reilly, J. Rowel, and J. Stone, "The Accumulation and Binding of Copper in Leaf Tissue of *Becium homblei* (De Wild.) Duvign. &Plancke," *New Phytologist* 69 (1970): 993–997, doi: 10.1111/j.1469–8137.1970.tb02478.x; A. Drew and C. Reilly, "Observations on Copper Tolerance in the Vegetation of a Zambezian Copper Clearing," *Journal of Ecology* 60 (1972): 439–444; J. J. Brummer and G. D. Woodward, "A History of the Zambian Copper Flower *Becium centrali-africanum* (B. homblei)," *Journal of Geochemical Exploration* 65 (1999): 133–140.
18 B. Leteinturier, A. J. M. Baker, L. Bock, J. Matera, and F. Malaisse, "Copper and Vegetation at the Kansanshi Hill (Zambia) Copper Mine," *Belgian Journal of Botany* 134:1 (2001): 41–50, doi: http://www.jstor.org/stable/20794476.
19 D. Selley et al., "A New Look at the Geology of the Zambian Copperbelt."
20 M.-P. Faucon, A. Meersseman, M. N. Shutcha, G. Mahy, M. N. Luhembwe, F. Malaisse, and P. Meerts, "Copper Endemism in the Congolese flora: A Database of Copper Affinity and Conservational Value of Cuprophytes," *Plant Ecology and Evolution* 143:1 (2010): 5–18.
21 N. M. Steven, "A Shaba-Type Cu-Co(-Ni) Deposit at Luamata, West of the Kabompo Dome, North Western Zambia," *Exploration and Mining Geology* 9:3–4 (2002): 277–287, doi: 10.2113/0090277.

22 Miombo is the name for *Brachystegia boehmii* and similar trees found across Southern Africa.
23 Chlorosis refers to the abnormal reduction or loss of the normal green coloration of plant leaves. This paragraph is based on P. Duvigneaud and S. Denaeyer-De Smet, "Cuivre et végétation au Katanga," *Bulletin de la Société Royale de Botanique de Belgique* 96:2 (1963): 93–231.
24 Ibid.
25 C. Reilly, "The Uptake and Accumulation of Copper by *Becium homblei* (De Wild.) Duvign. &Plancke," *New Phytologist* 68 (1969): 1081–1087, doi:10.1111/j.1469-8137.1969.tb06509.x; C. Reilly, J. Rowel, and J. Stone, "The Accumulation and Binding of Copper in Leaf Tissue of *Becium homblei* (De Wild.) Duvign. &Plancke," *New Phytologist* 69 (1970): 993–997, doi: 10.1111/j.1469-8137.1970.tb02478.x; C. Reilly, "Accumulation of Copper by Some Zambian Plants," *Nature* 215 (1967): 667–668, doi: 10.1038/215667a0.
26 S. N. Whiting, R. D. Reeves, and A. J. M. Baker, "Conserving Biodiversity: Mining, Metallophytes and Land Reclamation," *Mining Environmental Management* (2002); B. Leteinturier, A. J. M Baker, and F. Malaisse, "Early Stages of Natural Revegetation of Metalliferous Mine Workings in South Central Africa: A Preliminary Survey," *Biotechnologie, Agronomie, Société et Environnement* 3 (1999): 28–41; S. N. Whiting, R. D. Reeves, D. Richards, M. S. Johnson, J. A. Cooke, F. Malaisse, A. Paton, J. A. C. Smith, J. S. Angle, R. L. Chaney, R. Ginocchio, T. Jaffré, R. Johns, T. McIntyre, O. W. Purvis, D. E. Salt, H. Schat, F. J. Zhao, and A. J. M. Baker, "Research Priorities for Conservation of Metallophyte Biodiversity and their Potential for Restoration and Site Remediation," *Restoration Ecology* 12 (2004): 106–116, doi: 10.1111/j.1061-2971.2004.00367.x; S. Le Stradic, S. Boisson, M.-P. Faucon, M. Seleck, and G. Mahy, "Ten Years of Research on Copper-Cobalt Ecosystems in Southeastern D.R. Congo," in *Copper-Cobalt Flora of Upper Katanga and Copperbelt*, F. Malaisse, M. Schaijes and C. D'Outreligne (eds.) (Gembloux: Les Presses agronomiques de Gembloux, 2016), 27–37.
27 A. D. Bradshaw, "Populations of *Agrostis Tenuis* Resistant to Lead Zinc Poisoning," *Nature* 169:1098 (1952), doi:10.1038/1691098a0; A. D. Bradshaw, "Population Differentiation in *Agrostis Tenuis* sibth," *New Phytologist* 59 (1959): 92–103, doi: 10.1111/j.1469-8137.1960.tb06206.x.
28 W. H. O. Ernst, "Phytoextraction of Mine Wastes—Options and Impossibilities," *Chemie der Erde-Geochemistry-Interdisciplinary Journal for Chemical Problems of the Geosciences and Geoecology* 65 (2005): 29–42, doi: 10.1016/j.chemer.2005.06.001.
29 B. Leteinturier, A. J. M Baker, and F. Malaisse, "Early Stages of Natural Revegetation of Metalliferous Mine Workings in South Central Africa: A Preliminary Survey," *Biotechnologie, Agronomie, Société et Environnement* 3 (1999): 28–41.
30 J. Short and A. Milton, "Zambian Environmental Audits" (Mining Environmental Management, March 24–25, 1999); B. Leteinturier, "Evaluation du potentialphytocéno-tique des gisements cuprifères d'Afrique centro-australe en vue de la phytoremédiation des sites pollués par l'activité minière" (Gembloux: PhD Thesis, Faculty of Agricultural Sciences, 2002), 361.
31 R. R. Brooks, M. F. Chambers, L. J. Nicks and B. H. Robinson, "Phytomining," *Trends in Plant Science* 3 (1998): 359–362, doi: 10.1016/S1360–1385(98)01283–7; R. L. Chaney, J. S. Angle, C. L. Broadhurst, C. A. Peters, R. V. Tappero, and D. L. Sparks, "Improved Understanding of Hyperaccumulation Yields Commercial Phytoextraction and Phytomining Technologies," *Journal of Environment Quality* 36 (2007): 1429–1443, doi: 10.2134/jeq2006.0514; A. Van der Ent, A. J. M. Baker, R. D. Reeves, R. L. Chaney, C. Anderson, J. Meech, P. D. Erskine, M.-O. Simonnot, J. Vaughan, J.-L. Morel, G. Echevarria, B. Fogliani, and D. Mulligan, "'Agromining': Farming for Metals in the Future?" *Environmental Science and Technology* 49:8 (2015): 4773–4780, doi: 10.1021/es506031u
32 A. Van der Ent et al., "'Agromining': Farming for Metals in the Future?"
33 R. L. Chaney et al., "Improved Understanding of Hyperaccumulation Yields."
34 R. R. Brooks, M. F. Chambers, L. J. Nicks, and B. H. Robinson, "Phytomining."
35 A. Van der Ent et al., "'Agromining': Farming for Metals in the Future?"
36 B. Leteinturier et al., "Copper and Vegetation at the Kansanshi Hill (Zambia) copper mine."

37 A. Van der Ent et al. "'Agromining': Farming for Metals in the Future?"; A. Van der Ent, A. J. M Baker, M. M. J. van Balgooy, and A. Tjoa, "Ultramafic Nickel Laterites in Indonesia (Sulawesi, Halmahera): Mining, Nickel Hyperaccumulators and Opportunities for Phytomining," *Journal of Geochemical Exploration* 128 (2013): 72–79, doi: 10.1016/j.gexplo.2013.01.009.

38 R. S. Morrison, R. R. Brooks, R. D. Reeves, F. Malaisse, P. Horowitz, M. Aronson, and G. Merriam, "The Diverse Chemical Forms of Heavy Metals In Tissue Extracts of Some Metallophytes from Shaba Province, Zaire," *Phytochemistry* 20 (1981): 455–458; A. Paton and R. R. Brooks, "A Re-Evaluation of *Haumaniastrum* Species as Geobotanical Indicators of Copper and Cobalt," *Journal of Geochemical Exploration* 56 (1996): 37–45; F. K. Chipeng, C. Hermans, G. Colinet, and N. Verbruggen, "Copper Tolerance in the Cuprophyte *Haumaniastrum Katangense* (S. Moore) P.A. Duvign. &Plancke," *Plant and Soil* 328 (2009): 235–244, doi: 10.1007/s11104-009-0105-z; H. Peng, Q. Wang-Muller, T. Witt, F. Malaisse, and H. Küpper, "Differences in Copper Accumulation and Copper Stress Between Eight Populations of *Haumaniastrum Katangense*," *Environmental and Experimental Botany* 79 (2012): 58–65.

39 M.-P. Faucon, M. N. Shutcha, and P. Meerts, "Revisiting Copper and Cobalt Concentrations in Supposed Hyperaccumulators From SC Africa: Influence of Washing and Metal Concentrations in Soil," *Plant and Soil* 301:1–2 (2007): 29–36, doi: 10.1007/s11104-007-9405-3.

40 E. Lombi, K. G. Scheckel, and I. M. Kempson, "In Situ Analysis of Metal (Loid) S in Plants: State of the Art and Artefacts," *Environmental and Experimental Botany* 72:1 (2011): 3–17, doi: 10.1016/j.envexpbot.2010.04.005; W. J. Przybylowicz, J. Mesjasz-Przybylowicz, V. M. Prozesky, and C. A. Pineda, "Botanical Applications in Nuclear Microscopy," *Nuclear Instruments and Methods in Physics Research Section B: Beam Interactions with Materials and Atoms* 130:1 (1997): 335–345; J. Mesjasz-Przybyłowicz and W. J. Przybyłowicz, "Micro-PIXE in plant Sciences: Present Status and Perspectives," *Nuclear Instruments and Methods in Physics Research Section B: Beam Interactions with Materials and Atoms* 189:1 (2002): 47–481; J. Mesjasz-Przybyłowicz and W.J. Przybyłowicz, "PIXE and Metal Hyperaccumulation: From Soil to Plants and Insects," *X-Ray Spectrometry* 40:3 (2011): 181–185; E. Lombi and J. Susini, "Synchrotron-Based Techniques for Plant and Soil Science: Opportunities, Challenges and Future Perspectives," *Plant and Soil* 320:1 (2009): 1–35, doi: 10.1007/s11104-008-9876-x; T. Punshon, M. L. Guerinot, and A. Lanzirotti, "Using Synchrotron X-Ray Fluorescence Microprobes in the Study of Metal Homeostasis in Plants," *Annals of Botany* 103:5 (2009): 665–672, doi: 10.1093/aob/mcn264; G. Sarret, E. A. H. Pilon Smits, H. Castillo Michel, M. P. Isaure, F. J. Zhao, and R. V. Tappero, "Use of Synchrotron-Based Techniques to Elucidate Metal Uptake and Metabolism in Plants," *Advances in Agronomy* 119 (2013): 1–82, http://dx.doi.org/10.1 016/B978-0-12-407247-3.00001-9.

41 P. Koosaletse-Mswela, W. J. Przybyłowicz, K. J. Cloete, A. D. Barnabas, N. Torto, and J. Mesjasz-Przybyłowicz, "Quantitative Mapping Of Elemental Distribution in Leaves of the Metallophytes *Helichrysum Candolleanum, Blepharisaspera,* and *Blepharisdiversispina* from Selkirk Cu-Ni Mine, Botswana," *Nuclear Instruments and Methods in Physics Research Section B: Beam Interactions with Materials and Atoms* 363 (2015): 188–193; J. Mesjasz-Przybylowicz, W. J. Przybylowicz, A. Barnabas, and A. van der Ent, "Extreme Nickel Hyperaccumulation in the Vascular Tracts of the Tree *Phyllanthus Balgooyi* from Borneo," *New Phytologist* (2015), doi: 10.1111/nph.13712.

42 R. S. Morrison, "Aspects of the Accumulation," 287; R. S. Morrison, R. R. Brooks, R. D. Reeves, F. Malaisse, P. Horowitz, M. Aronson, and G. Merriam, "The Diverse Chemical Forms of Heavy Metals in Tissue Extracts of Some Metallophytes from Shaba Province, Zaire," *Phytochemistry* 20 (1981): 455–458.

43 R. V. Tappero, E. Peltier, M. Gräfe, K. Heidel, M. Ginder-Vogel, K. J. T. Livi, and D. L. Sparks, "Hyperaccumulator *Alyssum Murale* Relies on a Different Metal Storage Mechanism for Cobalt than for Nickel," *New Phytologist* 175:4 (2007): 641–654, doi:

10.1111/j.1469-8137.2007.02134.x; H. Küpper, B. Götz, A. Mijovilovich, F. Küpper, and W. Meyer-Klaucke, "Complexation and Toxicity of Copper in Higher Plants. I. Characterization of Copper Accumulation, Speciation, and Toxicity in *Crassula helmsii* as a New Copper Accumulator," *Plant Physiology* 151 (2009): 702–714.

44 A. J. M. Baker, W. H. O. Ernst, A. van der Ent, F. Malaisse, and R. Ginocchio, "Metallophytes: The Unique Biological Resource, its Ecology and Conservational Status in Europe, Central Africa and Latin America," in *Ecology of Industrial Pollution*, L. C. Batty and K. B. Hallberg (eds.) (Cambridge, UK: Cambridge University Press, 2010), 7–40; S. N. Whiting, R. D. Reeves, D. Richards, M. S. Johnson, J. A. Cooke, F. Malaisse, A. Paton, J. A. C. Smith, J. S. Angle, R. L. Chaney, R. Ginocchio, T. Jaffré, R. Johns, T. McIntyre, O. W. Purvis, D. E. Salt, H. Schat, F. J. Zhao, and A. J. M. Baker, "Research Priorities for Conservation of Metallophyte Biodiversity and their Potential for Restoration and Site Remediation," *Restoration Ecology* 12 (2004): 106–116, doi: 10.1111/j.1061-2971.2004.00367.x.

45 MMSD, "Breaking New Ground – Chapter 10: Mining, Minerals, and the Environment" (MMSD, 2002), 231–267, http://pubs.iied.org/pdfs/G00902.pdf.

46 B. Leteinturier et al., Copper and Vegetation at the Kansanshi Hill (Zambia) Copper Mine."

47 S. N. Whiting et al., "Research Priorities for Conservation of Metallophyte Biodiversity."

15 Field vignette
South Africa's underground women miners

Asanda Benya

Women's inclusion in South African mines, particularly in underground occupations, is a relatively new phenomenon, dating back to the early 2000s. Prior to that, legislation such as the 1911 Mines and Works Act No. 12 prohibited women from doing mine work. It categorically stated that "No person shall employ underground on any mine a boy apparently under the age of sixteen years or any female."[1] Internationally, Article 2 of the International Labour Organisation's (ILO) Convention 45 of 1935 reinforced this exclusion stating that "No female, whatever her age, shall be employed on underground work in any mine." Only females who occasionally went underground in non-manual occupations were exempted from this exclusion. More recently, the South African Minerals Act of 1991 also banned women from working underground.[2]

Since South Africa's transition to democracy in 1994, the prohibitive laws mentioned above have been repealed and replaced by the Mine Health and Safety Act of 1996 and the Mineral and Petroleum Resources Development Act (MPRDA) of 2002, thus opening underground occupations to women. Currently, South African mines have over 1.3 million employees and of these employees 52,000 (10.9 percent) are women.

The legislative shifts and increase in numbers of women in underground occupations since 2004 does not, however, mean that women's inclusion in mining has been easy. It has been received with both enthusiasm and hostility, precisely because on one hand it signifies a victory for women, but on another hand, it challenges the masculine occupational culture and "gender regimes" in mining.

Why then do women choose to work in the mines? The latest unemployment rate in South Africa when using the expanded definition, which includes people of working age who have stopped looking for work, is 35.6 percent. This rate is higher for young people and women. This means employment choices for women are limited. Most women reported that they could either work in the mines, or work as casual workers in retail, or as security guards with long working hours and pitiful wages, or be unemployed. Additionally, in platinum mining regions, Statistics South Africa reported that mines are the main employment providers, hence they become the first option for women. Unlike precarious part-time jobs, mines, women argued, usually hire them on a full-time basis with employment benefits such as the medical aid, maternity leave, pension/provident fund etc. By working in local mines instead of migrating to big cities like their parents did during apartheid, they can continue living and raising their children. Local mines, therefore, were their most practical option.

If women's inclusion in mining has not been easy, what challenges do they face? Just like men, women have to undergo several screening tests, such as the heat tolerance screening (HTS) and a medical examination, before they get slotted into mining jobs.[3] The HTS is a thirty minutes strenuous exercise which tests one's ability to withstand hot and humid conditions while doing physically demanding work. Women tend to complete this test with a body temperature that is higher than allowed thus a fail and are not hired. Those who pass the tests are hired, trained, and allocated jobs underground, mainly in entry level positions. The top structures in the mining hierarchy are still dominated by men and women are largely concentrated in low status entry level jobs.

While men live in hostels, women are not provided with accommodation. They are "encouraged" to live with their families because those in power believe that women are first obligated to their families. With their policies and procedures, hidden in protectionist discourse, they reinforce ideas about women and domesticity. With most women living far from work and with their families, to get to work on time for their shifts which sometimes start as early as four thirty in the morning, they have to leave home at early hours, walk long distances to bus stops, take public transport to work to make it on time for their shifts. This is not only a long, dangerous journey, but it is also expensive taking between 30–40 percent of their wages.[4]

Since they live with their families, unlike their male colleagues, that means they do not work only one shift but two and three shifts; first at home, then at work, and sometimes in their communities and their churches. With all the these "shifts" to do, by the time they get to work they are tired. Daily they have to choose between doing house chores, caring for their children or sleeping, and most choose the former. Others, however, relegate housework and child care to female relatives or outsource it to crèches and thus maintain a healthy balance between these responsibilities. Others, as noted above, wade the world of work sleep deprived and fatigued. At work, however, the sleep deprivation and fatigue are read as laziness. These women are seen as less productive and less committed than their male counterparts. While there are women who can do mine work, who are as productive as men, stereotypes about women's bodies as "naturally" unfit for mine work remain pervasive and on those grounds, they face hostility from males who see women in mining as an anomaly and as taking jobs from men.

Studies show that "fortunes" from minerals hardly trickle down to those who dig and are at the rock face, women are further down on this marginalization scale. While legislatively women in South Africa have been included in the mines, structurally the challenges they face render them outsiders.

Notes

1 P. Alexander, "Women and Coal Mining in India and South Africa, c1900–1940," *African Studies* 66:2–3 (2007): 201–222.
2 K. B. Simango, "An Investigation of the Factors Contributing to Failure of Heat Tolerance Screening by Women at Impala Platinum" (Unpublished Master's dissertation, The Da Vinci Institute for Technology Management, 2006).
3 Ibid.
4 Asanda Benya, "Gendered Labour: A Challenge to Labour as a Democratizing Force," *Rethinking Development and Inequality* 2 (special issue, 2013): 47–62.

16 Field vignette

Sapphire mining, water, and maternal health in Madagascar

Lynda Lawson

> *Before mining the river water was so clear you could see a needle in it—now I have three worms.*
> (Woman from Ilakaka south western Madagascar)

The above comments illustrate the close relationship between unregulated artisanal mining, water, and the health of a community. Some twenty years ago sapphires were discovered in south-west Madagascar near Ilakaka. Ilakaka with its formally crystal-clear stream had always been a favorite stop on the Route Nationale Sept, the highway that takes travelers from the capital of Antananarivo to the south. This tiny hamlet and its river was transformed by an influx of many thousands of miners in a just a few months and the rush continued along streams and river beds and the alluvial plains across the south-west around Sakaraha. At every stage of the sapphire rush—from the initial influx during the rush period to the quiet abandoned sites, women and their children are present: many are in the water sieving and washing gravel, others are providing services and, in some cases, selling the smaller stones. My research has investigated the impact of this activity on the daily lives of women. Detailed life histories were elicited from more than twenty women and each time the centrality of water in their work and health was evident.

In south-west Madagascar most sapphires are recovered from alluvial gravel deposits in ancient river beds. These are reached by underground tunnels and mostly by young male miners. The gem bearing gravel is transported—either by hand or in a cart or truck from the mine to a source of water, usually the river where the gravel is sieved. Women panning for gold are also present but to date there is no evidence of the use of mercury. Although no chemicals are used in the either gold panning or sapphire sieving, there are high levels of water turbidity from years of disturbance both to the river bed and further upstream where some are using light machinery for dredging and digging, generating large amounts of tailings. This can have devastating consequences for agriculture such as in 2016 when heavy rains flooded tailings onto rice fields and duck farms.

In the absence of any toilets or washing facilities, the river is used for all sanitation and most drinking water.

Siveran (Sieving)

Siveran involves washing the stones and shaking them in a sieve to locate sapphires which are heavier and are therefore found under the gravel. It is often done by women and children. Women may be sieving fresh gravel that has been brought down to the river from the quarries, but men usually have the first turn at this then women will dig up the tailings in the river bed and sieve again—this is called *tay siva*—"pooh mining." Sieving is best in the bright morning light making it easier to identify the stones, women typically work in the water from seven to three o'clock. Children help their mother with spotting sapphires. The sieves are made of solid wood that is usually water logged, they are heavy and the action of moving the stones back and forth in the water takes strength and skill.

Water and sapphire mining health impacts

> *We are sick because as you see, we stay in the water all day and the water gets inside our bodies, but we have no choice because if we find a sapphire, it can make a life.*
>
> *We are extremely exhausted, if you work too hard we have back aches and bloody stools.*[1]

In almost all interviews, when women were asked about water and their health, they reported stomach pains, intestinal problems, back aches, severe skin irritation, and worms. Many women reported miscarriages and losing children or young

Figure 16.1 South-west Madagascar, children accompany their family to the river to sieve.
Photo: Lynda Lawson

Figure 16.2 Women digging gravel from the river bed to sieve for sapphire.
Photo: Lynda Lawson

relatives to fever and malaria. The local doctor confirmed that women suffer from a wide range of illnesses especially urinary tract and genital infections.

> *All people on site are sick they drink water from the river we have stomach aches and it makes bloody stools.*

Sapphires are found in remote areas without infrastructure. There are few wells and women and children are forced to drink from the river; they cannot boil the water but try to drink the clear water from the top not the red muddy water below.

Despite considerable mineral wealth in these mining areas and the many thousands of informal miners working there, local government struggles to regulate the sector in any way. Infrastructure for water and sanitation is almost non-existent and the intensity of the mining activity is impacting streams and river and community health. Water is at the heart of the survival of these communities and women and children are often those primarily responsible for the collection and use of water. They are thus directly impacted by the negative impacts of resource extraction on water quality. It is thus vital to consult with and strengthen women's participation in water management.

Figure 16.3 Mother and child sieving for sapphires in south-west Madagascar
Photo: Lynda Lawson

Note

1 These comments come from women sapphire sievers at Ankaboke near Sakaraha, December 4, 2015.

Part IV
Reconciling scales of mining governance

Part IV

17 Strategies for working with artisanal and small-scale miners in sub-Saharan Africa

Nina Collins and Lynda Lawson

In brief

- Despite its importance as an economic activity and livelihood strategy in sub-Saharan Africa, artisanal and small-scale mining (ASM) is associated with many negative social, environmental and health impacts, and presents particular sustainable development challenges.
- Numerous universities, NGOs, associations, companies, and governments are working to resolve the many challenges created by ASM. These challenges include lack of formalization and regulation, poor environmental practices, health and safety risks, child and forced labor, inequitable distribution of benefits, security issues, and conflicts with large-scale mining, among many others.
- This chapter documents approaches to working with the ASM sector in sub-Saharan Africa to promote economic development, effective governance, protection and rehabilitation of the environment, and health and safety.
- The study draws primarily on an analysis of work plans focused on ASM developed by officials from developing countries who undertook capacity-building courses sponsored by the Australian Government through its Australia Awards program and the former International Mining for Development Centre. It also draws on interviews with key stakeholders in the ASM sector and is supplemented by a literature review.

Introduction

Artisanal and small-scale mining (ASM) represents a spectrum of mining activities ranging from individuals or small groups panning for gold or precious stones along riverbanks, to relatively large and organized operations involving excavators and mechanized plants.[1] In the literature and legislation, ASM is distinguished from industrial or large-scale mining by factors such as a low level of production;

relatively low degree of mechanization or technological development; a high degree of labor intensity; poor occupational health, safety, and environment conditions; lack of long-term planning; and little capital investment.[2] The types of minerals extracted by ASM in Africa range from high-value metals such as gold, coltan (columbite-tantalite), tungsten (wolframite), and tin (cassiterite); other metallic minerals such as copper, cobalt, and lead; precious and semiprecious stones such as diamonds, sapphires, emeralds, and other gemstones; and lower-value industrial and construction minerals such as limestone, marble, kaolin, clay, granites, sand, and salt.[3] Approximately 18 percent of Africa's gold and a high percentage of Africa's gemstones are produced by ASM.[4]

Scholars and international development institutions now widely agree that ASM is largely a poverty-driven activity.[5] It is significant that countries with a low Human Development Index have a high proportion of their population engaged in ASM.[6] While there has been no census of ASM activity across sub-Saharan Africa, in 2014 Hilson and McQuilken estimated that there were tens of millions of ASM miners, with at least 13 million in West Africa alone. Many more are involved in linked activities across the supply chain, such as in service activities.[7] The percentage of female artisanal miners in Africa exceeds the global average and is estimated to be 40–50 percent of the workforce and up to 75 percent of ASM workforce in countries like Guinea.[8] ASM activities are increasing due to a number of factors, including growing economic crises leading to unemployment, and natural and manmade disasters leading to decreased rural livelihood choices.

Despite its importance as an economic activity and livelihood strategy in sub-Saharan Africa, ASM is associated with many negative social, environmental, and health impacts, and presents particular sustainable development challenges. Numerous universities, NGOs, associations, companies and governments are working to resolve the many challenges created by ASM. These challenges include lack of formalization and regulation, poor environmental practices, health and safety risks, child and forced labor, inequitable distribution of benefits, security issues, and conflicts with large-scale mining, among many others.[9] This chapter focuses on strategies used in the ASM sector in sub-Saharan Africa to promote economic development, effective governance, protection and rehabilitation of the environment, and health and safety. ASM is a complex and multifaceted issue. The strategies and approaches discussed here at times overlap, reflecting this complexity.

Methodology

This study comprised a comprehensive review of the ASM literature, interviews with practitioners, and an analysis of work plans developed by officials working with ASM in sub-Saharan Africa.[10] It drew heavily on lessons learned from deep engagement with the sector through the Australian Government's capacity-building initiatives, such as the International Mining for Development Centre (IM4DC) education and training programs (2011–2015), and the Australia Awards Short Courses for the Extractives Industries (2012–2018). We specifically focus on the analysis of work plans and interviews with practitioners, and their reflections on new interventions to improve development outcomes of the sector.

In line with best practice in adult professional learning, an integral part of the pedagogical approach used in Australian Government capacity-building programs is to incorporate reflection about new knowledge acquired in relation to participants' home context.[11] This is facilitated by interactive classroom discussions, online forums such as IM4DC's alumni network (http://m4dlink.org/), written reflections, and, principally, the work plan on return (WPR).

Box 17.1   The work plan on return

Workplace-related training programs designed for adults are useful only to the extent that the skills taught are transferrable to the workplace.[12] This is particularly the case in work contexts that have few financial and material resources, such as many government ministries and civil society organizations in Africa. The objective of the WPR in the context of ASM is for participants to reflect on their learning and develop a tangible plan of action addressing a key issue related to mining in their home countries. The WPR is a tool for both personal development and training. It is designed to enable course participants to reflect on the complex issues raised by their work, to put in place a series of steps to enhance their capabilities to resolve these issues, and to transfer the competencies gained in capacity-building courses to their colleagues and external stakeholders.

The WPR has three focal points. First, before the course begins, the course coordinator asks participants and their line managers to identify a challenging issue in their work context that they wish to address. Second, during the capacity-building course, participants reflect on this issue in the light of new knowledge they are acquiring and craft a proposal for action to address the issue in the form of one project they are capable of completing. Finally, after the course, participants take action to implement the WPR in their home country. Critical to the success of the WPR is mentoring and peer-to-peer support for at least a year after the course finishes. Course designers should incorporate into the program ample time to plan, research, and workshop the WPRs and to anticipate barriers and facilitators to their achievement.[13] Mentoring and peer-to-peer support after participants return home is particularly important; for example, a follow-up call two months after the course significantly improved engagement and levels of completion with the WPRs in 2016.

A major issue raised in almost every capacity-building course assessed as part of this research was that of managing ASM; many WPRs chose to focus on this challenging issue. This chapter examines the themes chosen in thirty-four WPRs that focused on ASM. We also collected progress data for twenty of these WPRs via mail, telephone calls, and face-to-face interviews with participants.[14] In addition, we obtained information from post-course reports written by course coordinators following visits to Ethiopia and Madagascar to evaluate these courses, as well as participants' reported progress on their WPRs.

The course participants came from eighteen countries in Africa, Asia, and South America.[15] The WPRs were drawn from five Australian aid-funded courses that ran from 2012 to 2014.[16] Mining ministries in the various countries were represented by twenty-three participants (68 percent of the sample), universities or technical institutions were represented by five participants (14 percent), other government departments were represented by four participants (12 percent), and NGOs were represented by two participants (6 percent).

In addition to analysis of the WPRs, we carried out face-to-face interviews with eighteen practitioners working directly on initiatives dealing with ASM. Thirteen of these were officials from governments and universities from eleven developing countries, and five were academics and consultants working in developed countries on initiatives related to improving the governance of ASM in developing countries.[17]

We analyzed data (WPRs, progress data, interview transcripts) using NVivo software, and a simplified grounded theory approach whereby we analyzed the data according to categories and topics that emerged from the data rather than making a priori assumptions.[18] The WPRs considered ASM and the following challenges: economic development, effective governance, protection and rehabilitation of the environment, and health and safety.

Promoting economic development through ASM in Africa

ASM has been recognized in the Africa Mining Vision as an important factor for income and revenue generation that can increase local purchasing power, catalyze small and medium enterprises and foster local economic multipliers.[19] Box 17.2 provides examples of the WPRs reviewed for this study that focused on promoting economic development through ASM in Africa. These WPRs focused on improving miners' financial literacy and access to finance; the formation of cooperatives and associations; and certification schemes to enhance fairness and transparency in the mineral commodity chain. In this section we briefly discuss access to finance, and the formation of cooperatives and associations. Renzo Mori Junior critically examines certification schemes in this volume, Chapter 19, "Gauging the effectiveness of certification schemes and standards for responsible mining in Africa."

Box 17.2   Examples of WPRs focused on promoting economic development through ASM in Africa

Improving financial skills of ASM miners (Central African Republic)

The aim of this WPR was to reduce poverty and improve the livelihoods of artisanal miners by encouraging them to save their earnings and invest in small family businesses. The project involved working in cooperation with an artisanal mining cooperative and microcredit institution. The WPR was to have delivered a pilot workshop to the leader of an ASM cooperative in each targeted mining region, to be followed by several other workshops with other miners in the cooperative. Unfortunately, civil war in the CAR prevented progress on this WPR.

Upgrading mining technology in the small-scale sector (Ghana)

This project was developed by a not-for-profit ASM network in Ghana focused on organizing and improving the efficiency of the ASM sector. The WPR aimed to develop a training program focused on technology development and proposal writing to secure funding to implement the program. This WPR aimed to develop the training model and materials in such a way that miners could use them even if they were marginally literate and had limited time to attend training off site. This WPR also included plans for a longer-term project aimed at turning an abandoned mine into a tourist attraction.

Documentary on artisanal mining in West Africa (Senegal)

This WPR aimed to create a documentary on the benefits of fair trade gold in Senegal for consumers and the general public. Limited information was available on this WPR as the participant had taken leave from their position at the time of data collection.

Improvement of mining activities by consolidation of artisanal mining associations in cooperatives (Burundi)

This WPR aimed to increase the profitability and sustainability of ASM in Burundi by creating cooperatives. Its aims were broad, including components targeting the livelihood aspects of ASM, such as improving the livelihoods of miners by encouraging more profitable mineral extraction, processing techniques, and business practices through technical services and supervision; encouraging miners to use mining revenues to contribute to the sustainable development of surrounding communities; and raising local capacity to run tracking and certification schemes and to enforce bans on the transportation of non-compliant minerals.

Organizing ASM into cooperatives/enterprise group mining (Liberia):

This project aimed to promote compliance with environmental guidelines by encouraging ASM miners to form cooperatives or group mining enterprises

(continued)

(continued)

that would assist the government in monitoring their activities. As incentives to encourage miners to form cooperatives, the project would use financial assistance in the form of grants and loans, technology transfer to enhance productivity, and access to training and capacity building.

Developing a network of women working in ASM (Mozambique)

This WPR aimed to develop a network of women working in ASM in Mozambique. Through this network, the project would encourage women to form associations of women miners and would provide capacity building for alternative livelihoods, such as handicrafts and the production of ceramics.

Access to finance

A major challenge for African ASM miners is access to finance to scale up their business, to buy equipment, to access more valuable deposits, to improve productivity, and to enact reforms that could make their mines acceptable to certification schemes such as those developed by the Alliance for Responsible Mining and Fairtrade International. Yet empirical studies on strategies for economically empowering small-scale miners have been lacking.[20] Access to finance is often constrained by the legal status of ASM operations. Pit owners who operate in the informal sector are often able to secure financial support only through informal and inequitable channels, leading to a cycle of indebtedness.[21] Miners themselves also face financial entrapment, for example to local traders or buyers who lend them money for poor-quality equipment, medicines, and food.[22] Women pit owners and miners face additional challenges related to gender discrimination, including lack of access to finance, technical assistance and training, regardless of the registration status of the mine.[23] Women's work is also highly devalued and, as a result, they are frequently underpaid.

To avoid financial entrapment and non-repayment of loans, one approach taken by governments and donors has been to arrange the hire purchase of equipment by groups of miners. Multilateral institutions such as the United Nations Industrial Development Organization (UNIDO) and the World Bank have funded a number of small-scale grant schemes and equipment-leasing schemes in sub-Saharan Africa.[24] For example, from 2005 to 2012, under the Sustainable Management of Mineral Resources Project, the Ministry of Mines and Steel Development of Nigeria and the World Bank operated a gender-sensitive microgrant scheme open to mining cooperatives to formalize their activities.[25] Anene Nnamdi from the Nigerian Ministry of Mines and Steel noted that women miners were quick to take up the challenge, and when they formalized their activities they became eligible for small grants to purchase equipment.[26]

In 2009 Siegel and Veiga discussed the characteristics of successful user-friendly government loan facilities. In Namibia, for example, a government loan scheme financed through a Minerals Development Fund that included low interest rates, slow payment periods, and minimal bureaucratic overheads had high repayment rates. In Mozambique, a similarly successful scheme required miners to show their license and feasibility study, and to provide collateral and a payment plan. Siegel and Veiga suggest that rather than focusing on grants, donors could assist in developing such schemes and carrying the risk.[27]

Along with the need for government and donor-supported loans, miners also need business skills training with an emphasis on mobilizing micro-savings and careful needs analysis. They need to learn how to develop business plans to be able to move from artisanal mining to mining as a small-scale business so they can access credit. They also need training in the valuation of minerals, marketing, and upstream and downstream linkages, including business opportunities along the supply chain, service provision linked to ASM activity, and mineral beneficiation and manufacturing.

Formation of cooperatives and associations

During our interviews, key stakeholders frequently raised the idea of forming cooperatives and associations. A number of WPRs addressed this idea as a way to assist governments to regulate ASM by providing a formal body that is responsible for ensuring that a mining operation adheres to legislation and regulatory guidelines. Cooperatives could also support communication, cooperation, and coordination between miners; assist miners in sharing knowledge and resources; and contribute toward increasing local beneficiation.[28] Training and knowledge transfer or financial services for legal mining operations are often provided as incentives for miners to form cooperatives.

Holistic training built on the community's trust of trainers and extensive multistakeholder involvement inclusive of women were key success factors in a World Bank-funded initiative in Uganda.[29] The outreach, training, and small grants program resulted in increased profits for miners, enhanced production and marketing efforts, and improved mine safety conditions. In Rwanda the organization of ASM into cooperative companies has registered more than 10,000 miners and provided them with new opportunities to enter into joint ventures with overseas investors. The development of cooperatives in Rwanda was made possible by the government's support for a federation of cooperatives coupled with legislation that enabled easy land access. Findings from a largely unsuccessful USAID-funded project aimed at forming a cooperative scheme focused on diamond mining in Sierra Leone found that such schemes must be grounded in local best practice with clear accountability and entitlement guidelines and sources of contingency funding. Also critical is the provision of training for cooperative members in how to effectively manage a democratic cooperative and in effective and responsible mining.[30]

Despite these examples, there is a dearth of research reporting on the successes and failures of cooperatives in ASM. Further research into ASM cooperatives—drawing from experiences in mining and other industries—is needed to draw out important lessons for those looking to establish such ASM groupings in Africa and beyond.

Promoting effective governance of ASM in Africa

The governance of ASM in Africa presents fundamental challenges given that the majority of ASM activities are undertaken outside legal frameworks as part of the informal sector.[31] Key challenges exist not only in the creation of effective national legislation and policies that recognize the potential socioeconomic benefits of ASM, but also in terms of implementing such frameworks in locations where knowledge, political will, financial and human resources, and capacity are frequently lacking.[32]

The WPRs undertaken as part of this study identified lack of enforcement of legislation and regulations as a key issue in the governance of ASM. Approximately one-quarter of the WPRs reviewed for this study focused specifically on the governance or regulation of ASM. However, given that this category/topic overlaps significantly with other areas, the majority of the WPRs we reviewed were, in fact, indirectly aimed at assisting the implementation of legislation and regulations in the form of miner-friendly guidelines, checklists for inspectors, and capacity-building and sensitization programs. The WPRs more specifically focused on the area of governance in sub-Saharan Africa (Box 17.3) analyzed for this study fell into three broad categories: amending legislation, encouraging formalization through training and guidelines, and policy research.

Box 17.3   Examples of WPRs focused on promoting effective governance of ASM in Africa

Amending legislation

Improving the social dimension of the environmental specifications of small-scale mining operators (Madagascar)

This WPR aimed to review and modify existing environmental specifications for licensing of ASM operations in order to enhance the social responsibilities of small-scale mining operators. Activities included engaging with ASM operators to understand how they impact and are impacted by surrounding communities in order to develop strategies and specifications to enhance the welfare of communities, with the aim of including these in the environmental regulations and mining code of Madagascar.

Consideration of occupational health and safety (OH&S) in environmental texts for the Ministry of Mines (Madagascar)

While legislation governing mining in Madagascar discusses the environmental obligations of small-scale mining operators, it does not detail operators' obligations from a health and safety perspective. This WPR aimed to amend existing environmental legislation to include health and safety obligations for miners as a prerequisite for permits and mining authorizations, and to encourage the government to monitor and evaluate ASM activities from a health and safety perspective.

Guidelines for occupational health, safety, and protection of the environment in the extraction of aragonite and celestite (Madagascar)

This WPR aimed to apply knowledge about occupational health, safety, and environment (OHS&E) legislation and regulations gained through the course in Australia to establish guidelines, standards, and an OHS&E management system (enshrined in a code of conduct and commitment plan within existing legislation) for the mining of aragonite and celestite (celestine) in Madagascar. This project planned to conduct baseline studies to develop the guidelines, which would need to be approved by the Ministry of Mines. Mining administrators and inspectors would oversee and assist mine owners and operators in the implementation of the guidelines.

Encouraging formalization through training and guidelines

Understanding the views and needs of miners (Nigeria)

This WPR aimed to improve government and civil society's understanding of the views and the needs of miners and other stakeholders in order to formulate acceptable programs to enhance miners' compliance with regulations. It also focused on sensitizing miners on the benefits of obtaining mining licenses and the implications of not operating according to the requirements of legislation.

Guidelines for regulation of ASM (Madagascar)

This WPR aimed to assist in regulating gold rushes in northern Madagascar by developing guidelines for miners outlining the key legislation and regulations for ASM. The guidelines, which were published as a booklet in both French and Malagasy, would be accompanied by training and sensitization of mining communities and local authorities to encourage legal mining practices.

Conflict resolution between large-scale mining (LSM) companies and small-scale miners (Ghana)

This WPR, developed by a not-for-profit ASM network in Ghana, had a three-pronged objective: first, to foster trust and harmonious relationships between LSM companies and small-scale miners through consultative engagement; second, to negotiate with LSM companies to cede part of

(continued)

(continued)

their concessions to ASM through due process with regulators; and third, to prepare a simplified handbook for mining license acquisition and mining best practices to enable small-scale miners to understand license acquisition processes and best mining practices.

Policy research

Comparative desk study on ASM legal frameworks (Ghana)

This WPR aimed to compare legal frameworks for ASM in Ghana and Peru through a comparative desktop study in order to apply lessons from Ghana's attempt to formalize ASM to the development of a successful legal framework in Peru.

Formalization of ASM

Strategies to promote the effective governance of ASM in Africa tend to focus on formalization of the sector, which can refer specifically to the process of creating or amending legislative frameworks and policies to regulate ASM, as well as to the ongoing processes of implementing or adhering to these frameworks.[33] There are many barriers to formalization for ASM miners, including unavailability of untitled, mineralized land; limited tenure rights; and highly bureaucratic, centralized, or costly licensing processes.[34] While most countries in Africa legalized ASM in the 1990s, governments have tended to create legislation in a top-down manner, giving preferential treatment to LSM when granting concessions.[35] Legislation has also tended to misunderstand the nature of the ASM sector and the motivations of miners, and has not taken into account the precolonial history of ASM as a traditional livelihood activity deeply implicated in customary land tenure systems that often exist in parallel to common or civil legal systems in sub-Saharan Africa.[36] In some cases, classifications of ASM in legislation do not accurately represent the complexities miners actually face, which can also work as a disincentive for miners to formalize their activities.[37] In Ghana, for example, there have been recent discussions around amendments to legislation to include a medium-scale mining category in addition to the current large-scale and ASM categories, a proposal that has both detractors and supporters.[38] These discussions have been initiated due to the increasing mechanization of ASM sector.

In 2014 Fold, Jønsson, and Yankson argued that research and policies focused on the governance of ASM need to better understand how the formal and informal ASM sectors are intertwined, and how ASM labor markets function in Africa.[39] Because formalization relies on enforcement, it is important to involve local authorities and communities in policy-making processes, to ensure that policymakers consider local contexts.[40] In 2003 Hentschel, Hruschka, and Priester promoted the decentralization of governance of ASM as a way to ensure that legislation and regulation reflects the realities on the ground and that services are located close to where ASM activities are occurring.[41] In 2014, however, Hirons used the example of Ghana to show that decentralization that is not properly or fully implemented undermines the participation of traditional authorities and other political actors in the governance of ASM.[42] This undermining can intensify local conflicts and create additional complexities, including challenges associated with political rent seeking at the local level.

Prior studies have shown that formalization initiatives should include incentives to encourage miners to participate in the schemes.[43] For example, in Zimbabwe the government has raised its gold buying price and reduced royalty charges for artisanal miners to discourage the illegal mineral trade.[44] A major challenge for artisanal miners in many countries is the lack of geological information indicating mineralized areas for ASM concessions.[45] In Ghana, for example, the awarding of concessions for ASM with poor deposits of minerals or deposits that are not amenable to ASM has created a disincentive for miners to formalize their activities; they will often move beyond the boundaries outlined in their mining lease and onto the concessions of LSM companies.[46]

Education and training

Education and training are the cornerstone for promoting good governance of the ASM sector, by creating awareness of both the health and safety hazards and the environmental impacts of ASM.[47] Many scholars emphasize the importance of outreach and training activities with a participatory and multi-stakeholder focus that includes the communities surrounding ASM operations.[48] There has been a tendency for governments and donor bodies to promote education within more-established ASM regions, leaving a need for training and capacity-building programs in more-remote communities.[49] Formalized peer-to-peer learning can be a particularly effective method for training ASM miners. Education levels of the target audience need to be taken into account, including language and literacy levels.[50] For this reason, radio and other forms of multimedia can be effective for disseminating information to artisanal miners.[51] For example, the Nigerian government's ASM department has conducted a sensitization program that involves drama presentations at village squares, radio jingles, and billboards, to sensitize community members, and particularly women, about the dangers of children being involved in mining.[52]

Where possible, training should be delivered by local institutions that understand the culture- and context-specific issues that ASM miners face. Before beginning any training initiative, the training provider should conduct a thorough needs analysis with a representative pilot group, taking care to establish the literacy status of participants and their availability for such training. Externally delivered training and capacity-building programs should attempt to form partnerships with local institutions and build the capacity of local actors. They should also take into account local knowledge and attempt to build on previous initiatives and technologies that have been introduced. Too often, engagement and training programs in ASM have tried to start from scratch rather than building on existing initiatives that might have ceased due to lack of ongoing donor or government funding rather than because of a lack of potential for success.

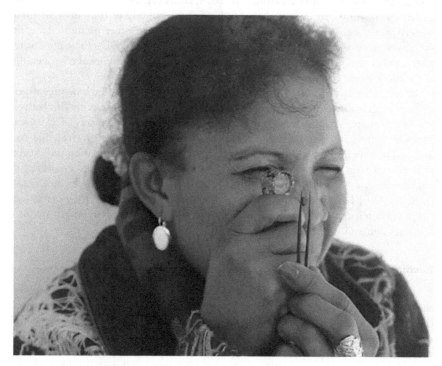

Figure 17.1 Malagasy gemstone miner leaning to use a loupe

Source: Lynda Lawson, 2016

An example of a successful program funded by Deutsche Gesellschaft für Internationale Zusammenarbeit (GIZ) was a field-based gemology training in Madagascar where local miners learned to identify stones, including the characteristics of more-valuable stones. Established relationships with local mayors and ASM community leaders enabled the training to take place in remote locations with a good representation of both men and women miners.[53]

Promoting protection and rehabilitation of the environment

The environmental impacts of ASM are a key issue facing the mining sector in Africa, and include land degradation and desertification, soil erosion, deforestation, biodiversity loss, and pollution of rivers and other water sources. Artisanal and small-scale gold mining (ASGM) is the largest source of anthropogenic mercury emissions globally.[54] While the use of mercury has garnered considerable attention from academics and international donor organizations, the reclamation of lands and other water-related issues have received less attention. Land degradation and the pollution of water bodies represent a serious sustainability issue for the sector given the importance of agriculture on the continent and increasing concern around the impacts of climate change. More research is required to understand the socioeconomic and environmental trade-offs and linkages in ASM to understand its broader development impacts.[55]

Eight of the thirty-four WPRs we included in the study focused primarily on the environment, but in total, over half (eighteen) of the WPRs touched on issues related to the environmental impacts of ASM.[56] Of the eight WPRs that specifically focused on the environment, three aimed to develop guidelines on environmental impacts; more specifically, two focused on promoting good practice in environmental management among miners and included training components, and one focused on environmental risk assessment aimed at building the capacity of inspectors. One WPR aimed to improve the monitoring of environmental impacts caused by ASM through a multi-stakeholder approach that targeted both government officials and miners. Two of the eight environment-focused WPRs aimed to better understand the impacts of mercury in specific regions; one WPR aimed to improve the rehabilitation of both small-scale and large-scale mines by identifying the challenges of doing so; and one WPR focused on alternative livelihoods as a way of protecting a World Heritage nominated site.

Box 17.4 Examples of WPRs focused on promoting protection and rehabilitation of the environment in Africa

Guidelines

Environmental management awareness in ASM (Malawi)

This WPR aimed to promote good practice and assist miners to gain knowledge about environmental management issues by developing environmental management guidelines on ASM in the local languages and providing training and sensitization to miners on these guidelines. The WPR proposed to monitor the effectiveness of the guidelines and training through inspections and to alter the guidelines iteratively, providing follow-up training and sensitization where necessary.

Proposal and popularization of a guide on the best environmental and social practices for ASM (Senegal)

The aim of this WPR was to develop a guide of good environmental and social practices to be adopted by miners to reduce the environmental and social impacts from ASM and improve the relationship between miners and local communities. After development of the guide, meetings would be held with different stakeholders to distribute the guide, including miners and their families, local communities, local authorities, and the government, via the regional committee of environmental monitoring.

(continued)

(continued)

Setting guidelines for environmental health risk assessment of large- and small-scale mining (Ethiopia)

The aim of this WPR was to contribute to the sustainable development of the mining sector and minimize and mitigate the environmental impacts of mining by developing a systematic environmental risk management framework and standardized guidelines of environmental health risk assessment for both ASM and LSM. The framework aimed to enhance the capacity of mining inspectors and would be based on risks identified through fieldwork.

Monitoring

System for monitoring environmental activities considering social responsibility (Madagascar)

This WPR aimed to promote community development through the mining sector and a sense of social responsibility among entities by establishing a pilot system for monitoring activities related to the environmental and social aspects of ASM and LSM. This pilot monitoring system would be developed through engagement with both government officials, miners and community members, through multi-stakeholder workshops and reflection groups.

Analysis on the effect of using mercury in ASGM in Chunya-Mbeya (Tanzania)

This project used risk assessment methodologies to assess the environmental and health impacts associated with the use of mercury in ASGM in the Chunya district of the Mbeya region of Tanzania. It also aimed to investigate options for mercury-free production and to promote best measures to reduce levels of mercury and address the issues identified.

Geochemical assessment of abandoned and active mine sites and the impacts on the health of mine workers in the south-west (Nigeria)

This WPR aimed to assess the type and level of pollution caused by ASM and the implications for the health of small-scale miners. It planned to assess this pollution through geological mapping of mine sites and identification of contaminated areas; geochemical analysis of groundwater, surface water, stream sediments, soil, and plants; and geochemical analysis of blood samples of mine workers, especially women and children.

Rehabilitation

Participatory evaluation of mine rehabilitation challenges (Ghana)

The aim of this WPR was to identify challenges facing post-mining landscape efforts for both industrial and artisanal gold mining companies in south-west Ghana through participatory assessment and evaluation of existing rehabilitation plans.

Alternative livelihoods

An attempt toward the control and legalization of artisanal mining within the Mwela Rock Art National Monument Area for sustainable community development (Zambia)

This WPR aimed to curb illegal artisanal mining in the Mwela Rock Art Protected Area by striking a balance between heritage conservation and the exploitation of artisanal mining products in the area. It would do this by demarcating specific areas for legalized artisanal mining and creating alternative livelihood opportunities.

Promoting cleaner technologies

Multilateral lending agencies and intergovernmental organizations in partnership with national governments have focused strongly on capacity-building and technology-transfer programs focused on more efficient and cleaner mining and processing methods to reduce environmental damage and mercury pollution.[57] Despite this, there is widespread lack of knowledge among miners of the benefits of adopting more efficient and environmentally sustainable technologies in their operations; in addition, miners lack financial incentives to do so.[58] There is a need to develop cleaner technologies that are firmly based in the local context.

Mercury retorts, centralized processing centers, and the promotion of mercury-free processing methods are three key strategies that have been used to reduce environmental emissions of mercury and associated health impacts. For example, a three-year project (2012–2015) in Burkina Faso, Mali, and Senegal, co-financed by a number of institutions including the Global Environment Facility and the UNIDO, focused on promoting affordable, simple, and intuitive mercury-free technologies that can be easily maintained and that take the needs of miners into account.[59] The program provided inclusive health education and technology training programs based on workshops, seminars, and trainings, as well as assistance to mining communities to obtain fair trade certification. Another project in Zamfara State in Nigeria (the Safer Mining Programme) introduced the iGoli process to miners.[60] This method, developed by Mintek, has been incorporated into a small plant in Tanzania suitable for groups of artisanal miners or very small commercial operators.[61] Unfortunately, the cost of iGoli has been prohibitive for some miners, and Sippl and Selin and Spiegel et al., among others, have suggested that improved technologies that are not produced locally might not be the most sustainable strategy for the ASGM sector.[62] This has been the case with mercury retorts—small devices that capture mercury vapors during the burning process, preventing inhalation and emissions to the environment. There has been some success with retorts that have been constructed locally using local materials, but there has been less success with externally constructed retorts.[63] In addition to facilitating uptake, the local manufacture of retorts provides an opportunity for local employment.[64] An example of a locally produced mercury-free method that has garnered some attention by the United Nations Environment Programme is the Ghanaian direct smelting kit, locally referred to by miners as *sika bukyia* (literally "money furnace").[65] It was developed at the University of Mines and Technology in Tarkwa, Ghana, and is based on what is often referred to as the borax method.[66] It is a promising approach, and has high recovery rates.[67] In addition, it processes gold quickly and consistently.[68] Dissemination of the technology will rely on subsidization and educational outreach, however.[69]

Centralized processing centers have had mixed success due to not taking miners' needs into account.[70] According to a 2007 study by Hilson, Hilson, and Pardie, centralized processing facilities are most effective in countries with localized gold deposits but not as effective where gold is widely dispersed.[71] Centralized facilities that require miners to haul tonnes of ore far from the mine might not be a viable solution, however.[72] While centralized processing facilities can serve the dual role of directly reducing environmental impacts and providing information about environmental management, in many cases in sub-Saharan Africa they have been installed before the project proponents ensured that miners needed them and are willing to use them.[73] For example, a World Bank-sponsored ore-processing center in Bolgatanga, Ghana, constructed in 2000, was underutilized because the equipment was not tailored to local geological conditions due to lack of research prior to establishment, and the processing center was located too far from the mining area, which would require funds for transport.[74] Other attempts to introduce processing centers in sub-Saharan Africa have run into issues with high demand superseding capacity, discouraging miners from using the centers due to lengthy waiting times, or the centers prioritizing miners with large amounts of ore, thus disadvantaging smaller producers.[75] Processing centers operating on an aid rather than business model could offer services at no cost to miners for a period of time, but then face the challenge of sustainability once the project finishes because they require resources for labor, equipment, maintenance, and administration.[76]

Reclamation of lands

Lack of geological information for ASM activities has led to substantial environmental and financial liabilities in sub-Saharan Africa: miners often operate in a haphazard manner in their attempt to find minerals, causing greater environmental destruction than would be the case if governments allocated mineralized concessions to ASM.[77] While most legislative frameworks provide provisions for the reclamation of land in ASM, in practice monitoring is largely absent, and reclamation is either ignored or undertaken superficially, such as by simply filling in mining pits. A pilot project focusing on the reclamation of ASM lands funded by the World Bank Mining Sector Development and Environment Project through the Ghana Minerals Commission was undertaken in Ghana in the late 1990s and early 2000s with the objective of demonstrating that mined land could be reclaimed and used for alternative economic activities and serving as a model for best practice and cost-effective methods for future reclamation programs.[78] This pilot project, undertaken by a Ghanaian contractor using labor from local villagers, reclaimed and revegetated three ASM sites covering 205 hectares in total, using economic trees (citrus and palm oil) and the participation and involvement of communities. Assessment of the pilot recommended that the Ghana Minerals Commission assess the pilot reclamation projects to determine the cost of a bigger program of environmental reclamation, which could potentially be funded through a levy on artisanal and small-scale miners.[79] A 2013 study looking at the process of reclamation of lands mined through ASM in Ghana estimates that the cost for reclamation of one acre of mined land excavated to a depth of 900 millimeters is approximately $52,419, including maintenance costs for a 5-year period of 2–3 percent.[80] Multilateral lending agencies such as the World Bank have considered establishing funds to assist in the organization of ASM groupings (e.g., cooperatives or associations) to cover the costs of environmental or social impacts. ASM producers can also raise funds to meet the cost of environmental sustainability through direct taxes, fees, or indirect costs incorporated into production cost functions.[81] The Asia Foundation, through its Engaging Stakeholders in Environmental Conservation Project, has introduced an innovative strategy for the reclamation of lands called frugal rehabilitation in Mongolia; this strategy might be suitable for application in Africa.[82] The strategy of frugal rehabilitation comprises comprehensive activities aimed to improve the value and productivity of degraded lands. It is structured around three components: technical rehabilitation, topsoil management, and biological rehabilitation targeting species that are native to the contextual environment. It is aimed specifically at ASM sites and acknowledges that ASM miners often lack the resources to achieve the level of land reclamation that large-scale mining companies can afford. For this reason, it focuses on achieving results at a reasonable cost that are accessible and affordable, and that address the concerns of both miners and communities.[83]

Promoting health and safety in ASM

Close to half of the thirty-four WPRs in this study attempted to address issues associated with health and safety, and all of these had a capacity-building focus.[84] Eight aimed to deliver sensitization or awareness training to miners, and four focused on developing guidelines, manuals, or checklists. Two WPRs included a train-the-trainer component, while four WPRs had a risk assessment focus, reflecting the content of the course in Australia and an attempt on the part of the participants to incorporate risk assessment into their work in their home countries. Over half of the health and safety-focused WPRs aimed to develop guidelines or informative materials as part of their program.

Box 17.5   Examples of WPRs focused on promoting health and safety in ASM in Africa

Sensitization and training initiatives for miners

Sensitization on OHS&E in ASM in Toamasina (Madagascar):

This project aimed to provide workshops and information packages based on research in priority areas around Toamasina.

(continued)

(continued)

Training for ASM mine managers/cooperative heads on health, safety, and environmental management (Rwanda)

These training workshops would draw on material and support provided by Rwandan police, the Rwanda Environmental Management Authority, the International Conference on the Great Lakes Region secretariat, and the German Federal Institute for Geosciences and Natural Resources Office in Rwanda.

Sensitization of stakeholders on the consequences of mercury use (Cameroon)

This WPR aimed to use a training program for mineworkers to sensitize stakeholders on the consequences of using mercury in gold mining activities. This train-the-trainer program would be implemented in partnership with three ASM cooperatives and with miners to develop a set of good practice guidelines for the reduction of the impacts of mercury and tools for a workplace health and safety management system. Its long-term aim was reduction in mercury in water in Bétaré-Oya and improving the livelihoods and living conditions for residents.

Job hazard analysis for small-scale miners (Malawi)

This project aimed to train mine inspectors and monitoring staff on the use of job hazard analysis. This analysis is a proactive safety management tool used to identify risks or hazards, such as environmental and human health hazards, that are associated with the use of chemicals in mining. It also aimed to promote a safety culture among miners, such as procedures for the proper handling, storage, and disposal of chemicals.

Education and awareness campaign on the health and safety of three small-scale mining operators in Kaduna State (Nigeria)

The aim of this WPR was to minimize, eliminate, substitute, or control OH&S hazards in the work environment of small-scale miners through a training and awareness building program with ASM operators. The project also envisaged training personnel at the Ministry of Mines and Steel Development on the importance of health and safety in mining to encourage stronger OH&S legislation and regulations in Nigeria.

The management and storage of chemicals used by artisanal miners (Sierra Leone)

Through a series of seminars and workshops with local miners, this WPR aimed to increase the awareness of artisanal miners of the environmental and human health hazards associated with the use of chemicals in mining and to promote a safety culture, including the procedures for proper handling, storage, and disposal of chemicals. The project would deliver seminars to mine inspectors and monitoring officers, who would subsequently deliver training to miners or communities.

Improving adoption of personal protective equipment through graphical illustrations for ASM in Nassarawa State (Nigeria)

This WPR focused on promoting the use of personal protective equipment through the use of graphical/pictorial information. It aimed to curtail injuries and common illnesses associated with using crude methods of mining by encouraging the use of masks, safety goggles, and gloves in mining operations. This was one of the few WPRs that recognized the importance of using appropriate graphics in communication materials for working with ASM miners.

Checklist/manuals for inspectorates

Enhanced mining inspections using checklists (Kenya)

This WPR aimed to improve the level of compliance with mining safety regulations in one small-scale mining operation and one artisanal gold mining site in Kenya by developing inspection checklists for the mining inspectorate to enhance inspections.

Implementation of a manual for OH&S inspection for mining activities (Madagascar)

The project aimed to develop a manual that would contain a series of essential inspection parameters for mining and environmental inspections of both small- and large-scale mines.

Guidelines for miners

Preparation of guidelines for OHS&E for ASGM (Madagascar)

This WPR aimed to prepare guidelines for inspectors that would incorporate risk management, monitoring and inspection, injury and illness management, and incident and accident reporting and investigation.

Guidelines for ASM OH&S (Burundi)

This WPR aimed to develop guidelines to help ASM operators prepare and implement management systems on OHS&E as a proactive strategy for addressing the OHS&E issues arising from mine operations. The focus would be on inspection (OHS&E assessment), prevention and remediation of mine hazards, mine damages repair, and mine conversion when required.

Conclusion

Table 17.1 Summary of key issues strategies and conditions of efficacy

Issue	Strategy	Conditions for Efficacy
Economic development	Access to finance Formation of cooperatives and associations	User-friendly loan facilities Donor-supported loans rather than grants Banks and lender financing of ASM activities, particularly those of women, at reasonable rates Incentives to form cooperatives Clear accountability and entitlement guidelines Training for members on managing a cooperative Contingency funding Research into what makes a cooperative successful
Governance	Amendment of legislation to better address needs of ASM miners Encouragement of formalization through training, guidelines, and facilitation of miner participation Policy research to assist the implementation of legislation and regulations Miner-friendly guidelines and checklists for inspectors Capacity building and sensitization programs	Government engagement with ASM to understand the needs and constraints that prevent uptake of formalization Engagement with and development of leaders from government, mining communities, and NGOs Engagement with ASM operators Establishment of training needs Political will and government capacity to enforce legislation and regulations
Protection and rehabilitation of the environment	Training and enabling of simple rehabilitation of mined land Focus on miners' responsibility to remediate areas they have disturbed, e.g., through levies on operators	Sufficient funding Miners' understanding of their rights and responsibilities Technologies that are firmly based in the local context, preferably manufactured locally using local materials Ensuring demand for new technologies and taking miners needs into account
Health and safety	Training in alternative methods for processing gold Expert advice on mine construction	Sufficient funding Build on existing initiatives with good understanding of context Needs analysis Acknowledgment of miners' potential lack of time, mobility, and literacy

Strategies for working with ASM 257

ASM in Africa provides a livelihood for millions of miners and their families. An extraordinary amount of the world's minerals is produced by ASM in Africa, from valuable commodities such as gold and precious stones through to the 3Ts (tungsten, titanium, and tin) and the so-called neglected development minerals. Much of this production is hidden from view and unregulated, and many miners—an increasing number of them women—are not benefiting from the mineral riches they are extracting. Development agencies are now taking seriously the potential to alleviate poverty and improve returns on ASM through changes in policy and practice. Further research is needed to understand more precisely the contribution of ASM to the economies of African states.

This chapter has reviewed the literature in four critical areas: promotion of economic development; effective governance; protection and rehabilitation of the environment; and health and safety. In addition to highlighting recent work in these areas from the literature, we incorporated the views of government and civil society representatives working with ASM issues in sub-Saharan Africa. The key learnings we focused on were in the areas of access to finance, formalization processes, including the formation of cooperatives and associations, cleaner technologies, reclamation of lands, and promotion of health and safety.

Many of the participants involved in this study discussed the need for more consultation and involvement of stakeholders—including those directly working with ASM—prior to the development of strategies aimed at addressing the many issues associated with the sector. They also emphasized the importance of communicating the findings of research to ASM communities themselves. One interviewee suggested that while many reports have been written about the issues surrounding ASM, little has been done in the way of implementing these findings. Commentators have supported this assertion, and speak of research fatigue in the ASM sector and the need for more action and less talk.

An ASM visioning workshop conducted by the International Institute for Environment and Development (IIED) in 2015 concluded that what was needed to tackle the complex challenges within the ASM sector was to "awaken that big, lazy, sleeping giant that is government."[85] The views shared by those interviewed here suggest that many in government are very much awake to the needs of ASM but require support to realize effective strategies. It is clear from this study that the strategies most likely to succeed are those that pay careful attention to the local context and are able to develop holistic approaches that reflect the issues as they play out in Africa.

Notes

1 J. Hinton, "Communities and Small Scale Mining: An Integrated Review for Development Planning. Report to the World Bank" (Washington, DC: Communities and Small-Scale Mining Initiative, 2006); Marcello M. Veiga, Peter A. Maxson, and Lars D. Hylander, "Origin and Consumption of Mercury in Small-Scale Gold Mining," *Journal of Cleaner Production* 14:3–4 (2006).
2 Miners often use picks, chisels, sluices, and pans, although the level of mechanization is increasing in some countries. J. Hinton, "Communities and Small Scale Mining"; Thomas Hentschel, Felix Hruschka, and Michael Priester, "Artisanal and Small-Scale Mining: Challenges and Opportunities" (London: International Institute for Environment and

Development and WBCSD, 2003); ILO, "Social and Labour Issues in Small-Scale Mines. Report for Discussion at the Tripartite Meeting on Social and Labour Issues in Small-Scale Mines" (Geneva: International Labour Organization, Sectoral Activities Program, International Labour Office, 1999).
3 J. Hinton, "Communities and Small Scale Mining"; Economic Commission for Africa, "Minerals and Africa's Development: The International Study Group Report on Africa's Mineral Regimes" (Addis Ababa, Ethiopia: Economic Commission for Africa, 2011); Cristina Villegas et al., "Artisanal and Small-Scale Mining in Protected Areas and Critical Ecosystems Programme (ASM-Pace): A Global Solutions Study" (London: Estelle Levin Limited and WWF, 2012).
4 African Union (AU), "Africa Mining Vision" (Addis Ababa: African Union and United Nations Economic Commission for Africa, February 2009).
5 Gavin Hilson and Chris Garforth, "'Everyone Now is Concentrating on the Mining': Drivers and Implications of Rural Economic Transition in the Eastern Region of Ghana," *Journal of Development Studies* 49:3 (2013); Benjamin N. A. Aryee, Bernard K. Ntibery, and Evans Atorkui, "Trends in the Small-Scale Mining of Precious Minerals in Ghana: A Perspective on Its Environmental Impact," *Journal of Cleaner Production* 11:2 (2003); Gavin Hilson and Sadia Mohammed Banchirigah, "Are Alternative Livelihood Projects Alleviating Poverty in Mining Communities? Experiences from Ghana," *The Journal of Development Studies* 45:2 (2009).
6 M. Hoadley and D. Limpitlaw, "The Artisanal and Small Scale Mining Sector & Sustainable Livelihoods," in *Mintek Small Scale Mining Conference, 9 September, 2004* (Nasrec: Book of Proceedings, 2004).
7 G. Hilson and J. McQuilken, "Four Decades of Support for Artisanal and Small Scale Mining in Sub Saharan Africa: a Critical Review," *The Extractive Industries and Society* 1:1 (2014).
8 Re: "40–50 percent of the workforce": K. C. Malpeli and P. G. Chirico, "The Influence of Geomorphology on the Role of Women at Artisanal and Small-Scale Mine Sites," *Natural Resources Forum* 37:1 (2013). Re: "countries like Guinea": K. Jenkins, "Women, Mining and Development: An Emerging Research Agenda," *The Extractive Industries and Society* 1:2 (2014).
9 CASM ICMM, IFC-CommDev, "Working Together – How Large-Scale Miners Can Engage with Artisanal and Small-Scale Miners" (International Council on Mining and Metals, Communities and Small-scale Mining, and IFC Oil, Gas and Mining Sustainable Community Development Fund, 2009); Thomas Hentschel et al., "Artisanal and Small-Scale Mining."
10 In this volume, Danellie Lynas, Gernelyn Logrosa, and Ben Fawcett in Chapter 18, "Community health, safety and sanitation: considering localized governance and monitoring of water quality" and Renzo Mori Junior in Chapter 19, "Gauging the effectiveness of certification schemes and standards for responsible mining in Africa," also cover various aspects of the literature on ASM.
11 A. Webster-Wright, "Reframing Professional Development through Understanding Authentic Professional Learning," *Review of Educational Research* 79:2 (2009).
12 S. Parker, D. Andrei, L. Wang, M. Pearce & A. Chapman, "Facilitating Learning and Development During IM4DC Short Courses," http://im4dc.org/wp-content/uploads/2013/09/Parker-Facil-FR-Completed-Report1.pdf (accessed February 20, 2017).
13 Ibid.
14 The twenty WPRs for which progress data were collected were in Burundi, Ghana, Ethiopia, Madagascar, Malawi, Mozambique, Nigeria, Rwanda, Peru, Tanzania, and Zambia.
15 The countries were Burundi, Cameroon, Central African Republic, Ethiopia, Ghana, Kenya, Liberia, Madagascar, Malawi, Mozambique, Nigeria, Peru, Philippines, Rwanda, Senegal, Sierra Leone, Tanzania, and Zambia.
16 The five programs were (1) 2012 Short Course in Managing Corporate Community Relations; (2) 2012 African Women in Mining and Development Study Tour; (3) 2012/

2013 Occupational Health, Safety and Environment Short Course; (4) 2012/2013 Community Aspects of Resource Developments (CARD) Short Course; and (5) 2014 Environmental Management in Mining.
17 The officials from governments and universities were from Burundi, Cameroon, Ethiopia, Ghana, Kenya, Liberia, Malawi, Nigeria, Philippines, Rwanda, and Zambia.
18 B. G. Glaser and A. L. Strauss, *The discovery of Grounded Theory: Strategies for Qualitative Research* (London: Weidenfeld and Nicolson, 1967).
19 Rodger Barnes, Kojo Busia, and Marit Kitaw discuss the Africa Mining Vision in this volume, Chapter 1, "Harmonizing African resource politics? The Africa Mining Vision and lessons from the African Mineral Development Centre."
20 Samuel J. Spiegel, "Microfinance Services, Poverty and Artisanal Mineworkers in Africa: In Search of Measures for Empowering Vulnerable Groups," *Journal of International Development* 24:4 (2012): 486.
21 Sadia Mohammed Banchirigah, "Challenges with Eradicating Illegal Mining in Ghana: A Perspective from the Grassroots," *Resources Policy* 33:1 (2008); Gavin Hilson, "Poverty Traps in Small-Scale Mining Communities: The Case of Sub-Saharan Africa," *Canadian Journal of Development Studies-Revue Canadienne D Etudes Du Developpement* 33:2 (2012).
22 R. Perks, "'Can I Go?'—Exiting the Artisanal Mining Sector in the Democratic Republic of Congo. *J. Int. Dev.* 23 (2011): 1115–1127; B. Verbrugge, "Artisanal and Small-Scale Mining: Protecting Those 'Doing The Dirty Work'" (IIED Briefing, October 2014), http://pubs.iied.org/17262IIED (accessed April 10, 2017); N. Garrett, "The Extractive Industries Transparency Initiative (EITI) & Artisanal and Small-Scale Mining (ASM): Preliminary Observations from the Democratic Republic of the Congo (DRC), EITI," https://eiti.org/sites/default/files/documents/Garrett_EITI_10_2007.pdf (accessed 10/04/2017).
23 Jennifer Hinton, Marcello M. Veiga, and Christian Beinhoff, "Women and Artisanal Mining: Gender Roles and the Road Ahead," in *The Socio-Economic Impacts of Artisanal and Small-Scale Mining in Developing Countries*, Gavin Hilson (ed.) (Netherlands: A. A. Balkema, Swets Publishers, 2003); S. E. Tallichet, M. M. Redlin, and R. P. Harris, "What's a Woman to Do? Globalized Gender Inequality in Small-Scale Mining," in *Socio-Economic Impacts of Artisanal and Small-Scale Mining in Developing Countries*, Gavin Hilson (ed.) (Netherlands: A. A. Balkema, Swets Publishers, 2003).
24 Samuel J. Spiegel, "Microfinance Services, Poverty and Artisanal Mineworkers in Africa." Marcello M. Veiga and Randy F. Baker, "Protocols for Environmental and Health Assessment of Mercury Released by Artisanal and Small-Scale Gold Miners," (Vienna: GEF/UNDP/UNIDO, 2004).
25 World Bank, "Nigeria - Sustainable Management of Mineral Resources Project: Restructuring: Main Report" (Washington, DC: World Bank, 2012).
26 Nnamdi Anene, personal communication, October 2014.
27 Shefa Siegel and Marcello M. Veiga, "Artisanal and Small-Scale Mining as an Extralegal Economy: De Soto and the Redefinition of "Formalization," *Resources Policy* 34:1–2 (2009).
28 E. Levin and A. B. Turay, "Artisanal Diamond Cooperatives in Sierra Leone: Success or Failure?" (Ottawa: Report Commissioned by Partnership Africa Canada, 2008); Thomas Hentschel et al., "Artisanal and Small-Scale Mining."
29 C. G. Sheldon et al., "Innovative Approaches for Multi-Stakeholder Engagement in the Extractive Industries, Oil, Gas, and Mining," *Unit Working Paper* (Washington DC: World Bank, 2013)
30 E. Levin and A. B. Turay, "Artisanal Diamond Cooperatives in Sierra Leone: Success or Failure?" (Ottawa: Report Commissioned by Partnership Africa Canada, 2008).
31 J. Hinton, "Communities and Small Scale Mining"; Shefa Siegel and Marcello M. Veiga, "Artisanal and Small-Scale Mining as an Extralegal Economy."
32 Thomas Hentschel et al., "Artisanal and Small-Scale Mining."; Roy Maconachie and Gavin Hilson, "Safeguarding Livelihoods or Exacerbating Poverty? Artisanal Mining and Formalization in West Africa," *Natural Resources Forum* 35:4 (2011).

33 Shefa Siegel and Marcello M. Veiga, "Artisanal and Small-Scale Mining as an Extralegal Economy"; Felix Hruschka, "'Illegal Mining'... A Factual or Conceptual Threat?" (ASM-Pace Guest Blog Series, 2013).
34 Shefa Siegel and Marcello M. Veiga, "Artisanal and Small-Scale Mining as an Extralegal Economy"; Gavin Hilson and Abigail Hilson, "Entrepreneurship, Poverty and Sustainability: Critical Reflections on the Formalisation of Small-Scale Mining in Ghana," *IGC Working Paper* (London: International Growth Centre, 2015); Eleanor Fisher, "Occupying the Margins: Labour Integration and Social Exclusion in Artisanal Mining in Tanzania," *Development and Change* 38:4 (2007); Thomas Hentschel et al., "Artisanal and Small-Scale Mining"; Gavin Hilson, Abigail Hilson, and Eunice Adu-Darko, "Chinese Participation in Ghana's Informal Gold Mining Economy: Drivers, Implications and Clarifications," *Journal of Rural Studies* 34:0 (2014); G. Hilson and C. Potter, "Why Is Illegal Gold Mining Activity So Ubiquitous in Rural Ghana?" *African Development Review-Revue Africaine De Developpement* 15:2–3 (2003).
35 G. Hilson et al., "Chinese Participation in Ghana's Informal Gold Mining Economy"; Eleanor Fisher, "Occupying the Margins."
36 Re: "motivations of miners": G. Hilson et al., "Chinese Participation in Ghana's Informal Gold Mining Economy." Re: "civil legal systems in sub-Saharan Africa": Frank K. Nyame and Joseph Blocher, "Influence of Land Tenure Practices on Artisanal Mining Activity in Ghana," *Resources Policy* 35:1 (2010); Samuel J. Spiegel and Marcello M. Veiga, "International Guidelines on Mercury Management in Small-Scale Gold Mining," *Journal of Cleaner Production* 18:4 (2010); S. Dondeyne et al., "Artisanal Mining in Central Mozambique: Policy and Environmental Issues of Concern," *Resources Policy* 34:1–2 (2009); CDS, "Livelihoods and Policy in the Artisanal and Small-Scale Mining Sector – an Overview" (Swansea: Centre for Development Studies, University of Wales, 2004).
37 G. Hinton, "Communities and Small Scale Mining."
38 E.g., Gavin Hilson and Abigail Hilson, "Entrepreneurship, Poverty and Sustainability."
39 N. Fold, J. B. Jønsson, and P. Yankson, "Buying into Formalization? State Institutions and Interlocked Markets in African Small-Scale Gold Mining," *Futures* 62 Part A (2014).
40 S. Lowe, "Consolidated Report: Small-Scale Gold Mining in the Guianas," (Guianas, Paramaribo: Report Prepared for the WWF, 2005) as cited by R. Maconachie and G. Hilson, "Safeguarding Livelihoods or Exacerbating Poverty?; and Samuel J. Spiegel, "Microfinance Services, Poverty and Artisanal Mineworkers."
41 Thomas Hentschel et al., "Artisanal and Small-Scale Mining."
42 Mark Hirons, "Decentralising Natural Resource Governance in Ghana: Critical Reflections on the Artisanal and Small-Scale Mining Sector," *Futures* 62 Part A (2014).
43 See, e.g., Abbi Buxton, "Responding to the Challenge of Artisanal and Small-Scale Mining: How Can Knowledge Networks Help?" (London: IIED, 2013); USAID, "Property Rights and Artisanal Diamond Development (Pradd) Project – Comparative Study: Legal and Fiscal Regimes for Artisanal Diamond Mining," (Washington, DC: USAID, 2010); Samuel J. Spiegel, "Formalisation Policies, Informal Resource Sectors and the De-/Re-Centralisation of Power: Geographies of Inequality in Africa and Asia" (Bogor, Indonesia: Center for International Forestry Research, 2012).
44 Veneranda Langa, "Govt Raises Gold-Buying Price," *NewsDay Zimbabwe*, September 28, 2015.
45 G. Hilson and O. Maponga, "How Has a Shortage of Census and Geological Information Impeded the Regularization of Artisanal and Small-Scale Mining?" *Natural Resources Forum* 28: 1 (2004); Benjamin N. A. Aryee et al., "Trends in the Small-Scale Mining of Precious Minerals in Ghana: A Perspective on Its Environmental Impact."; G. Hilson and P. Potter, "Why Is Illegal Gold Mining Activity So Ubiquitous in Rural Ghana?"
46 Re: "formalize their activities": G. Hilson and A. Hilson, "Entrepreneurship, Poverty and Sustainability"; G. Hilson and C. Potter, "Why Is Illegal Gold Mining Activity So Ubiquitous in Rural Ghana?" Re: "the concessions of large-scale mining companies":

FoN, "End of Project Report: Rights and Voices for Sustainable Small-Scale Mining" (Takoradi: Friends of the Nation, 2010).
47 Benjamin N. A. Aryee et al., "Trends in the Small-Scale Mining of Precious Minerals in Ghana."
48 E.g., Samuel J. Spiegel, "Microfinance Services, Poverty and Artisanal Mineworkers in Africa"; Gavin Hilson, Christopher J. Hilson, and Sandra Pardie, "Improving Awareness of Mercury Pollution in Small-Scale Gold Mining Communities: Challenges and Ways Forward in Rural Ghana," *Environmental Research* 103:2 (2007); Samuel J. Spiegel, "Socioeconomic Dimensions of Mercury Pollution Abatement: Engaging Artisanal Mining Communities in Sub-Saharan Africa," *Ecological Economics* 68:12 (2009); Rodolfo N. Sousa et al., "Strategies for Reducing the Environmental Impact of Reprocessing Mercury-Contaminated Tailings in the Artisanal and Small-Scale Gold Mining Sector: Insights from Tapajos River Basin, Brazil," *Journal of Cleaner Production* 18:16–17 (2010); Tara R. Zolnikov, "Limitations in Small Artisanal Gold Mining Addressed by Educational Components Paired with Alternative Mining Methods," *Science of the Total Environment* 419 (2012); Rebecca Adler Miserendino et al., "Challenges to Measuring, Monitoring, and Addressing the Cumulative Impacts of Artisanal and Small-Scale Gold Mining in Ecuador," *Resources Policy* 38:4 (2013).
49 Gavin Hilson, "Abatement of Mercury Pollution in the Small-Scale Gold Mining Industry: Restructuring the Policy and Research Agendas," *Science of the Total Environment* 362:1–3 (2006).
50 G. Hilson, C. J. Hilson, and S. Pardie, S "Improving Awareness of Mercury Pollution in Small-Scale Gold Mining Communities: Challenges and Ways Forward in Rural Ghana," *Environmental Research* 103: 2, (2007): 275–287.
51 Gavin Hilson, "Abatement of Mercury Pollution in the Small-Scale Gold Mining Industry"; G. Hilson et al., "Improving Awareness of Mercury Pollution in Small-Scale Gold Mining Communities."
52 Nnamdi Anene, personal communication, October 2014.
53 Lawson, Lynda, personal communication, April 2017.
54 UNEP, "Global Mercury Assessment: Sources, Emissions, Releases and Environmental Transport" (Geneva, Switzerland: Chemicals Branch, United Nations Environment Programme, 2013).
55 Vivian Schueler, Tobias Kuemmerle, and Hilmar Schröder, "Impacts of Surface Gold Mining on Land Use Systems in Western Ghana," *AMBIO* 40:5 (2011); Mark Hirons, "Trees for Development? Articulating the Ambiguities of Power, Authority and Legitimacy in Governing Ghana's Mineral Rich Forests," *The Extractive Industries and Society* 2:3 (2015); G. M. Hilson, "The Future of Small-Scale Mining: Environmental and Socioeconomic Perspectives," *Futures* 34:9–10 (2002); Adrián Saldarriaga-Isaza, Clara Villegas-Palacio, and Santiago Arango, "The Public Good Dilemma of a Non-Renewable Common Resource: A Look at the Facts of Artisanal Gold Mining," *Resources Policy* 38:2 (2013); Robin Bloch and George Owusu, "Linkages in Ghana's Gold Mining Industry: Challenging the Enclave Thesis," *Resources Policy* 37 (2012); Luke Danielson, "Artisanal and Small-Scale Mining from an Ngo Perspective," *Journal of Cleaner Production* 11:2 (2003).
56 More specifically, of the WPRs that were categorized as having a primary focus on OH&S, six also touched on environmental impacts, and of those WPRs categorized as having a primary focus on governance/regulation focus, four contained environmental aspects.
57 Benjamin N. A. Aryee et al., "Trends in the Small-Scale Mining of Precious Minerals in Ghana: A Perspective on Its Environmental Impact"; Kristin Sippl and Henrik Selin, "Global Policy for Local Livelihoods: Phasing out Mercury in Artisanal and Small-Scale Gold Mining," *Environment: Science and Policy for Sustainable Development* 54:3 (2012); Jennifer J. Hinton, Marcello M. Veiga, and A. Tadeu C. Veiga, "Clean Artisanal Gold Mining: A Utopian Approach?" *Journal of Cleaner Production* 11: 2 (2003).

58 Benjamin N. A. Aryee et al., "Trends in the Small-Scale Mining of Precious Minerals in Ghana."
59 UNIDO, "A Journey Towards Responsible Gold in West Africa" (United Nations Industrial Development Organization, 2015).
60 The iGoli process is a mercury-free method of extracting gold consisting of leaching gold concentrate with dilute hydrochloric acid and bleach and precipitating gold with sodium metabisulphite. Mintek, "IGoli," www.mintek.co.za/technical-divisions/small-scale-mining-beneficiation/technology-development/igoli/; ELI, "Artisanal and Small-Scale Gold Mining in Nigeria: Recommendations to Address Mercury and Lead Exposure" (Washington, DC: Environmental Law Institute, 2014).
61 Mintek," IGoli."
62 ELI, "Artisanal and Small-Scale Gold Mining in Nigeria: Recommendations to Address Mercury and Lead Exposure"; K. Sippl, and H. Selin, "Global Policy for Local Livelihoods: Phasing out Mercury in Artisanal and Small-Scale Gold Mining"; S. J. Spiegel, O. Savornin, D. Shoko, and M. M. Veiga, "Mercury Reduction in Munhena, Mozambique: Homemade Solutions and the Social Context for Change," *International Journal of Occupational and Environmental Health* 3 (2006): 215–221.
63 K. Sippl, and H. Selin, "Global Policy for Local Livelihoods"; G. Hilson et al., "Improving Awareness of Mercury Pollution in Small-Scale Gold Mining Communities"; S. J. Spiegel et al., "Mercury Reduction in Munhena, Mozambique."
64 K. Sippl and H. Selin, "Global Policy for Local Livelihoods."
65 C. E. Abbey et al., "Direct Smelting of Gold Concentrates, a Safer Alternative to Mercury Amalgamation in Small-Scale Gold Mining Operations," *American International Journal of Research in Science, Technology, Engineering & Mathematics* 7 (2014); UNEP, "Reducing Mercury Use in Artisanal and Small-Scale Gold Mining: A Practical Guide" (United Nations Environment Programme, 2012).
66 UNEP, "Reducing Mercury Use in Artisanal and Small-Scale Gold Mining."
67 Abbey et al., "Direct Smelting of Gold Concentrates, a Safer Alternative to Mercury Amalgamation."
68 UNEP, "Reducing Mercury Use in Artisanal and Small-Scale Gold Mining."
69 Abbey et al., "Direct Smelting of Gold Concentrates, a Safer Alternative to Mercury Amalgamation."
70 G. Hilson, "What Is Wrong with the Global Support Facility for Small-Scale Mining?"; Jesper Bosse Jønsson, Elias Charles, and Per Kalvig, "Toxic Mercury Versus Appropriate Technology: Artisanal Gold Miners' Retort Aversion," *Resources Policy* 38:1 (2013); Jennifer J. Hinton, Marcello M. Veiga, and A. Tadeu C. Veiga, "Clean Artisanal Gold Mining: A Utopian Approach?," *Journal of Cleaner Production* 11:2 (2003); Gavin Hilson, "Abatement of Mercury Pollution in the Small-Scale Gold Mining Industry."
71 G. Hilson et al., "Improving Awareness of Mercury Pollution in Small-Scale Gold Mining Communities."
72 UNIDO, "A Journey Towards Responsible Gold in West Africa."
73 Re: "about environmental management": J. Hinton et al., "Clean Artisanal Gold Mining."; G. Hilson, "Abatement of Mercury Pollution in the Small-Scale Gold Mining Industry." Re: "willingness to use them": G. Hilson, "What Is Wrong with the Global Support Facility for Small-Scale Mining?" G. Hilson, "Abatement of Mercury Pollution in the Small-Scale Gold Mining Industry."
74 G. Hilson et al., "Improving Awareness of Mercury Pollution in Small-Scale Gold Mining Communities."
75 Marcello M. Veiga et al., "Processing Centres in Artisanal Gold Mining," *Journal of Cleaner Production* 64:0 (2014); G. Hilson et al., "Improving Awareness of Mercury Pollution in Small-Scale Gold Mining Communities"; G. Hilson, "What Is Wrong with the Global Support Facility for Small-Scale Mining?"; G. Hilson, "Abatement of Mercury Pollution in the Small-Scale Gold Mining Industry."
76 UNIDO, "A Journey Towards Responsible Gold in West Africa."

77 B. N. Aryee et al., "Trends in the Small-Scale Mining of Precious Minerals in Ghana"; FoN, "End of Project Report."
78 B. N. Aryee et al., "Trends in the Small-Scale Mining of Precious Minerals in Ghana"; B. N. Aryee, "Small-Scale Mining in Ghana as a Sustainable Development Activity: Its Development and a Review of the Contemporary Issues and Challenges," in *The Socio-Economic Impacts of Artisanal and Small-Scale Mining in Developing Countries*, Gavin M Hilson (ed.) (Lisse: Swets & Zeitlinger B.V., 2003); World Bank, "Ghana - Mining Sector Development and Environment Project" (Washington, DC: World Bank, 1995).
79 World Bank, "Project Performance Assessment Report Ghana Mining Sector Rehabilitation Project (Credit 1921-GH), Mining Sector Development and Environment Project (CREDIT 2743-GH)," *Report No.: 26197*, (2003), http://documents.worldbank.org/curated/en/120891468749711502/pdf/multi0page.pdf.
80 J. B. K. Asiedu, "Technical Report on Reclamation of Small Scale Surface Mined Lands in Ghana: A Landscape Perspective," *American Journal of Environmental Protection* 1:2 (2013).
81 B. N. Aryee et al., "Trends in the Small-Scale Mining of Precious Minerals in Ghana."
82 Stacey, Jonathon, personal communication, November 2015; J. Stacey, "Developing a Frugal Rehabilitation Methodology for Artisanal Mining (ASM) and Applying it to the Rehabilitation of ASM Contexts Internationally," in *Between the Plough and the Pick: Informal Mining in the Contemporary World, 4–6 November 2015* (Canberra: Australian National University, 2015).
83 J. Stacey, "Developing a Frugal Rehabilitation Methodology for Artisanal Mining."
84 Danielle Lynas, Gernelyn Logrosa, and Ben Fawcett discuss the importance of occupational health and safety in ASM, focusing particularly on the connection of water and sanitation issues, in Chapter 18 of this volume, "Community health, safety and sanitation: considering localized governance and monitoring of water quality."
85 James McQuilken, "Waking the Lazy Sleepy Giant," (IIED, 2015), www.iied.org/waking-lazy-sleepy-giant.

18 Artisanal and small-scale mining community health, safety, and sanitation

A water focus

Danellie Lynas, Gernelyn Logrosa, and Ben Fawcett

In brief

- In order to provide successful water-related development and services within the African ASM context we need to understand the overlap between community and occupational health and safety, and the overlap of household and workplace.
- It is important to develop an integrated approach to water-related health, safety, and sanitation in Africa. Such an approach allows us to bridge the communication gap between local communities and donor organizations and development programs.
- Water plays an indispensable role in all aspects and in all stages of ASM and affected communities. In general, all mining and mineral processing operations require water, and maintaining life requires clean water.
- Wet and slippery workplace conditions cause many accidents, and many endemic diseases are waterborne. Water-related diseases can be caused by contamination of water from both surface water and groundwater sources.
- Water, sanitation, and hygiene, reach beyond the boundaries of traditional health care and public health sectors.

Introduction

The United Nations' shift from the Millennium Development Goals (MDGs) to the Sustainable Development Goals (SDGs), and the subsequent focus on equitable services, gives us the opportunity to review the artisanal and small-scale mining (ASM) poverty trap with fresh eyes. These goals, in turn, allow development programs to set meaningful targets. Because SDG 6 is committed to the universal provision of water and sanitation services, and SDG 5 addresses the specific needs of women and girls, they guide and stimulate support of ASM's community development. An integrated approach to water-related health, safety, and sanitation in

Africa allows us to bridge the communication gap between local communities and donor organizations and development programs. ASM is heavily reliant on water: preserving and protecting water for all users depends on careful program design and implementation.[1]

Water plays an indispensable role in all aspects and in all stages of ASM. Hence, its importance cannot be overemphasized when considering ASM community and occupational health, safety, and sanitation issues. Risk assessments of water-related health, safety, and sanitation conditions in ASM have been very limited and lack depth, however. Public health studies show that the majority of health problems are water related: the lack of access to clean water aggravates the increased risk of malaria, diarrhea, musculoskeletal pain, dehydration, schistosomiasis, cholera, nematode infections, headaches, and dermatological, visual, cardiac, and respiratory problems.[2] In Africa, miners and their families continue to be more vulnerable than those in other occupations to combined occupational and community health risks and diseases.[3]

A skin disease called Buruli ulcer is the second-most widespread mycobacterium infection in Ghana, with a reported higher prevalence in the Amansie West District.[4] Ghana is consistently second only to Côte d'Ivoire among six west African countries reporting the most cases of Buruli ulcer, worldwide, every year from 2010–2016. During this period, there has been a steady decline in global cases from nearly 5,000 to 1,864 cases in 2016.[5] Skin disease, transmitted through bacterially infected water, is just one aspect of a wide range of pressing challenges that result from lack of access to clean water and sanitation services and lack of capacity-building support services to improve water-related occupational health and safety practices.[6]

In order to provide successful water-related development and services within the African ASM context we need a holistic understanding of the overlap between community and occupational health and safety: if we separate household from workplace ASM conditions we will not be able to reach our goals of universal clean water and sanitation. Failure to take a holistic view explains why, despite a number of reliable biomarker studies that have been performed, health and safety problems persist in most African countries.[7] To address these problems we need risk assessment information in ASM sites that establish causal links to water sources.

We know that wet and slippery workplace conditions cause the most number of accidents, and that endemic diseases are waterborne. In addition, we know the open water sources used for gold panning and the abandoned mine pits filled with accumulated water are sites for mosquito breeding.[8] Standing water sources in most districts of Ghana were found within a 25-meter radius of 94 percent of ASM households, whereas only 54 percent of farming households were that close to open water. In fact, in 2011 the mosquito-borne disease malaria was the top self-reported disease, affecting 80 percent of all study participants in the ASM community.[9] It is only with a holistic understanding of water-related health and safety that we will be able to positively reframe the African sanitary discussions and offer suggestions to the existing epidemiologic analyses of human health.

In addition, we can draw a comparison from similar conditions shared by ASM communities and the urban slums. The most obvious parallel is the migration strategies driven by rural poverty and the reducing agricultural incomes. The use of temporary shelters is a characteristic pattern in ASM since miners tend to stay only as long as they find gold and to move on as soon as a better gold prospecting opportunity emerges somewhere else. Buxton suggests three key facets of ASM: vulnerability (economic, natural/physical, and sociopolitical), marginalization (geographical and political), and informality. These facets also apply to slums.[10] Given the parallels between the living conditions of both ASM miners and slum dwellers, we can undertake risk assessments to consider, in detail, similar water-related problems from their shared contexts. Flooding, for instance, is a water-related problem that often confronts both groups when they live in low-lying locations that are unsuitable for residential occupation.

Investigations considering water as the focus, therefore, can significantly assist in creating an integrated approach to developing practical management solutions. Building on previous knowledge gained from case study and field investigations in the Tarkwa district of Ghana, this chapter argues that we need to understand health, safety, and sanitation from a holistic point of view. Combined with the belated recognition by the United Nations of the universal human right to access to safe, clean, accessible, and affordable water and sanitation, the SDGs should provoke action to improve community and occupational conditions in ASM.

ASM community risks: integration of households and workplace

The first and most important step when assessing risks to individuals and communities is to establish the context. ASM is characterized by the overlap of household and workplace. For that reason, Basu et al. has used integrated and multidisciplinary strategy in dealing with ASM issues.[11] In this chapter we build on this approach to provide a holistic understanding of the effects that socioeconomic drivers impose on community health, and the safety risks in locations where water is the primary resource.

More than 1.1 million Ghanaians work in ASM.[12] Census data suggest that this number is growing.[13] As ASM activities occupy an increasingly large occupational footprint, it becomes even more important to improve workplace conditions.[14] Agriculture continues to be non-viable due to lack of irrigation systems and thus overdependence on rain-fed crops, which are limited by seasonality. Many households are being forced into mining as the only alternative.[15] The primary commodity in Ghana, among other countries in Africa, is gold, because it is profitable and stable in terms of market price. Gold remains attractive for the ASM community because of its fast and easy extraction from the raw mineral form. For many it holds the promise of immediate economic relief and is the mainstay of local, rural economies.[16] Despite the economic benefits, however, the ASM sector remains vulnerable to health and safety risks.

The social and political marginalization of the ASM sector translates into reduced access to workplace health and safety assistance and services for both miners and other affected groups, such as other downstream industries.[17] Poor mining practices in combination with the use of basic tools and equipment result in frequent accidents and significant environmental and health impacts, with the most well documented of health impacts being mercury exposure. Before addressing and redressing investigations into accidents, injuries, and fatalities, as well as environmental and health impacts, we need to take a holistic view of the community setting.

ASM miners vary their mining methods according to the occurrence of the gold deposit and its location. They use many mining methods, including alluvial mining along streams and riverbeds, open pit mining, and underground mining. All these methods pose safety risks to the miners involved.[18]

The seemingly easy methods of mining and processing alluvial gold captures the sense of inclusiveness in ASM; anybody can perform it: men, women, the elderly, and children. Shallow alluvial mining, which is popularly called "dig and wash," is characterized by ASM miners in valleys or low-lying areas who exploit deposits with depths not exceeding 3 meters.[19] Alluvial gold can naturally be liberated but is often partially attached to silts and clays when found along river banks. To recover the gold, ASM miners wash the silt or clay in a sluice box to disperse the clay. When the mineral deposit lies on or near the surface, miners instead use open-pit mining, excavating at depths of 7 to 12 meters. This method is usually more cost-effective and requires fewer miners to produce the same quantity of ore than its underground method counterpart. Both methods incur safety risks that require specific control measures. For instance, the risks associated with open-pit mine slopes include various forms of ground instability such as slumping, mudslides, toppling, or falling of material involving a part or the whole of a pit slope. Collectively, the International Labour Organization estimates indicate that ASM operations worldwide experience six to seven times more accidents than occur in large-scale mining operations.[20] Accidents and fatalities occurring in ASM contribute significantly to the bad reputation associated with the sector.[21] Publicly available reports from Ghana that include information from Tarkwa and a statement from a local nurse (see Box 18.1) support this perception.[22]

Box 18.1   Experience in a mining town in Tamale, Ghana

In a statement to the author in February 2017 by written interview, Nancy Waaley, registered nurse and ward administrator, revealed her experience in a mining town in Tamale, Ghana:

> The problems associated with mining and the consequences of the activities of the operators in Ghana is not only a problem for the sector but increasingly becoming a public health issue that needs a holistic and multi sector approach to address. Disease conditions emanating from the industry such as the case of Buruli ulcer and other water borne diseases is putting enormous pressure on health facilities, not only in the management but in health financing.
>
> The problem is further complicated by the fact that most of these activities are unregulated. Disregard for the use of appropriate personal protective clothing and use of harmful practices such as the use of potentially harmful chemicals such as mercury is increasing the disease

(continued)

(continued)

> burden. Minor cuts and abrasions that could be treated at primary health care levels finally land in the hospital as gangrenous wounds that will not heal but will once in a while end in amputation not only because of the delay in reporting to the health facility but also the ignorance of the operators and lack of knowledge of the effects of such chemicals and practices they engage in. Not only are these infections from chemical pollutions and harmful practices difficult to manage in terms of disease outcome, it significantly affects already scare resources available for clinical care and other activities of equal importance
>
> Public health departments continue to advocate for the safe handling of potentially dangerous chemical such as mercury, based on the situational reports of disease incidents and for regularization of these activities to ensure these operators of these dangerous activities benefit at least from some early interventional measures to at least prevent complications. Not only will lives be safe, our water bodies will be safe to support aquatic lives and safe for human consumption and activities.

Moreover, few studies exist on occupational injury or health in ASM in Ghana. Long, Sun, and Neitzel interviewed 173 participants from the upper east region in 2011 and re-interviewed 22 of them in 2013. They found that falls were the most common cause of injury. Legs and knees were the most common injury sites reported, with cuts, lacerations, burns, scalds, contusions, and abrasions the most common injuries reported. More-specific narratives might have allowed further data interrogation to ascertain the extent to which water was or was not involved in these accidents. In another study Kyeramateng-Amoah and Clarke examined hospital records from the Nkawkaw Hospital in the eastern region of Ghana between 2006 and 2013 and found that the most frequent cause of injury was pit collapse, and that fractures, contusions, and spinal cord injuries were the most common types of injury reported.[23] A 2014 study undertaken in Tarkwa interviewed 404 small-scale miners about their occupational injury experiences over the previous 10 years. The study indicated that almost a quarter of those interviewed reported being injured in the previous 10 years, with the overall rate calculated to 5.39 injuries per 100 person years. The rate was significantly higher for women (11.93 per 100 person years) and for those with little mining experience (25.32 per 100 person years for those with less than 1 year of work experience). Most of the reported injuries (71 percent) resulted from miners being hit by objects, with 17 percent resulting from mishandling of tools or machinery, and 5.8 percent from falls. Lack of reliable records on accidents among those involved in ASM presents an ongoing challenge in implementing proactive preventative or mitigating controls to lessen the personal and community impacts of injuries.

In general, all mining and mineral processing operations require water. Water is used in deep open-pit and hard rock mining methods during their downstream grinding and washing of the discharge, generating contaminated wastewater.[24] Water contamination due to mining has been widely discussed in the literature.[25] Specific water quality problems that are highlighted in the South Africa Environment Outlook Report include salinity and acidification.[26] Furthermore, the resulting seepage of contaminated water and potential migration into the groundwater environment can compromise the health and safety of both miners and local communities.

Water, sanitation, and hygiene-related health risks in ASM

The World Health Organization (WHO) defines health as "a state of complete physical, mental and social well-being and not merely the absence of disease or infirmity".[27] Determinants of health, such as water, basic sanitation, and hygiene, reach beyond the boundaries of traditional health care and public health sectors. Water-related decisions extend to include education, transportation, agriculture, and the environment. In the developing world, where more than 660 million people in 2015 lacked an improved water supply and 2.3 billion had no access to basic sanitation and other necessities such as access to safe water and improved sanitation are universally accepted as being essential not only for human life, but also for dignity and human development as well.[28]

The lack of access to clean water and sanitation services and to capacity-building support services to improve water-related occupational health and safety practices causes a wide range of diseases, which result in diminished livelihoods for artisanal miners and their communities.[29] Lack of proper sanitation facilities on mining sites and in many residential areas means that miners and surrounding communities are faced not only with the risk of accidents, but also with the risk of water-related diseases due to contamination of water from both surface water and groundwater sources.[30] Such interplay of the environmental, economic, and social dimensions of water-related risks makes ASM a perfect candidate for the integration of sustainable development and sustainable urban water management principles.[31]

As rapidly growing population relates ASM to urbanized communities, parallels in socioeconomic, institutional, and political perspectives associate informal ASM to informal urban settlements, called slums.[32] Slums can provide lessons about improving conditions, particularly in health, water, sanitation, and hygiene that can be applied to ASM communities. The physical environments are not necessarily similar, although in some aspects they might be: what brings people into both communities, and their circumstances while they live there, indicate some potentially useful similarities, however. The most obvious parallel is that both involvement in informal, small-scale mining and migration to slums are strategies people use to cope with rural poverty and declining agricultural incomes, and to help to sustain individual and household livelihoods. Indeed, Sinding has suggested, "ASM may serve as an important supplement to other rural economic activities, thus mitigating the growth of urban areas and particularly urban slums."[33] Both lifestyle choices result from a combination of pull factors, toward economic possibilities, and push factors, from unproductive or overcrowded land. Such choices often involve only one or a few members of a household moving, in order to supplement a rural household's income.

Drawing from the indicative parallels between ASM and slum communities, we explore water-related risks in the context of setting meaningful targets to achieve the new equity-focused SDGs. It is with this view that we present a case study of Tarkwa, Ghana, to better understand the various water-related health, safety, and sanitation issues encountered in ASM. Building on case study research findings, we make recommendations that could contribute to improving the livelihoods of communities that are dependent on a sustainable mining future.

Tarkwa district: a case study

Tarkwa is the oldest and main mining district in Ghana; it is located within the Tarkwa-Nsuaem administration district of the western region, where alluvial mining is predominant along the region's streams and riverbeds. In 2012 the district had a population of 90,477 and the highest concentration of both large-scale and small-scale mines in the region.[34] Tarkwa contributes significantly to the overall gold production in the country.[35] The following section combines collective works of two projects funded through the International Mining for Development Centre (IM4DC), which is supported by the Australian government. The work included desktop review, interviews, and in-field observations.

In 2014 a project funded by IM4DC was undertaken to identify the major water-related safety concerns for ASM operators in the Tarkwa region. It found that the top three water-related accident causes were drowning, falls, and mudslides.[36] One-quarter of the accidents reported by interviewees were directly water related; in the remaining accidents, water played a somewhat indirect causal link (e.g., injuries resulting from the use of manual machines in wet conditions). The continuous use of water in gold processing creates wet and slippery ground. The most common accidents are trips or falls due to wet conditions created by the continuous use of water in gold mining and processing.[37] Significantly, continuous use of water can contribute to pit collapses, which account for the majority of deaths in ASM, especially in the rainy season.[38] While these findings are supported by a number of authors who highlight and acknowledge the occurrence of accidents in ASM, however interview participants suggested the lack of causal exploration could be linked to underreporting: miners avoid accident investigations by regulatory authorities by not reporting accidents.

It is noteworthy that cumulative effects of water-related risks to communities are not limited to water contamination alone. Topic experts suggest that the exploitation of marginal areas such as riverbanks, protected forests, and weak strata often cause accidents in ASM; this in turn has implications for overall mine worker and community safety.[39] Miners can misguidedly excavate low-strength soil at varying depths, leading to mine collapse. Furthermore, the fine waste material resulting from the series of mining washing activities is sometimes left near riverbanks, increasing the safety risks of slipping and falling. Lack of knowledge and inappropriate use of available technology contribute substantially to the level of risk undertaken by ASM miners.

In 2015 UNICEF and WHO reported that 84 percent of the rural population of Ghana had access to an improved water supply.[40] Our extensive literature review in this case study confirmed that water resources in most Ghanaian ASM communities such as the district of Tarkwa are under pressure because the town of Tarkwa is a hub for increased mining activities.[41] Tarkwa district predominantly relies on the Bonsa River as a water source, with a piped supply delivered through the Bonsa Treatment Plant. Water use from this piped supply, which is managed by Ghana Water Company Limited, was reportedly reduced to only 40 liters per person each day in 2008, due in part to increased population. Many residents use open wells, either as their primary water source or to supplement that from the piped, municipal supply. Such wells are prone to contamination through poor construction and are at risk of contamination by flooding, which can be exacerbated by mining activities.[42]

Most of the diseases identified in Tarkwa were water related. The main sanitation issues we identified from the research were lack of proper waste disposal mechanisms and inadequate sanitation facilities.[43] Diseases that were commonly reported by miners and surrounding communities include malaria, diarrhea, and various skin conditions, with many of the ASM operators interviewed indicating that they regularly suffered from recurrent episodes of malaria. They also reported that bouts of diarrhea were common, particularly among younger children. Miners reported various skin conditions, many of which reportedly worsened with exposure to stagnant water while they were mining and washing minerals. Some interviewees had weeping skin lesions medically diagnosed as Buruli ulcer, which is a common condition in the Tarkwa area. While Buruli ulcer is prevalent in areas subject to rapid environmental change such as those subject to logging, irrigation, agriculture, mining, and dam construction, it is also prevalent in environments with poor water quality and in arsenic-rich aquatic environments.[44] Tuberculosis and mercury poisoning are also prevalent within ASM communities.[45] Our interviews highlighted the commonality of disease incidence between surrounding communities and the miners themselves.

Water is polluted by ASM activities, mainly through mercury use, uneconomical mineralized rocks (waste rocks), and the slurry that remains after gold extraction (tailings). Most interview participants acknowledged that no proper means of waste disposal were available, meaning that they dumped waste anywhere at ASM sites and in the nearby communities, with this waste later finding its way into local water bodies. The waste material contains heavy metals such as mercury and cyanide. In addition to contaminating the water bodies through bio-accumulation, these metals become health hazards when consumed through water or fish from these water bodies.[46] While the effects of direct mercury ingestion and exposure, such as neurological disorders, kidney dysfunction, and immune-toxicity/autoimmune dysfunction, are well documented, along with its ability to cross the placenta causing significant birth defects, many ASM operators remain ignorant of the health and safety implications of mercury use and its disposal.[47] The widespread ASM practice of using mercury in the gold amalgamation process, in combination with lack of access to health and safety information regarding mercury use, and general reluctance among those involved in ASM to utilize either retorts or mercury-free extraction processes, increases the vulnerability of women and children within local communities.

Desktop findings of this research revealed that while availability and access to natural water bodies, including surface water and groundwater, is relatively high, access to clean water is low. Some township residents are unable to access treated and piped water, so they rely on streams and rivers for domestic use. It is common practice for users to buy water from vendors, including water to drink. Water is a vector of diarrhea, malaria, cholera, and Buruli ulcer, which are diseases prevalent in Tarkwa. Notwithstanding the threats, access to water is essential to maintain the livelihoods of ASM miners. Coupled with the poor sanitation in Tarkwa and its surrounding informal mining sites, this indicates that there is a clear opportunity to create awareness on the use of clean water, improved sanitation, and better hygiene practices to prevent outbreaks of water-related diseases. For example, a reported outbreak of cholera in Tarkwa in November 2012, infecting 140 cases in 2 weeks and resulting from poor sanitation in the district, was said to be a recurrence of an annual event.[48] This outbreak reinforces the need to improve water, sanitation, and hygiene support services and facilities, which would promote the health of all miners, including women, and thereby increase their productivity and reduce their poverty and vulnerability.

We confirmed other studies that have found that the same water is used for washing and drinking at the mine sites and in the communities.[49] Miners combine mercury with a substantial amount of water to extract the gold; this water source is also used in the communities. In addition, the miners work partly submerged in dirty water that they have used for mineral washing, and their clothes are wet the entire time they are on a mining site. Asamoah found that most mining communities in Ghana are established close to water bodies, forests, and places where water is accessible, with streams and rivers providing communities with livelihood essentials such as water for cooking, drinking, farming, building, recreation, and aesthetics.[50] In combination, these factors highlight the importance of a focus on water when considering health, safety, and sanitation in ASM, and underpin the links between occupational health and safety and water supply, sanitation, and hygiene.[51]

Understanding community culture, attitudes, and belief systems might help to clarify the behavior of those involved in ASM and surrounding communities and highlight what might motivate them to change their behavior, especially with regard to sanitation and hygiene. Understanding and meeting local demand as well as behavioral change are key elements to bringing about sustainable water, sanitation, and hygiene improvements in any community, including ASM communities and slums, and enable promotion of services and facilities that are acceptable, affordable, and appropriate to the local conditions. Women are widely acknowledged by those working in the water, sanitation, and hygiene sector as being particularly disadvantaged and at particular risk, and therefore in need of special consideration in planning and implementing improvements.

More water- and gender-related health and safety issues: fieldwork findings

Building on the IM4DC action research findings, fieldwork conducted on an ASM site in 2015 provided us with the opportunity to better understand some of the water-related health and safety issues faced by women miners in the Tarkwa region. Typical of gender-based ASM activities, the women interviewed were engaged in heavy manual work, often carrying up to 30 kilograms of ore materials in metal pans on their heads. The men on site were largely responsible for operating the machinery, including crushing machines and sieves, and mercury-based amalgamation. The women reported that this work was in addition to their domestic responsibilities. These overlapping duties created additional exposure to water-related health and safety issues. Despite minimal earnings, their financial contribution was essential to household income for payment of items such as school fees and medical expenses.

A number of the women indicated that they worked throughout their pregnancies and subsequent breast feeding (meaning that small children were often on site), usually undertaking the same workload as pre-pregnancy, including gold amalgamation off site. The women reported suffering miscarriages, but there are no data that establish a link between miscarriage and mining activity. All of the women displayed slow healing and infected cuts and abrasions on their feet and their hands; most of the injuries were the result of digging with their bare hands in mud and dirt on the mine site, and working in often-stagnant pools of water. Many reported itching skin and skin rashes that worsened when they had to work in water, with Buruli ulcer being a very common water-related skin condition. The women also reported eye and throat irritations from contact with dust, mud, and contaminated water on site. With no sanitation or washing facilities on site, the women indicated they avoided drinking water. When they needed to urinate or defecate, they went into the surrounding bushes.

Many of the women interviewed indicated that they knew that their work practices were damaging to their health and compromising their safety, but admitted they did not know the extent of the damage they were exposing themselves to, or the longer-term implications of their work, nor did they know where to go to get reliable advice on how to look after their health. They had a strong desire to know more about the health and safety issues they were exposed to and the implications these issues had for their health and well-being and that of their families, in particular the effects on their children and the communities in which they lived. Many reported a feeling of marginalization, with the associated implication that they were not worthy of receiving appropriate health and safety advice and interventions.

Discussion

Like many developing countries, Ghana faces governance issues that endanger water-related health, safety, and sanitation in ASM. Although Ghana has been known for its mining legislation that is inclusive of the ASM sector, legal provisions have been often criticized as being insufficient in addressing the needs of the community, protecting the environment, and legalizing the mining operations.[52] Small mining companies can obtain a license that allows them to divert and impound water from rivers and streams as outlined in Article 17 of the Minerals and Mining Act of 2006. This can compromise both social and environmental sustainability of valuable water resources and, in turn, could lead to communities and the environment deprived of access to water. Our review of Ghanaian policies and regulations highlights the decentralized and cumbersome processes involved in obtaining a small-scale license. This difficulty drives miners toward illegal mining operations, contributing to the existing adverse environmental effects such as water pollution and unsafe work practices.

As noted earlier in this chapter, several socioeconomic, institutional, and political parallels exist between ASM communities and those living in slums. Buxton suggests three key facets of ASM: vulnerability (economic, natural/physical, and sociopolitical), marginalization (geographical and political), and informality; these facets also apply to slums.[53] Urban slums are usually characterized by the absence of formal tenure for residents, inadequate infrastructure and services, overcrowded and substandard dwellings, location on land that is unsuitable for occupation (e.g., low-lying and prone to flooding, or on steep slopes prone to landslips), and lack of formal recognition by local authorities. Most slum dwellers are considered squatters, without the formal right to occupy the land where they live. In a similar way, mineral tenure—the right to mine a particular piece of land—is a key issue in ASM. Without such tenure, in both cases, the potential tenure holder is unlikely to invest in improvements, for example in equipment or in upskilling in ASM, or in improved housing and related infrastructure (including water supply, sanitation, and electrical connections) in both ASM and slums. As Sinding writes, informal miners live "on the fringes of or completely outside the law, ignoring factors such as existing property rights, environmental regulations and taxation systems."[54] The same applies to slum dwellers. Miners in the informal sector might be in conflict with large-scale miners for the same resources.[55] Likewise, slum dwellers are often in conflict with landowners, who might be government or private, for rights to the land they are living on. Both those involved in ASM and slum dwellers live unstable lives, with the possibility of eviction. Small-scale miners cannot exist without protection of some sort in their living and working conditions; similarly, protection systems, in the form of influential people, who are often connected with local politicians, exist in slums, even though they are extralegal.[56] Informality, in both ASM and slums, implies possibilities for corruption and bribes, especially with officials who sometimes act as powerful brokers that overlook the illegal status.

ASM remains largely ignored by the international development sector. According to Hilson and McQuilken, "More than four decades after entering the international development lexicon, ASM remains on the periphery of poverty alleviation and local economic development policy."[57] Recognition of the significance of urban slums and interest in their development is growing among international development agencies and professionals, however. Slums house nearly 1 billion occupants worldwide, and are expected to be the center of the highest population growth in the near future. This is particularly the case in Africa, where the United Nations Human Settlements Programme estimates that 62 percent of urban dwellers (213 million people) were living in slums in 2012, and that this number will probably grow to at least 300 million by 2030.[58] Given this case, what can we learn from development activities and processes in slums, in particular in the water, sanitation, and hygiene sector? In particular, which lessons would improve health and working conditions in ASM communities?

As Hilson and McQuilken conclude in their review of the situation of ASM in sub-Saharan Africa, ASM as a people-centered activity needs to be much better understood in order to help those people to improve both their livelihoods and the environment in which they live and work.[59] There is also a growing understanding that slum development must be based on a thorough understanding of the local context: Kennedy-Walker and colleagues, who undertook studies in slums in Lusaka, Zambia, with a focus on sanitation, stress that a good understanding of power, politics, and history is essential to creating effective strategies for work in the slums.[60] Similarly, McGranahan has summarized and discussed three key and related challenges in making improvements in sanitation in slum areas: collective action, coproduction, and tenure.[61]

Many authors on water and sanitation improvement, including Kamal Kar and Robert Chambers, have acknowledged that collective action is key to achieving widespread sanitation improvements in villages, for example through community-led total sanitation to eliminate defecation in the open and thus to achieve significant improvements in public health for all. Even though the public health imperative is even stronger in densely populated slums, and there is a strong demand for better facilities and services—especially among women and girls who suffer enormous indignity, as well as abuse and sometimes violence, from the need to defecate in the open—such collective action is generally more difficult in the heterogeneous communities found in slums than in traditionally based rural communities.[62] McGranahan cites two successful examples, in informal settlements in Karachi, Pakistan; and in Mumbai and Pune, India.[63] The first example involved concerted attempts to organize the local community, led by a charismatic community development worker, to build and manage low-cost sewers, which the local authority eventually allowed to be connected to the city's main sewers. In the second example, in Mumbai and Pune, an alliance of organizations working with slum dwellers organized the residents to build and manage community toilet blocks. Both examples demonstrated that the collective nature of the problem could become part of the justification for collective action. The same principle applies to developing services and facilities to deliver better health, water supplies, and sanitation to ASM communities. Kwesi and colleagues give useful photographic evidence that sanitation conditions in some ASM communities in Tarkwa, the case study that we have considered in this chapter, are all too similar to those in many slums, with widely spread heaps of solid waste acting as sites for open defecation and attracting scavenging animals.[64] The health implications are all too obvious, as is the need for collective action.

Coproduction is a term and a concept suggested by Elinor Ostrom, the economist who was awarded the Nobel Prize in Economic Sciences in 2009, to show how synergies between, for example, local government and civil society groups, can be used for more-effective development of services, including water and sanitation.[65] Local authorities in relatively poor countries often do not have the resources nor do they recognize the incentive to improve services in informal settlements. If slum residents work collectively with the local authorities, however, better services for residents should result. In return, public agencies should be able to improve public behavior from some of their most impoverished citizens. Coproduction in the example of community toilet blocks in India discussed above involves the community managing the day-to-day functioning of the toilets, while the authorities take responsibility for regular, safe disposal of the resulting fecal waste. A conscious process of coproduction in such contexts should create social capital and improve relations between public service users and their governments.

This would be particularly beneficial where residents often see the authorities as a threat and authorities see residents as a nuisance or worse. Similarly, those involved in informal ASM, or *galamsey* as it is known in Ghana, are viewed as illegal by the authorities. Breaking down the barriers of suspicion, fear, and disapproval between authorities and informal miners should lead to valuable developments based on coproduction.

As noted above, tenure is a significant challenge to investments in services, such as sanitation. In some cases, governmental restrictions might not allow authorities to provide services to informal settlements, and in other cases they are allowed but not obliged to do so. Private utilities are very reluctant to provide services to poorer customers from whom they expect difficulties in recovering fees. Residents, whether tenants or landlords, who fear eviction are reluctant to make capital investments in facilities unless they are confident that they will be able to use them for a worthwhile period of time. Based on their research in the slums of Dakar, Senegal, Scott, Cotton, and Sohail suggest that informal and de facto tenure arrangements and not formal, legal tenure (de jure) are all that are necessary to encourage household investment in sanitation infrastructure.[66] Perhaps similar, flexible approaches to mineral tenure could be found to facilitate acceptance of ASM activities and provision of services to improve health and safety.

Osumanu and colleagues provide a useful Ghanaian example of how water supply and sanitation improvements can be made in slums when they discuss the work of the local organization People's Dialogue Ghana (PDG).[67] PDG is a local affiliate of Shack/Slum Dwellers International, a global network of community organizations of the urban poor that aims to build the voice and collective capacity of those communities. PDG's five key strategies in supporting slum dwellers are (1) building and organizing communities; (2) facilitating savings; (3) strengthening negotiating power; (4) establishing and providing a support base; and (5) bringing together communities and local government. Notice that many of these strategies align with the issues discussed by McGranahan.[68] Martin Mulenga, in the preface to Osumanu and colleagues, summarizes the key findings of five similar studies in different countries, including a study of the PDG, and stresses the importance of collaborative working—within each community, between communities, among government and service providers, and between local and international members of Shack/Slum Dwellers International; of creative, adaptable, and appropriate financing mechanisms, including collective loans; of collecting and using data, and of monitoring and sharing information; and, finally, of finding ways to scale up successful interventions.[69] All of these aspects of work with ASM communities need improvement in order to ameliorate health and safety. Authorities need to work collaboratively with NGOs and civil society, including ASM representatives, to develop workable and sustainable solutions to environmental health problems and dangerous working conditions in order to improve the social and economic lives of ASM communities and slum dwellers.

Conclusion

A water-related focus is key to identifying opportunities for considering and tackling issues related to health, sanitation, and safety faced by ASM and affected communities.

The principles behind management solutions and pathways presented in this chapter can help frame the sustainable development context according to the relevant needs of the resource sector today. In the shift from the MDGs to the SDGs, short-term deliverables can be converted into meaningful targets that prioritize human welfare.

Safe and healthy working conditions come at a certain price for ASM miners and their families. It is only when our knowledge network is shaped by an understanding of the vulnerable, marginalized, and informal nature of the sector that we can collaboratively set targets that are meaningful, achievable, and sustainable.

Notes

1 Peter J. Ashton, David Love, Harriet Mahachi, and P. H. G. M. Dirks, "An Overview of the Impact of Mining and Mineral Processing Operations on Water Resources and Water Quality in the Zambezi, Limpopo and Olifants Catchments in Southern Africa," *Contract Report to the Mining, Minerals and Sustainable Development (Southern Africa) ENV-PC 42* (Project, by CSIR-Environmentek, Pretoria and Geology Department, University of Zimbabwe-Harare, 2001).

2 K. Obiri-Danso, C. A. A. Weobong, and Keith Jones, "Aspects of Health-Related Microbiology of the Subin, an Urban River in Kumasi, Ghana," *Journal of Water and Health* 3:1 (2005): 69–76; Joyce Yaa Avotri and Vivienne Walters, "'You Just Look at our Work and See If You Have Any Freedom on Earth': Ghanaian Women's Accounts of Their Work and Their Health," *Social Science & Medicine* 48:9 (1999): 1123–1133; Nana O. B. Ackerson and Esi Awuah, "Urban Agriculture Practices and Health Problems Among Farmers Operating on a University Campus in Kumasi, Ghana," *Field Actions Science Reports. The Journal of Field Actions* Special Issue 1 (2010).

3 Afedzi Abdullah, "NGO Calls for Occupational Health Policy in Ghana," *Ghana News Agency*, April 30, 2016, www.ghananewsagency.org/health/ngo-calls-for-occupational-health-policy-in-ghana-103300 (accessed February 2017); Pius Agbenorku, Margaret Agbenorku, Adela Amankwa, Lawrence Tuuli, and Paul Saunderson, "Factors Enhancing the Control of Buruli Ulcer in the Bomfa Communities, Ghana," *Transactions of the Royal Society of Tropical Medicine and Hygiene* 105:8 (2011): 459–465; Joe-Steve Annan, Emmanuel K. Addai, and Samuel K. Tulashie, "A Call for Action to Improve Occupational Health and Safety in Ghana and a Critical Look at the Existing Legal Requirement and Legislation," *Safety and Health at Work* 6:2 (2015): 146–150; Geoffrey K. Amedofu, "Hearing-Impairment Among Workers in a Surface Gold Mining Company in Ghana," *African Journal of Health Sciences* 9:1 (2002): 91–97.

4 Buruli ulcer is caused by a necrotizing bacterium of uncertain aetiology; it is found in higher concentration in mining areas with stagnant water pools and mining pits. Alfred A. Duker, Alfred Stein, and Martin Hale, "A Statistical Model for Spatial Patterns of Buruli Ulcer in the Amansie West District, Ghana," *International Journal of Applied Earth Observation and Geoinformation* 8:2 (2006): 126–136.

5 World Health Organization (WHO), "Buruli Ulcer," (WHO, 2017), www.who.int/mediacentre/factsheets/fs199/en/ (accessed February 2017).

6 The mechanism that makes diseases such as Buruli ulcer thrive and become resistant to existing epidemiologic interventions is still unknown, however; Thomas Akabzaa, and Abdulai Darimani, "Impact of Mining Sector Investment in Ghana: A Study of the Tarkwa Mining Region," *Third World Network* (2001); Samuel J. Spiegel and Marcello M. Veiga, "Building Capacity in Small-Scale Mining Communities: Health, Ecosystem Sustainability, and the Global Mercury Project," *EcoHealth* 2:4 (2005): 361–369.

7 Niladri Basu, Edith Clarke, Allyson Green, Benedict Calys-Tagoe, Laurie Chan, Mawuli Dzodzomenyo, Julius Fobil et al., "Integrated Assessment of Artisanal and Small-Scale Gold

Mining in Ghana—Part 1: Human Health Review," *International Journal of Environmental Research and Public Health* 12:5 (2015): 5143–5176.
8. Gavin Hilson, "A Contextual Review of the Ghanaian Small-Scale Mining Industry," *Mining, Minerals and Sustainable Development* 76 (2001).
9. K. Lu, R. Long, M. Rajaee, N. Basu, J. Akizili, T. Robins, A. Yee, C. Sharp, E. Renne, and M. L. Wilson, "An Exploratory Study on the Effects of Social Resources, Environmental Exposures and Malaria Prevention Practices on the Prevalence of Malaria-Like Symptoms in a Gold-Mining Community in Ghana" (Ann Arbor, MI: University of Michigan, 2011).
10. Abbi Buxton, "Responding to the Challenge of Artisanal and Small-Scale Mining" (London: IIED, 2013).
11. Niladri Basu et al., "Integrated Assessment of Artisanal and Small-Scale Gold Mining in Ghana"; Marcello M Veiga, Malcolm Scoble, and Mary Louise McAllister, "Mining with Communities," *Natural Resources Forum* 25:3 (2001): 191–202.
12. Gavin Hilson and James McQuilken, "Four Decades of Support for Artisanal and Small-Scale Mining in Sub-Saharan Africa: A Critical Review," *The Extractive Industries and Society* 1:1 (2014): 104–118.
13. Punam Chuhan-Pole, Andrew L. Dabalen, Andreas Kotsadam, Aly Sanoh, and Anja K. Tolonen. "The Local Socioeconomic Effects of Gold Mining: Evidence from Ghana," *Policy Research Working Paper 7250* (World Bank Group, 2015).
14. Bernd Dreschler, "Small-Scale Mining and Sustainable Development Within the SADC Region," *Mining, Minerals and Sustainable Development* 84 (2001); Nellie Mutemeri, and Francis W. Petersen, "Small-Scale Mining in South Africa: Past, Present and Future," *Natural Resources Forum* 26:4 (2002): 286–292.
15. Gavin Hilson, "Poverty Traps in Small-Scale Mining Communities: The Case of Sub-Saharan Africa," *Canadian Journal of Development Studies/Revue canadienne d'études du développement* 33:2 (2012): 180–197.
16. Gavin Hilson, "Small-Scale Mining, Poverty and Economic Development in sub-Saharan Africa: An Overview." *Resources Policy* 34:1 (2009): 1–5.
17. Herman Gibb and Keri Grace O'Leary, "Mercury Exposure and Health Impacts Among Individuals in the Artisanal and Small-Scale Gold Mining Community: A Comprehensive Review," *Environmental Health Perspectives (Online)* 122:7 (2014): 667; Yasaswi Paruchuri, Amanda Siuniak, Nicole Johnson, Elena Levin, Katherine Mitchell, Jaclyn M. Goodrich, Elisha P. Renne, and Niladri Basu. "Occupational and Environmental Mercury Exposure Among Small-Scale Gold Miners in the Talensi–Nabdam District of Ghana's Upper East Region," *Science of the Total Environment* 408:24 (2010): 6079–6085.
18. Thomas Akabzaa and Abdulai Darimani, "Impact of Mining Sector Investment in Ghana."
19. Benjamin N. A. Aryee, Bernard, K. Ntibery, and Evans Atorkui, "Trends in the Small-Scale Mining of Precious Minerals in Ghana: A Perspective on its Environmental Impact," *Journal of Cleaner Production* 11:2 (2003): 131–140.
20. International Labour Organisation, "Small-Scale Mining on the Increase in Developing Countries" (News Release, 1999), www.ilo.org/global/about-the-ilo/media-centre/press-releases/WCMS_007929/lang--en/index.htm (accessed May 24, 2014).
21. Abbi Buxton, "Responding to the Challenge of Artisanal and Small-Scale Mining"; Felix Hruschka and Christina Echavarria, "Rock-Solid Chances for Responsible Artisanal Mining," *Arm Series on Responsible ASM* 3 (2011).
22. Niladri Basu et al., "Integrated Assessment of Artisanal and Small-Scale Gold Mining in Ghana."
23. Rachel N. Long, Kan Sun, and Richard L. Neitzel. "Injury Risk Factors in a Small-Scale Gold Mining Community in Ghana's Upper East Region," *International Journal of Environmental Research and Public Health* 12:8 (2015): 8744–8761; E. Kyeremateng-Amoah, and Edith E. Clarke, "Injuries among Artisanal and Small-Scale Gold Miners in Ghana," *International Journal of Environmental research and Public health* 12:9 (2015): 10886–10896.

24 M.A. Hermanus, "Occupational Health and Safety in Mining-Status, New Developments, and Concerns," *Journal of the Southern African Institute of Mining and Metallurgy* 107:8 (2007): 531–538.
25 David Banks, Paul L. Younger, Rolf-Tore Arnesen, Egil R. Iversen, and Sheila B. Banks, "Mine-Water Chemistry: The Good, the Bad and the Ugly," *Environmental Geology* 32:3 (1997): 157–174; S. H. H. Oelofse, P. J. Hobbs, J. Rascher, and J. E. Cobbing, "The Pollution and Destruction Threat of Gold Mining Waste on the Witwatersrand: A West Rand Case Study," in *10th International Symposium on Environmental Issues and Waste Management in Energy and Mineral Production* (Bangkok: SWEMP, 2007), 11–13.
26 Department of Environmental Affairs and Tourism (DEAT), "South Africa Environment Outlook. A Report on The State of The Environment. Executive Summary and Key Findings" (Pretoria: Department of Environmental Affairs and Tourism, 2006).
27 World Health Organization, "Preventing Disease Through Healthy Environments," 2013, www.who.int/ipcs/assessment/public_health/mercury_asgm.pdf (accessed May 16, 2013).
28 UNICEF, "Water, Sanitation and Hygiene, About WASH," (UNICEF, n.d.), www.unicef.org/wash/3942_3952.html; World Health Organization (WHO), "Progress on Sanitation and Drinking Water: 2015 Update and MDG Assessment," (WHO, 2015), www.who.int/water_sanitation_health/monitoring/jmp-2015-update/en/ (accessed February 2017).
29 Thomas Akabzaa, and Abdulai Darimani, "Impact of Mining Sector Investment in Ghana"; Samuel J. Spiegel and Marcello M. Veiga, "Building Capacity in Small-Scale Mining Communities."
30 J. S. Kuma, and E. Ewusi, "Water Resources Issues in Tarkwa Municipality, South-West Ghana," *Ghana Mining Journal* 11:1 (2010).
31 R. K. Amankwah, and C. Anim-Sackey, "Strategies for Sustainable Development of the Small-Scale Gold and Diamond Mining Industry of Ghana," *Resources Policy* 29:3 (2003): 131–138; Michael Redclift, "Sustainable Development and Popular Participation: A Framework for Analysis," *Revisiting Sustainable Development* (1992): 43; R. R. Brown, N. Keath, and T. H. F. Wong, "Urban Water Management in Cities: Historical, Current and Future Regimes," *Water Science and Technology* 59:5 (2009): 847–855.
32 "The term 'slum' usually has derogatory connotations and can ... legitimate the eviction of its residents. However, it is a difficult term to avoid for at least three reasons. First, some networks of neighbourhood organizations choose to identify themselves with a positive use of the term, partly to neutralize these negative connotations, eg the National Slum Dwellers Federation in India. Second the only global estimates for housing deficiencies, collected by the United Nations, are for what they term 'slums.' And third, in some nations, there are advantages for residents of informal settlements if their settlement is recognised officially as a 'slum'; indeed their residents may lobby to get their settlement classified as a 'notified slum.'" Diana Mitlin, "Will Urban Sanitation 'Leave No One Behind'?" *Environment and Urbanization* 27: 2 (2015): 365–370, fn3.
33 Knud Sinding, "The Dynamics of Artisanal and Small-Scale Mining Reform," *Natural Resources Forum* 29: (2005): 243–52.
34 Ghana Statistical Service, "2010 Population and Housing Census: Final Results" (Ghana Statistical Service, 2012), www.statsghana.gov.gh/docfiles/2010phc/2010_POPULATION_AND_HOUSING_CENSUS_FINAL_RESULTS.pdf (accessed March 19, 2014).
35 Thomas Akabzaa and Abdulai Darimani, "Impact of Mining Sector Investment in Ghana."
36 Danellie Lynas, Sarah Goater, Mark Griffin, Alycia Moore, Gernelyn Logrosa, Evelyn Kamanga, and Melissa Pearce, "Water-Related Safety, Health and Sanitation of Artisanal and Small-Scale Miners and the Affected Communities" *Action Research Report Project* CO22 (International Mining for Development Centre).
37 International Labour Organisation, "Small-Scale Mining on the Increase in Developing Countries."

38 E. Kyeremateng-Amoah and Edith E. Clarke, "Injuries Among Artisanal and Small-Scale Gold Miners in Ghana."
39 Evelyn Kamanga, "Water-Related Health, Safety and Sanitation of Artisanal and Small-Scale Miners and the Affected Communities: Tarkwa Case Study (Ghana)" (Master's thesis, The International Water Centre, 2014).
40 World Health Organization. 2014. "Progress on Drinking Water and Sanitation" (Joint Monitoring Programme, 2014), www.who.int/water_sanitation_health/publications/2014/jmp-report/en/ (accessed May 24, 2014).
41 J. S. Kuma and E. Ewusi, "Water Resources Issues in Tarkwa Municipality, South-West Ghana."
42 Ibid.
43 Evelyn Kamanga, "Water-Related Health, Safety and Sanitation of Artisanal and Small-Scale Miners."
44 Richard W. Merritt, M. Eric Benbow, and Pamela L. C. Small, "Unraveling an Emerging Disease Associated with Disturbed Aquatic Environments: The Case of Buruli Ulcer," *Frontiers in Ecology and the Environment* 3:6 (2005): 323–331.
45 Gavin Hilson, "Abatement of Mercury Pollution in the Small-Scale Gold Mining Industry: Restructuring the Policy and Research Agendas," *Science of the Total Environment* 362:1 (2006): 1–14; Sumanth G. Reddy, "A Comparative Analysis of Diseases Associated with Mining and Non-Mining Communities: A Case Study of Obusai and Asankrangwa, Ghana" (Master's thesis, University of North Texas, 2005); Samuel J. Spiegel and Marcello M. Veiga, "Building Capacity in Small-Scale Mining Communities: Health, Ecosystem Sustainability, and the Global Mercury Project," *EcoHealth* 2:4 (2005): 361–369.
46 F. A. Armah, S. Obiri, D. O. Yawson, A. N. M. Pappoe, and Bismark Akoto, "Mining and Heavy Metal Pollution: Assessment of Aquatic Environments in Tarkwa (Ghana) Using Multivariate Statistical Analysis," *Journal of Environmental Statistics* 1:4 (2010): 1–13; R. K. Amankwah and C. Anim-Sackey, "Strategies for Sustainable Development of the Small-Scale Gold and Diamond Mining Industry of Ghana," *Resources Policy* 29:3 (2003): 131–138; Benjamin A. Teschner, "Small-Scale Mining in Ghana: The Government and the Galamsey," *Resources Policy* 37:3 (2012): 308–314.
47 WHO, "Progress on Drinking Water and Sanitation."
48 Ghana Mining, "Minerals Commission," 2012, www.ghana-mining.org/ghanaims/Institutions/MineralsCommissionMC/tabid/155/Default.aspx (accessed March 30, 2014).
49 Thomas Akabzaa, and Abdulai Darimani. "Impact of mining sector investment in Ghana."
50 Evans Asamoah, "The Impact of Small Scale Gold Mining Activities on the Water Quality of River Birim in the Kibi Traditional Area" (PhD dissertation, 2012).
51 Peter J. Ashton et al., "An Overview of the Impact of Mining and Mineral Processing Operations."
52 Frederick A. Armah, "Artisanal Gold Mining and Mercury Contamination of Surface Water as a Wicked Socio-Environmental Problem: A Sustainability Challenge?" (paper presented to the proceedings of 3rd World Sustainability Forum, sciforum.net platform, November 1–30, 2013); Benjamin A. Teschner, "Small-scale Mining in Ghana: The Government and the Galamsey."
53 Abbi Buxton, "Responding to the Challenge of Artisanal and Small-Scale Mining."
54 Knud Sinding, "The Dynamics of Artisanal and Small-Scale Mining Reform."
55 Bannock Consulting, "The Impact of Price Fluctuations on Livelihood Strategies in Artisanal and Small-Scale Mining Communities Compared with Other Non-Financial Shocks" (Rugby, UK: Paper prepared in association with Intermediate Technology Consultants, n.d.).
56 Knud. Sinding, "The Dynamics of Artisanal and Small-Scale Mining Reform"; Deepa Joshi, Ben Fawcett, and Fouzia Mannan. "Health, Hygiene and Appropriate Sanitation: Experiences and Perceptions of the Urban Poor," *Environment and Urbanization* 23:1 (2011): 91–111.

57 Gavin Hilson and James McQuilken, "Four Decades of Support for Artisanal and Small-Scale Mining in Sub-Saharan Africa: A Critical Review," *The Extractive Industries and Society* 1:1 (2014): 104–118.
58 UN Habitat, "Issue Papers and Policy Units of the Habitat III Conference" (UN, 2015), https://unhabitat.org/wp-content/uploads/2015/04/Habitat-III-Issue-Papers-and-Policy-Units_11-April.pdf (accessed February 2017).
59 Gavin Hilson, and James McQuilken, "Four Decades of Support for Artisanal and Small-Scale Mining in Sub-Saharan Africa."
60 Ruth Kennedy-Walker, Jaime M. Amezaga, and Charlotte A. Paterson, "The Role of Power, Politics and History in Achieving Sanitation Service Provision in Informal Urban Environments: A Case Study of Lusaka, Zambia," *Environment and Urbanization* 27:2 (2015): 489–504.
61 Gordon McGranahan, "Realizing the Right to Sanitation in Deprived Urban Communities: Meeting the Challenges of Collective Action, Coproduction, Affordability, and Housing Tenure," *World Development* 68 (2015): 242–253.
62 Kamal Kar and Robert Chambers, "Handbook on Community-Led Total Sanitation" (Institute of Development Studies, University of Sussex, 2008); Deepa Joshi et al., "Health, Hygiene and Appropriate Sanitation."
63 McGranahan, Gordon. "Realizing the Right to Sanitation in Deprived Urban Communities."
64 Edward Kwesi, Attimo Amyiah, Appiah Sampson, Isaac Borsah, Taggoe Naa Dei, and Tinadu Kwame, "Impacts of 'Galamsey' on Drainage and Sanitation in The Mining Communities of Tarkwa, Ghana" (Sofia: Paper presented at the FIG Working Week, May 17–21, 2015).
65 Elinor Ostrom, "Crossing the Great Divide: Coproduction, Synergy, and Development," *World Development* 24:6 (1996): 1073–1087.
66 Pippa Scott, Andrew Cotton, and M. Sohail. "Using Tenure to Build a 'Sanitation Cityscape': Narrowing Decisions for Targeted Sanitation Interventions," *Environment and Urbanization* 27:2 (2015): 389–406.
67 K. Osumanu, L. Abdul-Rahim, J. Songsore, F. Braimah, and M. Mulenga, "Urban Water and Sanitation in Ghana: How Local Action is Making a Difference," *Human Settlements Working Paper, Water and Sanitation—25* (London: International Institute for Environment and Development, 2010).
68 Gordon McGranahan, "Realizing the Right to Sanitation in Deprived Urban Communities."
69 K. Osumanu et al., "Urban Water and Sanitation in Ghana."

19 Gauging the effectiveness of certification schemes and standards for responsible mining in Africa

Renzo Mori Junior

In brief

- African countries have an abundance of natural resources but low levels of human development. To address substantially unregulated mining practices, sustainability certification schemes and standards, most of them voluntary, are emerging organically and rapidly. In the wake of this fast growth are questions about their effectiveness in delivering sustainable economic, social, and environmental outcomes.
- Sustainability certification schemes and standards are valuable instruments with potential not only to deliver positive outcomes, but also to drive change and share the wealth of the mining sector equitably beyond the mining life cycle.
- As global interest has grown in how the extractive industries both fail to deliver sustainable growth and fuel conflicts, development of different governance initiatives, such as sustainability certification schemes and standards, has also grown.
- This chapter provides a literature review to identify and assess the challenges and to reflect on how sustainability certification schemes and standards can be used as instruments to deliver positive outcomes in Africa.

Introduction

Africa is well known for having abundant economically valuable mineral resources. The continent is believed to contain a significant percentage of the world's mineral reserves and has the largest or second-largest reserve of bauxite, chromite, cobalt, industrial diamonds, manganese, phosphate rock, platinum-group metals, soda ash, vermiculite, and zirconium.[1] Mineral resources can have different implications for and impacts on the communities, regions, and countries where mining activities take place, however. Although the abundance and exploitation of mineral resources with high economic value could in theory contribute positively to Africa's development, in practice these mineral resources have not been successfully converted into human development and improvements.[2]

The African Union and United Nations Economic Commission for Africa uses the term *paradox of Africa's mineral wealth* to refer to the mineral wealth of the continent and the persistent poverty of the majority of its people.[3] Haglund provides a similar statement, and uses the term *resource curse* to explain how extractive industries often fail to deliver sustainable growth in Africa.[4] Instead, those industries have negative impacts, such as drawing skilled workers away from other industries into the mining sector, fewer jobs than expected, jobs that are less suited to local communities, and the unpredictable nature of governments' revenues from the industries, which increases the risk of corruption. Various studies have adopted different approaches to document this correlation between mineral wealth and negative outcomes, associating the term *resource curse* with the corruption, poverty, slow growth, inequalities, and violence that beleaguer African countries.[5]

This contrast between countries that are rich in natural resources yet have low levels of per capita GDP and human development has been demonstrated through different international rankings and indexes. For instance, of the twenty countries with the lowest per capita GDP in 2016 eighteen were in Africa; of the twenty countries with the lowest Human Development Index rankings in 2016, nineteen were in Africa.[6] Some of those countries have substantial mineral resources, including the Democratic Republic of the Congo (DRC), Sierra Leone, and Mozambique.[7]

The mining sector plays a significant role for much of Africa, but despite the boom of mineral commodity prices and investments that began in 2003, African countries continue to export primary commodities; in doing so, they are supplying developed and industrialized countries without providing adequate returns to the continent.[8] A significant share of the mines in Africa are owned and operated by foreign companies and most of the minerals are exported in raw form; in addition, the industry imports most of its inputs from abroad.[9]

It is important to highlight the importance and relevance of artisanal and small-scale mining (ASM) in the African context, including its economic, social, and environmental impacts. Usually, ASM miners turn to mining either as their sole source of income or as part of a diversified strategy to earn a livelihood. These miners use rudimentary exploration and extraction techniques, thereby limiting their earnings. Millions of Africans depend directly or indirectly on ASM for their livelihood, but it is often associated with significant adverse social and environmental impacts and with low contribution to national revenues.[10] In this context, the importance of the ASM sector is frequently underestimated.[11]

As global interest has grown in how the extractive industries both fail to deliver sustainable growth and fuel conflicts, development of different governance initiatives, such as sustainability certification schemes—and standards, has also grown.[12] These governance initiatives aim to build collective responsibility for clean global mineral trade to provide accountability for civil society actors and to enable companies and governments to demonstrate that they are operating responsibly. Sustainability certification schemes and standards are a relatively new practice that emerged in the early 1980s and has grown fast since then. The International Trade

Centre—Standards Map, an online platform that enables users to explore and compare sustainability certification schemes, provides information on more than 170 schemes addressing sustainability in global supply chains.[13] These certification schemes address a wide range of aspects, activities, products, objectives, and sectors, including minerals and mining activities. In the mining sector, this practice emphasizes transparency and accountability of natural resource exploration practices, and addresses increased societal concerns about environmental destruction, human rights, conflicts, pollution, and social inequalities.[14]

These initiatives that use standards to define practices and assess sustainability performance could improve the performance of mining activities and fill the gap created by the absence of regulation. In addition, such initiatives have the capacity to foster local development and capacity building.

Several authors have explored and discussed the benefits and challenges faced by sustainability certification schemes and standards and how these initiatives can deliver positive economic, social, and environmental outcomes. Although these initiatives have grown quickly since the early 2000s, to date there have been few attempts to comprehensively synthesize the literature regarding sustainability certification schemes and standards in mining and their impacts, in particular considering the challenges in the African context. This chapter addresses the need for synthesis, and provides a literature review to identify and assess the challenges and to reflect on how sustainability certification schemes and standards can be used as instruments to deliver positive outcomes in Africa. I identify four main challenges to improve the effectiveness of certification schemes and standards: ASM formalization, interoperability, local development, and consequences of non-compliance and sanctions.

Artisanal and small-scale mining formalization

Most difficulties in controlling the mineral trade in Africa are associated with the illegality of ASM. Many proposals and interventions for regulation have been suggested and developed but with limited success. Different authors have provided reasons for these failures: Childs mentions the poverty cycle that traps ASM miners who use inadequate technologies that cause negative social and environmental impacts and low productivity levels, resulting in low levels of income. Garrett, Mitchell, and Levin studied internal and external trade mechanisms in artisanally mined diamonds in the DRC and Sierra Leone, and point out that the main reason these proposals and interventions fail to deliver positive outcomes is that they have not taken into account existing trade mechanisms and systems and have not provided sufficient resources to the institutions that must impose the new systems over those already in place.[15] Childs, Garrett, Mitchell, and Levin suggest that a different approach that seeks to remove penalties for the informal sector and at the same time incentivizes formalization should be adopted.

After assessing mineral resources and Africa's development strategies, the African Union and United Nations Economic Commission for Africa provided an interesting recommendation to improve interoperability between ASM and large-scale mining. These authors recommend that cooperation between ASM and large-scale mining operations should be encouraged in order to convert

ASM into viable operating enterprises through a pragmatic approach that would distinguish marginal from potentially viable ASM operations.[16]

After studying Certified Trading Chains in the Great Lakes Region in Rwanda, Franken and colleagues assert that fostering ASM formalization of sustainability certification schemes can be a fundamental tool for development in ASM communities.[17] Such schemes have the capacity to foster good governance and progressively transform and formalize informal mining and trading. Indeed, according to Franken and colleagues, formalization is a prerequisite for achieving transparent recording of production and trade, improving governance, and reducing conflicts associated with the mining sector.

Some researchers have pointed to the development of mining cooperatives as an instrument with potential to reduce illegality and deliver positive outcomes for local mining communities. After studying the Peace Diamond Alliance integrated management initiative in Sierra Leone, Maconachie and Binns find that mining cooperatives play an important role in addressing the problems of equity by improving benefits for the diggers, creating transparency in mining, promoting realistic trading agreements in the industry, enhancing realistic prices, and promoting wealth creation.[18] Cooperatives also have the capacity to engage with the community and increase their participation in minerals management, mobilize local surveillance, minimize corruption, enhance government capacity, and reform policies.

In this context, Garrett, Mitchell, and Levin recommend that the process of formalizing ASM should be progressive, proceeding through an understanding of the sector with the goal of increasing the regulatory capacities of the state. Trading structures should adopt an approach that facilitates and improves existing trading structures to make them more desirable and to make trading structures more cost-effective for operators to work legally, rather than pursuing a control approach. Garrett, Mitchell, and Levin also suggest that governments should implement alternatives to connect miners with markets, and that Governments should work with financial institutions to should develop initiatives to encourage the use of banking systems, which can provide appropriate and reliable services, and to formalize financial transactions. Perks, after reviewing the mineral dynamic of conflict minerals in the Great Lakes Region of Central Africa, similarly suggests that two of the more important aspects of the solution are improved organization of ASM and active political will.[19]

Regarding the formalization of ASM and the use of sustainability certification schemes and standards as an instrument to deliver positive outcomes in Africa, Childs recommends that, to succeed, these initiatives need to articulate a better understanding of ASM demographics, and emphasize a participatory approach.[20] This participatory approach should consider the complexity of ASM relationships with different stakeholders, attempt to understand the motivations of illegal miners, consider the involvement of socially and economically marginalized stakeholders in making decisions over their lives, and be based on more-accurate and more-reliable data. In fact, according to Childs, without those considerations, initiatives will continue to fail.

Hilson suggests a different approach.[21] Examining the fair trade gold certification scheme in sub-Saharan Africa, Hilson suggests that rather than just working toward putting artisanal miners into direct contact with Western jewelers and/or retailers, the fair-trade gold scheme should emphasize the lack of good governance and transparency and the informality of and complications with purchasing arrangements, among other aspects. Hilson suggests that host governments are the true end consumers and rejects the idea that the solution to the ASM mining problems in sub-Saharan Africa should be based on increased interactions between miners and retailers. Among other things, Hilson recommends that fair trade gold certification schemes should help miners to access assistance programs and vital support services such as technologies and legitimate credit, as well as to enable governments and donor bodies to address social and environmental problems in mining communities and to foster development of a transparent, credible, and functional network of ASM cooperatives.

Attention should also be paid to the side effects of sustainability certification schemes in the ASM sector. Exploring the sourcing of tin, tantalum, tungsten, and gold from the DRC, the Business for Social Responsibility (BSR) points out that certification schemes could exacerbate inequalities by allowing only those participants with the financial and technical capacity to be certified to reap the benefits of being certified, thus reducing local development opportunities.[22] BSR states that because of the complexities and difficulties in tracing minerals through the supply chain of the ASM sector in the DRC, smelters might choose to source only from larger mines to guarantee their compliance with international conflict-free supply chain regulations (e.g., the Dodd-Frank Act), thus excluding ASM sites from the global supply chain.[23] BSR recommends improved understanding of the regional context of the ASM sector to ensure not only that the trade of conflict minerals is avoided, but also that the ASM sector continues to be included in the global supply chain economy and continues to improve its sustainability performance.

Schure and colleagues examined the institutional framework of artisanal mining in the Congo basin and came to similar conclusions.[24] They recommend that, in order to develop effective mechanisms to foster local and regional development, institutions need to take into account the local organizations, vulnerable groups, interests of multiple actors, and the fact that the majority of the miners combine mining with other activities (e.g., agriculture, fishing, or harvesting forest products) to earn a livelihood. They also recommend an integrated approach, in both the forestry and mining sectors, to maximize positive outcomes.

It is important to highlight the role governments, mining companies, and consumers have in ensuring that compliance and concerns with conflict-free supply chain regulation or certification schemes and standards do not contribute to the exclusion of vulnerable stakeholders—in this case ASM miners—by limiting their access to the global market. Equally important, sustainability certification schemes and standards should address the complexity of relationships in the ASM sector and their causes and effects in the local context in order to be effective in delivering sustainable and fair local development in the long term.

Governments, financial institutions, schemes, and business associations should use a progressive understanding approach considering the local context and exploring sustainable development opportunities rather than an immediate command-and-control approach. For example, policymakers and schemes could identify and foster collaboration between large-scale mining companies and ASM communities, provide formalization and traceability incentives of minerals, develop and improve relevant legislation and institutions, develop and implement capacity-building programs, and adopt a more-transparent and more-accountable approach.

Interoperability

The term *interoperability* refers to whether a sustainability certification scheme or standard recognizes or references any other certification schemes or standards within its own processes, as well as legal requirements, guidelines, international protocols, and regulations. Interoperability has the potential to reduce costs and amplify the outcomes achieved by individual schemes through coordination and sharing of knowledge and practices.[25]

According to Vlassenroot, sustainability certification schemes and standards should explore opportunities for coordination and cross-recognition to maximize synergies, reduce costs and bureaucracy, and increase legitimacy and reach. In a similar vein, Cuvelier and colleagues state that there is a lack of coordination between different schemes and suggest that schemes need to coordinate activities, share resources and data, and develop joint efforts for a more transparent mining sector in the DRC.[26]

Better coordination and harmonization of data among different certification schemes and standards is the key to minimizing overlaps and unproductive information overflow. Brockmyer and Fox made recommendations for the representatives of sustainability certification schemes and standards and participating governments to improve their relationships with each other in order to facilitate communication and knowledge sharing.[27]

Bleischwitz, Dittrich, and Pierdicca studied implications for certification schemes in Central Africa, and provide a similar comment. They highlight the importance of coordination and harmonization of data for minimizing overlaps and unproductive information overflow among certification schemes and standards. They also state that interoperability between governance initiatives and international laws and regulations is important for promotion of transparency, certification, and accountability. Similarly, Bone, after assessing the Kimberley Process Certification Scheme (KPCS), advocates more-effective cooperation and collaboration within and between governments to improve the effectiveness of the KPCS.[28] Bone highlights the important role governments play in the KPCS context and states that cooperation is fundamental for addressing current and future risks to the credibility of the KPCS.

Indeed, Young, studying certification approaches for conflict minerals and conflict-free minerals, states that by working together manufacturing companies can avoid the need to maintain independent compliance programs and thus are able to share costs and efforts.[29] The same principle could be applied to certification schemes and initiatives: by working together these they can improve their own effectiveness by reducing the number and duplication of requests to participants, which will directly affect compliance costs and effort.

In conclusion, the existence of different sustainability certification schemes or standards does not guarantee good outcomes unless sustainability certification schemes and standards cover the entire value chain. The lack of interoperability among these initiatives could lead to lower levels of stakeholder confidence, reducing acceptance and increasing confusion. In addition, the existence of numerous schemes with different requirements could lead to high compliance costs, greenwashing, schemes with loose performance standards, and lack of credibility.

Previous studies have alerted us to the overlap and low level of interoperability between sustainability certification schemes and standards; and between sustainability certification schemes and standards and existing legal requirements, guidelines, international protocols, and regulations in the mining sector.[30] One of the main challenges that sustainability certification schemes and standards face in terms of interoperability lies in the different design characteristics, scope, participants, and aims among them. Without common understanding, compatible assurance processes, and equivalent requirements, cross-recognition and referencing is compromised.

Local development

Some sustainability certification schemes and standards in the mining sector state that one of their priorities is fostering and delivering local development. Stimulating local development in mining communities is also one of the motivations for ASM certification schemes. The development and implementation of mechanisms to identify and support vulnerable stakeholders (in this case local communities affected by mining activities) are regarded as good practice by the literature on the effectiveness of sustainability certification schemes and standards. In her assessment of mineral development and trade strategies in the Great Lakes Region of Central Africa, Perks identified that many strategies have focused more closely on security aspects than on an economic development agenda. Bone and Amnesty International point out that the sole focus of the KPCS is on avoiding conflict diamonds, rather than on incorporating other important aspects such as local development, environmental protection, and human rights for both diamond miners and local communities.[31]

Sustainability certification schemes and standards should play an important role in delivering local development. Sustainability certification schemes and standards should develop and implement programs to provide finance and technical support, deliver capacity building, and improve accessibility. Perks states that mineral permits, financial services, and technological assistance are necessary to foster local development. Pact also made an important statement about the importance of initiatives reinvesting mining revenues in improved income-generating activities.[32] Such initiatives and programs are fundamental to delivering local development not only during mining booms, but also during mining busts or downturns.

To illustrate this point, in their study of technical assistance initiatives and the conflict-free mineral chain certification scheme of the International Conference on the Great Lakes Region in the eastern DRC, Partnership Africa Canada determined that technical assistance, in this case an improved sluice, increased gold yields by 25–30 percent. Similarly, Haglund points out that skills training and customized advice on topics related to implementing schemes and self-assessments are crucial for producers to shape their own destinies and improve their competitiveness by developing the quality of their products as well as their efficiency and sustainability. ISEAL Alliance suggests that capacity building for data collection and management, for example, could be used as an instrument to deliver local development.[33] Capacity-building programs allow producers to develop their production, processing, marketing, and trading methods and therefore take ownership of their own livelihoods. Such programs leverage and multiply the impact of local community development initiatives by creating inclusive processes that improve commitment, build good relationships, and strengthen trust.

In relation to the role that sustainability certification schemes and standards can play to deliver positive outcomes, Maconachie points out that the KPCS has forced a large volume of illicit diamonds into official channels, thus increasing export earnings that are vital for the reconstruction of community structures and war-damaged infrastructure. Smillie considered better knowledge of actual production and trade capacities, improved tax revenue, reduced illegality, reduced likelihood of conflict diamonds, and improved formalization to be benefits of the KPCS.[34]

Hilson assessed the case of Malawi's Nyala rubies, which its suppliers describe as a fair trade gem, to illustrate how fair trade mineral programs can potentially be misbranded.[35] Hilson concludes that the certification scheme in place, delivering Nyala rubies to markets, prioritizes the alleviation of consumers' concerns about conflict minerals and human rights abuses rather than delivering benefits to local producers, such as small-scale gemstone miners in Malawi. Hilson recommends that sustainability certification schemes and standards in the mining sector should consider the empowerment of small producers and be developed based on universal guidelines and baseline criteria debated and agreed on through a multi-stakeholder approach.

From this perspective, it is important to emphasize the importance of a multi-stakeholder approach. Positive engagement with different stakeholders allows sustainability certification schemes and initiatives to be more effective in developing solutions and to contribute in positive ways to development by considering regions' contexts and singular particularities. For instance, after assessing the KPCS Bone states that a multi-stakeholder approach creates a sense of ownership and accountability that enables different stakeholders to believe that they have made an investment in something important and lasting.[36] Furthermore, a multi-stakeholder approach increases stakeholders' confidence in the scheme and improves legitimacy and credibility.

A multi-stakeholder approach could present some challenges, however. Inclusion of divergent stakeholder concerns and attempts to reach consensus or a simple majority vote can be challenging and create delays. ISEAL Alliance suggests that it is important to determine the most appropriate occasion to engage with stakeholders so as not to engage stakeholders unnecessarily at the expense of efficiency. Global Witness defends the idea that due diligence processes of supply chains, conducted by downstream organizations, are quicker than certification schemes at tackling the trade of conflict minerals in Africa.[37]

The lack of transparency around the reporting and decision-making processes of sustainability certification schemes and standards in the mining sector is another issue that some authors criticized, saying that it affects the schemes' capacity to foster local development: Bone argues for greater transparency of the KPCS with regard to reports and decision-making processes. Mori Junior, Franks, and

Ali argue that transparency is one of the key issues for an effective sustainability certification scheme. Indeed, Bleischwitz, Dittrich, and Pierdicca say that the importance of initiatives to foster and improve transparency should be supported because those initiatives have a positive impact on markets and against corruption. Maconachie and Binns say that to be effective in Africa, to deliver positive outcomes, and to be more accountable, however, such initiatives must consider the local context and must develop instruments that support governments with financial, technological, and skilled human capacity constraints. Without such support, the development of good governance and accountability is likely to take many years, especially in countries emerging from a long period of conflict.[38]

Although the literature has provided several guidelines and recommendations for improving practices, in some circumstances specific regional contexts could represent real challenges. For example, Maconachie states that it is a significant challenge to include civil society actors in Sierra Leone in equitable participation because the power structures of Sierra Leonean society remain hierarchical and highly inequitable.[39] Maconachie highlights the existence of many powerful groups with a vested interest in continuing to exploit diamond reserves illegally as an additional challenge that sustainability certification schemes and standards have to overcome.

Sustainability certification schemes and standards must now rise to the challenge of responding to stakeholder pressure to deliver positive outcomes and to foster local development in a transparent and accountable manner. The specific challenges lie in how sustainability certification schemes and standards can understand and recognize community values, how different stakeholders should be involved and participate, how programs can address regional contexts, and how sustainability certification schemes and standards representatives can monitor and improve the efficiency and effectiveness of these programs, such that they provide lasting tangible positive impacts for local communities affected by mining in a transparent and accountable manner.

Consequences of non-compliance and sanctions

Consequences and sanctions for non-compliance offer an incentive for participants to be in compliance and are important for guaranteeing the credibility of sustainability certification schemes and standards. The ISEAL Alliance states that sanctions can be seen as an incentive to conform rather than an attempt to penalize participants.[40] Furthermore, the lack of consequences and sanctions in situations of non-compliance can affect stakeholders' perceptions and their levels of trust, as well as participants' enthusiasm and interest in compliance.[41]

After assessing 15 sustainability certification schemes in mining, Mori Junior, Franks, and Ali found that only 60 percent of the schemes analyzed have consequences and sanctions for situations of non-compliance in place and publicly available, although they are an important aspect of good practice. Similarly, Partnership Africa Canada has criticized the KPCS for having situations of non-compliance that did not result in consequences, affecting stakeholders' level of trust in the KPCS and the transparency of its processes. Regarding this situation where the credibility and the level of stakeholders' trust in the KPCS were affected, Smillie recommends that the KPCS should go back to basics and reexamine its purpose and the methodologies that KPCS has chosen to achieve it.[42]

In this context, Pact, in its assessment of the ITRI Tin Supply Chain Initiative, suggested that governments in the DRC, Rwanda, and Burundi should make use of penalizing infractions in a timely and appropriate manner as a way to maintain market confidence. Acosta provided a similar recommendation.[43]

Although the use of sanctions in a timely and appropriate manner could help to maintain market confidence and credibility, governments should assess the negative impacts of such sanctions beforehand to reduce any negative side effects. Hilson and Clifford, after assessing the temporary suspension of diamond exports from Akwatia, Ghana, in 2006 and 2007 under the KPCS requirements, provide an interesting example. They argue that even though the suspension facilitated improvements in the diamond monitoring system in Ghana (i.e., the system of paperwork to track diamonds to their source) it had catastrophic economic impacts in Akwatia: "The city that was once a vibrant diamond-trading centre populated by hundreds of buyers, merchants, miners, and traders is now an economically depressed township."[44] Hilson and Clifford also mention that confidence in Ghana's small-scale diamond sector as a whole has suffered in response to this sanction.

Cuvelier and colleagues, after assessing the impacts of the Dodd-Frank Act in the DRC, present similar findings.[45] They point out that the Kabila, DRC, mining embargo in 2010 provided some positive impacts, such as improving awareness of the urgent need to address negative aspects of the mining industry, speeding up the process of mining reform, and stimulating involvement in due diligence initiatives. It had a paralyzing effect on the regional economy and a negative impact on living conditions in eastern DRC's mining sites and urban centers, however.

Sustainability certification schemes and standards should define and make publicly available how sustainability certification schemes and standards will address situations of non-compliance and the corresponding sanctions that will accompany continued non-compliance in order to ensure credibility. Sustainability certification schemes and standards should clearly inform all participants and key stakeholders of what constitutes non-compliance and the associated sanctions; sustainability certification schemes and standards should also confirm stakeholders understand non-compliance. In addition, compliance assessments and the application of sanctions should be unbiased and based on procedures and processes that participants have predetermined and preapproved. The challenge here lies in how consequences of non-compliance and sanctions could be defined and applied to guarantee credibility without affecting accessibility and penalizing the more vulnerable stakeholders.

Conclusion

Although it has been difficult to convert the mineral resource wealth of the African continent into human development and local improvements, governance initiatives such as sustainability certification schemes and standards can play an important role in improving this situation. This chapter states that such governance initiatives are valuable instruments with the potential not only to deliver positive outcomes and pay living wages for local communities, but also to drive change and share the real wealth of the mining sector with the more vulnerable players through a local development model that goes beyond the mining life cycle. These initiatives still need to improve their effectiveness and overcome challenges, however, to fully achieve their goals and bring about sustainable local development in Africa.

In addition to the four main challenges (ASM formalization, interoperability, local development, and consequences of non-compliance and sanctions) identified by this chapter through a literature review, it is important to highlight that sustainability certification schemes and initiatives should extend the scope of sustainability certification schemes and initiatives beyond limited commodities, specific sectors of the trading chain, and large-scale mining.

Finally, the chapter emphasizes that societal awareness, of both sustainability and corporate social responsibility, is one of the key factors affecting the existence and effectiveness of sustainability certification schemes and standards. Without this market pressure for better outcomes and improvements, the success of such initiatives in converting the abundance and exploitation of mineral resources into human development in Africa would be seriously compromised. Therefore, consumers, clients and business partners should pressure organizations operating in markets with weak governmental performance and low levels of transparency to adopt ethical approaches, and to avoid illegal or non-certified products in their supply chain, and/or to prevent illegal or non-certified products mixing with legal products along their supply chain.

To conclude, the success of a sustainability certification scheme or standards depends on the extent to which the initiative can understand and recognize the local context, manage stakeholders' expectations, and deliver sustainable positive impacts with transparency and accountability. Here, "deliver sustainable positive impacts" means not only delivering positive outcomes for certified entities, but also to truly contribute to local development in the long term and beyond mining cycles.

Notes

1 KPMG, "Mining in Africa Towards 2020" (South Africa: KPMG, 2013); U.S. Geological Survey, "Mineral Commodity Summaries 2015" (Virginia: KPMG, 2015); Thomas R. Yager et al., "The Mineral Industries of Africa," in *2010 Minerals Yearbook* (U.S. Department of the Interior U.S. Geological Survey, 2010).
2 African Union and United Nations Economic Commission for Africa, "Minerals and Africa's Development. An Overview of the Report of the International Study Group on Africa's Minerals Regimes" (Addis Ababa: African Union and United Nations Economic Commission for Africa, 2011); Dan Haglund, "How Can African Economies Turn the Resource Curse into a Blessing?" *Bridges Africa* 1:3 (2012); Scott Pegg, "Poverty Reduction or Poverty Exacerbation?" (Indiana: Department of Political Science. Indiana University Purdue University Indianapolis, 2003).
3 African Union and United Nations Economic Commission for Africa, "Minerals and Africa's Development."
4 Dan Haglund, "How Can African Economies Turn the Resource Curse into a Blessing?"
5 Democratic Republic of the Congo (DRC): Nadira Lalji, "The Resource Curse Revised: Conflict and Coltan in the Congo," *Harvard International Review* 29:3 (2007);

Nigeria: Xavier Sala-i-Martin and Arvind Subramanian, *Addressing the Natural Resource Curse: An Illustration from Nigeria* (Berlin: Springer, 2008); sub-Saharan Africa: Matthias Basedau, "Context Matters – Rethinking the Resource Curse in Sub-Saharan Africa," *GIGA Working Paper No. 1* (German Institute for Global and Area Studies, May 1, 2005); Angola: John L Hammond, "The Resource Curse and Oil Revenues in Angola and Venezuela," *Science & Society* 75:3 (2011); Chad and Cameroon: Scott Pegg, "Can Policy Intervention Beat the Resource Curse? Evidence from the Chad–Cameroon Pipeline Project," *African Affairs* 105:418 (2006); South Africa: Ainsley D. Elbra, "The Forgotten Resource Curse: South Africa's Poor Experience with Mineral Extraction," *Resources Policy* 38:4 (2013); Ghana: Heikki Holmås and Joe Oteng-Adjei, "Breaking the Mineral and Fuel Resource Curse in Ghana" (Development Co-Operation Report, 2012).
6 http://data.worldbank.org/indicator/NY.GDP.PCAP.CD (accessed September 1, 2015); http://hdr.undp.org/en/composite/HDI.
7 Dan Haglund, "How Can African Economies Turn the Resource Curse into a Blessing?"; Scott Pegg, "Poverty Reduction or Poverty Exacerbation?"; United Nations Development Programme, "Human Development Report 2014. Sustaining Human Progress: Reducing Vulnerabilities and Building Resilience," (New York: United Nations Development Programme, 2014).
8 African Union and United Nations Economic Commission for Africa, "Minerals and Africa's Development."
9 Ibid.; Punam Chuhan-Pole et al., "Africa's Pulse. An Analysis of Issues Shaping Africa's Economic Future" (World Bank Group, 2015).
10 African Union and United Nations Economic Commission for Africa, "Minerals and Africa's Development"; Marieke Heemskerk, "Collecting Data in Artisanal and Small-Scale Mining Communities: Measuring Progress Towards More Sustainable Livelihoods" (Paper presented at the Natural Resources Forum, 2005); Gavin Hilson, "'Constructing' Ethical Mineral Supply Chains in Sub-Saharan Africa: The Case of Malawian Fair Trade Rubies," *Development and Change* 45:1 (2014).
11 Gudrun Franken et al., "Certified Trading Chains in Mineral Production: A Way to Improve Responsibility in Mining," in *Non-Renewable Resource Issues* (Berlin: Springer, 2012).
12 This paper uses the term "sustainability certification schemes," "certification schemes," or simply "schemes" to describe all schemes that address governance, social, and/or environmental issues and provide a certificate, label, or claim attesting compliance to their standards. Compliance is determined by an assurance, audit, or verification process; ISEAL Alliance (2010, page 5) defines standard as "document that provides, for common and repeated use, rules, guidelines or characteristics for products or related processes and production methods, with which compliance is not mandatory."
13 The International Trade Centre (ITC) is a joint agency of the World Trade Organization and the United Nations. Its objective is for businesses in developing countries to become more competitive in global markets, speeding economic development and contributing to the achievement of the United Nations' Millennium Development Goals; The ITC Standards Map, www.standardsmap.org/ (accessed October 13, 2015).
14 Graeme Auld, Lars H Gulbrandsen, and Constance L McDermott, "Certification Schemes and the Impacts on Forests and Forestry," *Annual Review of Environment and Resources* 33:1 (2008); Mike Barry et al., "Toward Sustainability: The Roles and Limitations of Certification" (Washington: RESOLVE Inc, 2012); Allen Blackman and Jorge Rivera, "Producer-Level Benefits of Sustainability Certification," *Conservation Biology* 25:6 (2011); Boudewijn Derkx and Pieter Glasbergen, "Elaborating Global Private Meta-Governance: An Inventory in the Realm of Voluntary Sustainability Standards," *Global Environmental Change* 27:0 (2014); Renzo Mori Junior, Daniel M. Franks, and Saleem. H. Ali (eds.), "Designing Sustainability Certification for Impact: Analysis of the Design Characteristics of 15 Sustainability Standards in the Mining Industry" (Brisbane: University of Queensland, 2015); Rachel Perks, "Digging into the Past: Critical

Reflections on Rwanda's Pursuit for a Domestic Mineral Economy," *Journal of Eastern African Studies* 7:4 (2013); Petrina Schiavi and Fiona Solomon, "Voluntary Initiatives in the Mining Industry: Do They Work?" *Greener Management International* 53 (2007).

15 John Childs, "Reforming Small-Scale Mining in Sub-Saharan Africa: Political and Ideological Challenges to a Fair Trade Gold Initiative," *Resources Policy* 33:4 (2008); N. Garrett, H. Mitchell, and E. Levin, "Regulating Reality: Reconfiguring Approaches to the Regulation of Trading Artisanally Mined Diamonds," *Artisanal Diamond Mining: Perspectives and Challenges* (2008).

16 African Union and United Nations Economic Commission for Africa, "Minerals and Africa's Development."

17 G. Franken et al., "Certified Trading Chains in Mineral Production."

18 Roy Maconachie and Tony Binns, "'Farming Miners' or 'Mining Farmers'?: Diamond Mining and Rural Development in Post-Conflict Sierra Leone," *Journal of Rural Studies* 23:3 (2007).

19 Garrett, Mitchell, and Levin, "Regulating Reality: Reconfiguring Approaches to the Regulation of Trading Artisanally Mined Diamonds; Rachel Perks, "Digging into the Past: Critical Reflections on Rwanda's Pursuit for a Domestic Mineral Economy."

20 John Childs, "Reforming Small-Scale Mining in Sub-Saharan Africa."

21 Gavin Hilson, "'Fair Trade Gold': Antecedents, Prospects and Challenges," *Geoforum* 39:1 (2008).

22 Business for Social Responsibility, "How Can Business Contribute to the Ethical Mining of Conflict Minerals? Addressing Risks and Creating Benefits Locally in the Artisanal and Smallscale Mining Sector in the Democratic Republic of the Congo," 2014.

23 In 2010 the United States of America Congress passed the Dodd-Frank Act, which directs the Securities and Exchange Commission (SEC) to issue rules requiring certain companies to disclose their use of conflict minerals if those minerals are "necessary to the functionality or production of a product" manufactured by those companies. Under the Act, those minerals include tantalum, tin, gold or tungsten. Section 1502 of the Act mandates the Securities and Exchange Commission (SEC) to issue rules requiring the disclosure by publicly traded companies of the origins of listed conflict minerals (http://www.sec.gov/News/Article/Detail/Article/1365171562058, accessed October 1, 2015).

24 Jolien Schure et al., "Is the God of Diamonds Alone? The Role of Institutions in Artisanal Mining in Forest Landscapes, Congo Basin," *Resources Policy* 36:4 (2011).

25 Barry et al., "Toward Sustainability: The Roles and Limitations of Certification"; International Organization for Standardization and International Electrotechnical Commission, "Guide 2: Standardization and Related Activities," in *Eight Edition* (Genève: International Organization for Standardization and International Electrotechnical Commission, 2004); ISEAL Alliance, "Principles for Credible and Effective Sustainability Standards Systems: Iseal Credibility Principles" (London: ISEAL Alliance, 2013); R. Mori Junior et al., "Designing Sustainability Certification for Impact"; WWF, "Searching for Sustainability – Comparative Analysis of Certification Schemes for Biomass Used for the Production of Biofuels" (Berlin: WWF, 2013); Steven B. Young, Alberto Fonseca, and Goretty Dias, "Principles for Responsible Metals Supply to Electronics," *Social Responsibility Journal* 6:1 (2010).

26 Koen Vlassenroot, *Artisanal Diamond Mining: Perspectives and Challenges* (Cambridge, MA: Academia Press, 2008); Jeroen Cuvelier et al., "Analyzing the Impact of the Dodd-Frank Act on Congolese Livelihoods," (DRC Affinity Group, 2014).

27 Brandon Brockmyer and Jonathan Fox, "Assessing the Evidence: The Effectiveness and Impact of Public Governance-Oriented Multi-Stakeholder Initiatives" (London: The Transparency and Accountability Initiative, 2015).

28 Raimund Bleischwitz, Monika Dittrich, and Chiara Pierdicca, "Coltan from Central Africa, International Trade and Implications for Any Certification," *Resources Policy* 37:1 (2012); Andrew Bone, "The Kimberley Process Certification Scheme: The Primary Safeguard

for the Diamond Industry," in *High-Value Natural Resources and Post-Conflict Peacebuilding*, P. Lujala and S. A. Rustad (eds.) (London: Earthscan Publications Ltd, 2012); The KPCS is a scheme that came into force in 2003 with the endorsement from the United Nations General Assembly to stem the flow of conflict diamonds used by rebel movements to finance wars against legitimate governments (www.kimberleyprocess.com/).

29 Steven B. Young, "Responsible Sourcing of Metals: Certification Approaches for Conflict Minerals and Conflict-Free Metals," *The International Journal of Life Cycle Assessment* (2015).

30 Tom Campbell, "A Human Rights Approach to Developing Voluntary Codes of Conduct for Multinational Corporations," *Business Ethics Quarterly* 16: 2 (2006); B. Derkx and P. Glasbergen, "Elaborating Global Private Meta-Governance: An Inventory in the Realm of Voluntary Sustainability Standards"; Renzo Mori Junior, Daniel Franks, and Saleem Ali, "Sustainability Certification Schemes: Evaluating Their Effectiveness and Adaptability," *Corporate Governance: The International Journal Of Business in Society* 16: 3 (2016).

31 Estelle Levin, "Certification and Artisanal and Smal-Scale Mining: An Emerging Opportunity for Sustainable Development" (Communities and Small-Mining, 2008); ISEAL Alliance, "Setting Social and Environmental Standards V5.0" (London: ISEAL Alliance, 2010); "Principles for Credible and Effective Sustainability Standards Systems: Iseal Credibility Principles; Kristin Komives and Amy Jackson, "Introduction to Voluntary Sustainability Standard Systems," in *Voluntary Standard Systems* (Berlin: Springer, 2014); Renzo Mori Junior et al., "Designing Sustainability Certification for Impact"; Rachel Perks, "Digging into the Past"; Andrew Bone, "The Kimberley Process Certification Scheme"; Amnesty International, "Chains of Abuse: The Global Diamond Supply Chain and the Case of the Central African Republic" (London: Amnesty International, 2015); The term "conflict diamonds" means rough diamonds that are used by rebel movements to finance their military activities, including attempts to undermine or overthrow legitimate governments. United Nations, *Resolution Adopted by the General Assembly 55/56* (2001).

32 Rachel Perks, "Digging into the Past:"; Pact, "Unconflicted: Making Conflict-Free Mining a Reality in the DRC, Rwanda and Burundi" (Washington, DC: Pact, 2015).

33 Partnership Africa Canada, "L'or Juste – Just Gold – Artisanal Gold Pilot Project: Democratic Republic of Congo," in *Intergovernmental Forum on Mining, Minerals, Metals and Sustainable Development* (Geneva: Partnership Africa Canada, 2014); Dan Haglund, "How Can African Economies Turn the Resource Curse into a Blessing?"; ISEAL Alliance, "Assessing the Impacts of Social and Environmental Standards Systems," in *ISEAL Code of Good Practice* (London: ISEAL Alliance, 2014).

34 Roy Maconachie, "Diamonds, Governance and 'Local' development in Post-Conflict Sierra Leone: Lessons for Artisanal and Small-Scale Mining in Sub-Saharan Africa?" *Resources Policy* 34:1 (2009); Ian Smillie, "Assessment of the Kimberley Process in Enhancing Formalization and Certification in the Diamond Industry–Problems and Opportunities," *Deutsch Gesellschaft für Internationale Zusammenarbeit (GIZ) GmbH* (2011).

35 G. Hilson, "'Constructing' Ethical Mineral Supply Chains in Sub-Saharan Africa."

36 Andrew Bone, "The Kimberley Process Certification Scheme."

37 ISEAL Alliance, "Principles for Credible and Effective Sustainability Standards Systems: Iseal Credibility Principles"; Global Witness, "Do No Harm: Excluding Conflict Minerals from the Supply Chain" (London: Partnership Africa Canada, 2012).

38 Andrew Bone, "The Kimberley Process Certification Scheme"; Renzo Mori Junior et al., "Designing Sustainability Certification for Impact"; R. Bleischwitz, M. Dittrich, and C. Pierdicca, "Coltan from Central Africa, International Trade and Implications for Any Certification"; Roy Maconachie and Tony Binns, "Beyond the Resource Curse? Diamond Mining, Development and Post-Conflict Reconstruction in Sierra Leone," *Resources Policy* 32:3 (2007).

39 Roy Maconachie, "Diamonds, Governance and 'Local' Development in Post-Conflict Sierra Leone: Lessons for Artisanal and Small-Scale Mining in Sub-Saharan Africa?," *Resources Policy* 34:1 (2009).

40 ISEAL Alliance, "Assuring Compliance with Social and Environment Standards: Code of Good Practice" (London, ISEAL Alliance, 2011).
41 Renzo Mori Junior et al., "Designing Sustainability Certification for Impact"; Partnership Africa Canada, "Diamonds and Human Security: Annual Review 2009" (Ontario: Partnership Africa Canada, 2009); Khadija Sharife and John Grobler, "Kimberley's Illicit Process," *World Policy Journal* 30:4 (2013); Amanda Stark and Estelle Levin, "Benchmark Study of Environmental and Social Standards in Industrialised Precious Metals Mining," (Solidaridad, 2011), www.cocoasolidaridad.org/sites/solidaridadnetwork.org/files/Revised%20Solidaridad_Benchmark_Study_Revised_Final%20_Dec_2011.pdf (accessed April 22, 2011).
42 Renzo Mori Junior et al., "Designing Sustainability Certification for Impact"; Partnership Africa Canada, "Diamonds and Human Security: Annual Review 2009"; Khadija Sharife and John Grobler, "Kimberley's Illicit Process"; Ian Smillie, "Assessment of the Kimberley Process in Enhancing Formalization and Certification in the Diamond Industry–Problems and Opportunities."
43 Pact, "Unconflicted: Making Conflict-Free Mining a Reality in the DRC, Rwanda and Burundi"; iTSCi (ITRI Tin Supply Chain Initiative) is a joint initiative that assists upstream companies (from mine to the smelter) to institute the actions, structures, and processes necessary to conform with the OECD Due Diligence Guidance (DDG) at a very practical level, including small and medium size enterprises, cooperatives and artisanal mine sites, www.itri.co.uk/index.php?option=com_zoo&task=item&item_id=2192&Itemid=189; Andres Mejia Acosta, "The Extractive Industries Transparency Initiative: Impact, Effectiveness, and Where Next for Expanding Natural Resource Governance? A," *U4 Brief* 6 (2014).
44 Gavin Hilson and Martin J Clifford, "A 'Kimberley Protest': Diamond Mining, Export Sanctions, and Poverty in Akwatia, Ghana," *African Affairs* 109:436 (2010): 449.
45 Cuvelier et al., "Analyzing the Impact of the Dodd-Frank Act on Congolese Livelihoods."

20 Field vignette

The Australia-Africa Minerals and Energy Group (AAMEG)

Trish O'Reilly

The footprint of the Australian resources sector on the African continent is significant. Australia is one of the top investors in mineral exploration in Africa with over 170 Australian companies developing or operating over 400 projects across 35 African countries (AAMEG Research 2016). Australia's two-way goods and merchandise trade with Africa has increased over the past decade from around $6 billion in 2004–2005 to $8.5 billion in 2014–2015. It is estimated that 60 percent of Mining Equipment, Technology and Services ("METS") companies in Australia export their services to African countries and these services are worth approximately $15 billion annually. Africa is important to Australia.

The Australia-Africa Minerals & Energy Group (AAMEG) is the leading member association supporting the Australian extractives industry in dealing with non-technical issues in their operations in the countries of Africa. The organization was formally incorporated in April 2011 with strong support from the Australian Federal Government.

AAMEG supports members operating in Africa by:

1 **Being the voice of industry on member-driven issues:**
   AAMEG advocates publicly on behalf of its members interests and stresses the importance of Australia continuing its strategic engagement with Africa and changing the conversation from one of aid and dependency, to one of economic diplomacy and shared value impact.

2 **Building the Australian government's understanding of and support for the Australian resources sector operating in Africa, and assisting with engagement between the resources sector and government:**
   AAMEG provides a forum with a collaborative approach, that brings together industry, governments, academia and NGOs in order to more effectively deal with issues such as political and social risk, anti-bribery and corruption, health issues, security and human rights issues, capacity building initiatives, community development and social-licence-to-operate matters, meaningful engagement of communities and a range of other issues.

3 **Building relationships with African governments and providing venues for collaboration on issues of common interest:**
   In order to increase engagement and opportunity for positive discussion, AAMEG has established Associate Membership for African Governments

and regularly meets with Australian and African Governments to provide the opportunity to facilitate strategic partnerships and dialogue including: facilitating discussion between Associate Members and Mining companies over operational issues; determine capacity-building opportunities in host countries that members can engage in; and determine opportunities for host countries to engage with Australian Governments, academia and NGOs.

4   **Creating a forum for members to network and share operational experiences in order to strengthen the Australian resources industry operational capability in Africa;**

AAMEG organizes a range of member forums including presentations, roundtables and workshops through which members can share their individual experiences and be better prepared to deal with what are often confronting issues. Below is a selection of subjects dealt with and the AAMEG members who led discussions:

- Impact of HIV on the mining industry, Murdoch University, Perth.
- Medical challenges you may face when doing business in Africa, International SOS, Perth.
- On the take. . . the rise of fraud and corruption, PwC, Perth.
- Management of Financial Crime Risk in Africa, PwC, Melbourne.
- Dispute Resolution in Africa, Herbert Smith Freehills, Perth.
- The Underlying Business Case for CSR, ACCSR Masterclass, Curtin University, Perth.
- Development of GIS database related to west African Geoscience, SNL Metals & Mining, Perth.

A key focus of the organization is to actively supports international best practice. In particular AAMEG strongly supports Australia's approach to the Extractive Industries Transparency Initiative, the OECD Guidelines for Multinational Enterprises and the Voluntary Principles on Security and Human Rights. AAMEG also supports the worthy aspirations of African leaders as expressed in the Africa Mining Vision (2009). In response to need we have developed a number of toolkits and handbooks that promote international best practice, including, for example; A Social Aspects Management Handbook – Increasing Shared Value through better Business Practices, 2016.

AAMEG has a clear focus on community engagement and development. In terms of community engagement and benefits, mining can be something of a paradox. The establishment of a mining project in a remote jurisdiction can bring jobs, training, education and economic flow-on to a community otherwise estranged from the material benefits of big business. On the other side, however, mining can pose a threat to the local environment, displace residents, create wealth inequalities and disrupt traditional livelihoods and culture.

Resource companies operate in some of the most underdeveloped regions on earth. Many of the countries and communities in which they operate face significant challenges in health, education, economic development, and basic infrastructure. Meanwhile, the extractives sectors present an opportunity for social change on a massive scale: A number of the world's twenty largest companies operate in these sectors. While the social imperative is clear, business too is imperative in improving interactions with host communities. Companies can lose billions each year to community strife.

Noting this, all sectors have a role to play and must look at the best way of achieving positive results for all and sharing the benefits. Shared value is an important management strategy that has gathered significant momentum. Companies are increasingly recognizing the importance of business opportunities to respond more effectively to social problems. While philanthropy and aid focus on "giving back" or minimizing the harm business has on society, shared value helps focus company leaders on maximizing the competitive value of solving social problems in new customers and markets, cost savings, talent retention, and more.

More companies are now building and rebuilding business models around social good, which sets them apart from the competition. With the help of NGOs, governments and other stakeholders, business has the power of scale to create real change on monumental social problems.

AAMEG provides the opportunity for companies to showcase good practice and to share these experiences for future learning and positive growth. We are committed to working collaboratively with industry, federal, and state governments, international organizations, NGOs and the education sector, with a key aim to enhance the capability and credibility of the Australian resources sector in Africa and to position Australian companies as partners of choice and employers of choice in the development of Africa's resource potential and associated regional development initiatives.

21 Field vignette

Sourcing "conflict-free" minerals from Central Africa

Steven B. Young

Illegal mining and smuggling of tin, tantalum, tungsten, and gold has been rampant in and around the eastern Democratic Republic of the Congo (DRC) since at least the 1990s. The trade of these four raw materials—collectively labelled the "conflict minerals"—is associated with significant loss of life, severe humanitarian problems, and financing of armed conflict.

In the absence of conventional government-led solutions, a complex international response has developed that relies on certification. Minerals sourced from government-confirmed conflict-free areas in the region are "bagged-and-tagged" with unique labels. A chain-of-custody record follows bags along transportation routes to local sorting, through international shipping ports, and ultimately to overseas smelters where tags are confirmed as conflict-free, and minerals are processed and refining to metals and other useful forms. Minerals entering smelters from other regions of the world are also covered under conflict-free programs to provide due diligence that smuggled materials from the DRC are not flowing through other countries outside the high-risk areas in central Africa.

Two overarching governance mechanisms support this system. First, since 2010, the OECD Due Diligence Guidance for Responsible Supply Chains of Minerals from Conflict-Affected and High-Risk Areas has provided a backdrop for governments and companies concerned about social and human rights in the DRC. Secondly, a United States regulation arising from the Dodd-Frank Financial Reform Act has, since 2012, required corporations listed on U.S. stock exchanges to report on their sources of tin, tantalum, tungsten, and gold that are used in manufactured products. The government-based mechanisms mesh with the private-sector certification approach provided by the "bag-and-tag" program in central Africa, and are further connected into private sector programs that link minerals to smelters to users of metals.

As a result of the network of management systems for conflict-free sourcing, tin, tantalum, tungsten, and gold from central Africa enter global supply-chains destined for high-tech electronics and other consumer goods. Companies in the multinational electronics-industry are the main members of the Conflict Free Sourcing Initiative, whereas gold refiners in the London Bullion Metals Association follow their Responsible Gold Guidance, and the Responsible Jewellery Council serves manufactures and retailers of gold and gemstone products.

The benefits of responsible sourcing are diverse, but are most apparent downstream. Brand name end-user companies gain confidence about the provenance of their materials, and use this information to meet regulatory obligations in the United States and to communicate about their responsibility to consumers and other stakeholders, for example, in corporate "Supplier Responsibility" reports. Through signals in the supply-chain, customer confidence and sales can be maintained. For example, the majority of smelters have signed-up to voluntarily participate in programs (including subjecting themselves to annual audits): almost 100 percent of tantalum producers, more than 60 percent of tin refiners, and most of the largest gold refiners are engaged.

Benefits upstream, on-the-ground in the DRC, are not as easily characterized. NGO reports suggest that conflict incidents at tantalum and tin mines have been reduced, whereas tungsten production is too small to judge. The United Nations reports that illicit gold mining and smuggling remains a challenge and, according the UN experts group as of 2016, continues to finance armed groups. Informal data suggest that non-certified minerals in the DRC and Rwanda have a price disadvantage of about 30 percent compared to minerals from confirmed suppliers that are part of the certification scheme. Better assessments of community and local economic outcomes need to be done, nonetheless, mining continues to be important in the region and certification is part of their future.

Challenges and criticisms are notable. Numerous political and economic concerns have been raised by companies, industry associations and politicians (especially in the USA where Dodd-Frank is being challenged). Observers point to the difficulty of administering due diligence systems and the costs of auditing. This is particularly apparent in the DRC, where conflict minerals mining is mostly done by small-scale informal artisanal producers. The integrity of chain-of-custody of minerals from remote regions, and through multiple transport nodes, has also been a concern. Tags can be lost, stolen or counterfeit, and non-compliant materials may leak into the "conflict-free" supply chain. Thus, controls are not perfect. Moreover, the scope of coverage has been limited. Due to the US regulation the focus has been limited to only the four conflict minerals and only to risks in the eastern DRC, where mining and mineral trade contributes to financing armed conflict. Other regions and materials are overlooked. For example, there are parallel risks and concerns for other metals, like copper and cobalt, and in other regions, like the southern DRC, where mining is linked to human rights violations and forced labor.

Despite challenges and limitations, it is expected that certification and assurance approaches will grow in the future: additional regions, more materials and a broader range of sustainable development issues are being considered. The European Commission has announced a requirement expected to be in effect in 2020, which is similar to the U.S. Dodd-Frank rule. Starting in 2016, the third edition of the OECD guidance on minerals expanded the scope of risks and regions, and now includes a range of economic and social (but not environmental) issues associated with sourcing of minerals. Perhaps most powerfully, a number of large electronics and automotive firms have expanded their public commitments to using ethical raw materials, both in corporate statements and via the new Responsible Minerals Initiative. Multinational companies and their suppliers that use minerals and other material resources are thus expected to expand efforts accordingly. Thus, market drivers and expectations for mineral certification in Africa, and responsible sourcing of raw materials in general, are developing steadfastly.

Conclusion
A multifaceted fortune

Kathryn Sturman and Saleem H. Ali

"Africa is rich!" read the T-shirts worn by protestors in recent years outside the Mining Indaba, an annual investment conference attended by mining companies and African government delegations in Cape Town, South Africa. The slogan is often used by civil society organizations to decry the paradox of abundant mineral resources in the context of enduring poverty in Africa. For example, the Africa Progress Panel's findings on Illicit Financial Flows in 2015 have been interpreted by Global Justice Network, a non-governmental organization coalition, as evidence that "multinational companies are stealing Africa's wealth."[1] Africa, in the political imagination of its citizens and foreigners alike, remains locked in controversy over what its natural resources are worth and who benefits from them.

This book has explored the value of Africa's mineral fortune from a diverse range of perspectives and disciplines. It has brought together environmental and political scientists, economists and geographers, geologists and mining engineers, to consider the worth of minerals to the continent and ways to derive greater value from them for sustainable development. The Sustainable Development Goals (SDGs) provide a consensus-based framework for determining this value in its fullest sense, well beyond the ever-fluctuating global commodity prices. Mineral economics is frustratingly unpredictable for miners and lawmakers alike. There are many policy decisions, however, that could enhance the contribution of mining to sustainable development if those decisions are well informed by science and not derailed by politics. The book takes stock of the value of Africa's mineral endowment in all its facets, and relative to the value of the other natural resources of land, water, biodiversity, and clean air to Africa's people and animals, and to the whole planet.

Part I has shown that mining stakeholders in Africa are ahead of those in other parts of the world in having articulated a broad and detailed policy for mineral development. The African Mining Vision (AMV) has inspired mining ministers within the Association of Southeast Asian Nations (ASEAN) to agree on a regional plan, called the ASEAN minerals cooperation action plan (AMCAP). Industry-led groups in Latin American countries have also prepared vision statements for mining, such as Peru's draft Mining Vision 2030 and Chile's Valor Minero. There is global recognition that resource-rich countries are in competition for foreign direct investment in this sector. At the same time, governments can learn from each other for their mutual benefit. Regional policy harmonization, such as the AMV process, mitigates the worsening of social and environmental regulation of mining that is one of the negative aspects of globalization.

At the other end of the spectrum, resource nationalism is a concern voiced by actors in the mining industry. The EY reports on business risks in the mining and metals industry have ranked resource nationalism as one of the top ten business risks for several years running during the mining boom. This is undoubtedly part of the political game played between companies and governments over their respective share of profits and revenues from mining. It is occasionally a clear risk, however, to the viability of the industry at all in some jurisdictions. For example, in July 2017 newly elected president John Magufuli of Tanzania accused Acacia gold mining company, which is majority owned by Barrick Gold, of owing $190 billion in unpaid taxes for understating gold production and export figures.[2] The amount claimed by a government-appointed committee that audited the company reflects highly inflated popular perceptions of the value of minerals produced by companies in Africa. Mine nationalization has become a political rallying point for disaffected and marginalized constituencies in post-apartheid South Africa, although the mining sector itself has shrunk from about 20 percent of GDP in 1980 to only 7.3 percent in 2016.[3]

Acute conflict over mining is clearly in no one's interests, and may be ameliorated through multi-stakeholder dialogue and transparency. The Extractive Industries Transparency Initiative (EITI) has been implemented in twenty-five African countries, accounting for almost half of all implementing countries worldwide. Originally focusing on the polarizing issue of taxes and royalties, the initiative has broadened its scope in the 2016 EITI Standard. The national EITI multi-stakeholder groups are now mandated to report on a wide range of valuable data, including mineral production figures, the beneficial ownership of extractive companies operating in their jurisdiction, and public expenditure of mineral revenues. For a continent that is resource rich but data poor, this expanded process is likely to inform and enhance multi-stakeholder resource governance.

The geostrategic interests of old and new global powers continue to be played out in Africa—by the United States, the European Union and bilateral European trade and development partners, the BRICs (Brazil, Russia, India, and China), and new mining investors in Africa from Australia and Canada. This brings acute challenges, but also new opportunities for the champions of state sovereignty over Africa's resources. In some cases, African governments have become adept at negotiating to their advantage by playing off one potential investor against another during the global commodity price boom. In other cases, there has been a lack of capacity and political will of government officials to exercise due diligence over foreign companies and to negotiate the best deal for the country, local communities, and the environment. In this case too information is power. African governments can gain only from knowing and understanding the value of their own mineral assets, as well as from the profile and track record of potential investors.

Part II of the book turned our attention to the importance of data for informing policy choices. Accurate, publicly available geological data about the value of mineral resources in the ground is a vital first step in governance, according to the World Bank and the Natural Resource Governance Institute's extractive industries decision chain. Geological data sharing is gaining some tepid traction across Africa but far more needs to be done in this regard to ensure there is efficient exploration investment. Economic data also needs more-effective interface with policy action, and the use of new rapid economic assessment techniques holds promise for remote mineral-rich parts of the continent. The case study from Madagascar exemplifies how such methods can be applied to measure non-fiscal impacts. A key conclusion of this part is that governments in Africa will need to refine their data acquisition methods and also calibrate them better to both private sector investment needs as well as public indicators of well-being.

Part III of the book demonstrated that the value of Africa's mineral endowment cannot be assessed in isolation from that of other natural resources. The environmental health and safety dimension of resource extraction needs to be given utmost priority to prevent a proverbial "race to the bottom," whereby worse-performing companies are attracted to lax regulations or enforcement in Africa. The incorporation of ecosystem service methodologies, phytomining techniques are hopeful signs that innovations in Africa's mining sector can prevent such a negative outcome or remediate past harms. Risk management is at the heart of this conversation and specific ways of integrating risk analysis methodologies within the African context deserves prioritization by policymakers.

Finally, Part IV of the book considered scales of mining and the trade-offs between large-scale and small-scale operations. There is broad-based value in the form of livelihoods for many in low-tech, labor-intensive mining, which is slowly coming to the attention of policymakers. However, the promotion of artisanal and small-scale mining (ASM) formalization competes with the easier route to valuable resource rents African governments can derive from encouraging large-scale mining. The African Development Bank and OECD's African Economic Outlook for 2017 has focused on the theme of entrepreneurship and industrialization as drivers of sustained growth and prosperity for the continent.

A seminal lesson of this book is the importance of harmonizing natural science and social science research in a useful praxis that can be applied by policymakers most efficiently. Scientific rigor should always inform policymaking around minerals because they are ultimately the fundamental primary natural resource. Africa has an emerging research sector but is the continent with the most growth potential in this arena. We hope that this book will have a positive impact to ensure a more constructive role in African mineral economies achieving their sustainable development goal targets through a confluence of scientific research being applied to effective policymaking.

Notes

1 T. Jones and M. Curtis, *Honest Accounts: How the World Profits from Africa's Wealth* (UK: Global Justice Network, 2017), www.globaljustice.org.uk/sites/default/files/files/resources/honest_accounts_2017_web_final.pdf.
2 www.independent.co.uk/news/business/news/tanzania-acacia-mining-tax-bill-190-billion-africa-unpaid-taxes-penalty-interest-gold-mining.
3 "Deep trouble: South African mining is in Trouble," *The Economist,* July 8, 2017.

Index

Note: Page references in **bold** refer to tables; page references in *italics* refer to figures.

African Development Bank 15–17, 24, 26n35, 26n39, 149, 304
African Minerals Development Centre (AMDC): 9–10, 19; business plan 17; establishment 26n38; related to UNDP 26n39; workstreams 26n41
African Mining Vision (AMV): 9–11, 147, 149–50, 302; action plan 16–17, 28, 41; addressed challenges 17–18; Africa rising narrative 28; formation of plan 15–17, 26n35; and market volatility 22–3; *see also* artisanal and small-scale mining
African Union Assembly of Heads of State and Government (AU) 2, 10, 24, 41, 149, 283
Ambatovy Local Business Initiative (ALBI) 138 *see also* Madagascar
Ambatovy Projects *see* Madagascar
Anosy Regional Affairs Centre of QMM (CARA) 138
apartheid 228, 303
artisanal and small-scale mining: 5, 26n41, 237–8; access to clean water 269–70; African Mining Vision 11; community and environment 273–6; community health 264–6; community risks 266–8; economic development 240, **256**, 257, 293n13; formalization 93, 284–7, 304; gender 272; Ghana 65–7, 267–8, 270–2; local development 274, 288; Madagascar 123–4; mineral resources 282; Rwanda 36, 96–101; and work plan on return 239–46, 249–51, 253–5, 258n14, 261n56
Association of Southwest Asian Nations (ASEAN) 302
Australia-Africa Minerals and Energy Group (AAMEG) 297–9

biological diversity (biodiversity) 155, 163; hotspots 156–7, 160, 166, 185
Brazil, Russia, India, and China (BRICs) 303
Bretton Woods 15
Buhovac, Adriana Rejc 202
Burkina Faso **72**, 86, *95*, 148–50, 268

cash-flow: discounted approach 196; *see also* Madagascar
Central African Copper-Cobalt Belt 71, 209
Centre for Social Responsibility in Mining (CSRM) 121
Chad *see* stabilization funds
Chambishi mine 48, **54**
Chatham House Centre 2, 188, 193n19
Chinese Mining: Human Rights Watch report 51–6; China Non-Ferrous Metal Mining Co. (CNMC) 48–57; investment in medium- to large-scale African mining 47; investment in the world and in Africa 46–7; Konkola mine 56; major mining projects locations *49*, *50*; minerals 48; Non-Ferrous Company Africa (NFCA) 48–50; non-neoliberal practice 56–7; pay at copper mines in Zambia 53–5; safety at copper mines in Zambia 53; and Zambia 45, 51; *see also* Zambia's Copper-Cobalt Belt
Commonwealth Secretariat United Kingdom 15
community-based natural resource management 39, 157
community development agreements 39
Conference of Ministers Responsible for Mineral Resources Development 16
conflict prevention: climate change 158–9; conservation priority areas 157–61,

163; current priorities 154; Democratic Republic of Congo 300–1; ecosystem service assessments 159; Madagascar 168–73; mining potential impact 161–3; Mozambique 165–8; *see also* biological diversity
conflict risks: economic development 40; extractive industries 29–30, 38; governance mechanisms 37–40; Kenya 35–6; Mozambique 33–5; Rwanda 36–7; types of 41–2
conflicts: local conflict 12, 30, 33; resource conflict 28–31, 37; social conflict 30
Conservation International 2, 169, 176n38
conservation planning 155, 157, 159
corporate social responsibility (CSR) 202; *see also* sustainability certification schemes
Country Mining Vision (CMV) 9, 19–24; guidebook 19
crony capitalism 35

Democratic Republic of Congo (DRC) 72; *see also* emerging infectious diseases; conflict prevention
Deutsche Gesellschaft für Internationale Zusammenarbeit (GIZ) 138, 248
Dodd-Frank Act 37, 286, 291, 294n23, 300–1

Ebola 2, 182–91
Economic Community of West African States 38
Emerging infectious diseases (EIDs): 183–4, 192; Katanga Province 188–9; managing risk for the extractive industry 185–8; *see also* Ebola
Environmental and social impact assessments (ESIAs) 162
Europe's financial crisis 51
EUROTRADE International 99
exploration activity 78–80
Extractive Dependence Index (EDI): indicators 146–7; and various country scores 147
Extractive Industries Transparency Initiative (EITI) 21, 39, 124, 131, 166, 303

foreign direct investment (FDI): 13–14, 302; China 90–2; and Mozambique 168–9; outbound foreign direct investment (OFDI) 46–8, 58n21; *see also* Chinese Mining

Geological Survey of Namibia (GSN) *see* Namibia
geological surveys: national level 82–3; *see also* artisanal and small-scale mining; Namibia; Q-sort methodology
German Federal Institute for Geosciences and Natural Resources (BGR) 100, 254
Ghana: waste management AKOBEN program 63–4; economic linkages 92–3; human development index ratings **95**; microeconomic indicators **94**; policy on artisanal and small-scale mining 65–7; *see also* artisanal and small-scale mining; mineral investment decision making
Ghana Institute of Public Management and Administration (GIMPA) 111
Global Justice Network 302
global mineral reserves 11, 160
global steel cycles 13–14
globally important reserves 12
Great Rift Valley 29
Greece *see* Europe's financial crisis
grievance mechanisms 39
Gross Domestic Product (GDP) 11; Ghana 197; macroeconomic indicators **94**; Madagascar 119, 122, 125, 127, 129, 133, 169; Mozambique 165–6; Namibia 81–2; Rwanda 96; Tanzania 198

Haumaniastrum katangense 214
Haumaniastrum robertii 215–16
high-risk venture *see* mineral investment decision making
Human Development Index 11–12, 90, 121, 169, 283
Human Rights Watch 45–6; *see also* Chinese Mining

Illicit Financial Flows 302
inductively coupled plasma-atomic emission spectroscopy (ICP-MS) 82, 215
International Conference on Mineral Exploration 79, 84n13
International Council on Mining and Metals (ICMM) 14, 84n16, 164, 186, 216
International Health Regulations (IHR) 186; core capacities 186–92; *see also* emerging infectious diseases
International Institute for Environment and Development (IIED) 257; *see also* artisanal and small-scale mining
International Labour Organization 38, 267
International Mining for Development

Centre (IM4DC) 41n2, 217, 238; *see also* Tarkwa district
International Monetary Fund 26n46; Zambia 55–6; *see also* Chinese Mining
International Study Group 15–16, 292n2
International Tin Research Institute (ITRI) 37
International Trade Centre (ITC) 283–4, 293n13
interoperability 287
invisible hand 9
IUCN World Conservation Congress 163, 172–4

Japan *see* metals and materials
Junior, Renzo Mori 240

Katanga Province *see* emerging infectious diseases
Kenya 41n2; *see also* conflict risks
Kimberley Process Certification Scheme 287

Leontief, Wassily 88
Lesotho 19–21
life cycle: of mineral extraction 2, *3*
linkages: changes in the global development landscape 89; current approaches to measuring resource-sector 91; economic linkages: 3, 11, **17**, 18, 38; extractive industries context 88; global value chain analysis (GVC) 92–3; input-output analysis 87–8, *89*; linkage effects 92; *see also* Ghana; Madagascar
Lowi, Miriam 5
Luanshya mine 50–1, **54**, **55**

Madagascar: Ambatovy Projects 122–8, 130, 133, 135–9, 142–3, 144n1; cash-flow model 125–7; GDP *129*; economic linkages 136–44; fiscal contributions 130; large-scale mining development 119–21, 129; mineral investments 136; mining overview 121–3; sapphire mining 230–3; total estimated fiscal income *134*, 135; total fiscal income from large-scale mining *132–3*; Wuhan Iron and Steel Corporation (WISCO) 123, 126–8, 131, 133, 135–7, 139–45; *see also* conflict prevention
Madagascar's Ankeniheny-Zahamena Forestry Corridor (CAZ) 169
McKinsey Global Institute 12
metals and materials: challenges to the development of mineral resources 80; critical minerals for African production and reserves **77**, 83n2; exploration activity 78–9, 84n14; global supply 12; partnership with Japan 77; selected African countries by commodities mined **72–4**, 83n1; worldwide demand 76; *see also* Namibia
Millennium Development Goals 10
mineral investment decision making: commodity price and environmental uncertainties 195; disaster risk management strategies 190–1; key mining regulations in Ghana 197, key mining regulations in Tanzania 198; the model 202–6; mining in South Africa 198–9; modeling environmental uncertainty 200–2; modeling gold price uncertainty 199; value and operating policy 196
Mineral Resources Policy and Strategy *see* Mozambique
mineral rights 38
Mining, Minerals and Sustainable Development project 14
Ministry of Mineral Resources 21
Mozambican Liberation Front (FRELIMO) 33
Mozambican National Resistance (RENAMO) 33
Mozambique 21–2, 41n2; *see also* conflict prevention; conflict risks
Multi-sectoral Mining Integrated Project (MMIP) 67 *see also* Ghana

Namibia: challenges 80; commodities **74**, 172, 242; geoscientific data 71, 81–2, 85n22; successful development of mineral resources 80–2
Natural Resource Governance Institute 37
Netherland's Dutch Disease 13
New Partnerships for Africa's Development 15, 25n30; *see also* African Union
1962 Mining and Mineral Policy 19–20
Non-renewable resources 1, 3, 14

Organisation for Economic Cooperation and Development 15

paradox of Africa's mineral wealth 283
People's Dialogue Ghana (PDG) 276
Policy and Strategy of Mineral Resources 22
political settlements 5, 28–9, 31, *32*, 33, 36, 41
Protected Area downgrading, downsizing, and degazettement (PADDD) 164

public participation 37–8; participatory planning methods 40
public policy 3, 4, 51, 87, 92; themes linking science and public policy 4

QIT Madagascar Minerals (QMM) 122, 131, 133, 144n1
Q-sort methodology 87, 111; analysis 114–15; Ghana and Zambia 111–14; Rwanda 111

real options technique 196–7
renewable resources 1, 39, 76–7
resource curse 2, 12, 29, 41, 87, 90, 92, 143, 283
resource endowment 11–12, 29, 90
resources boom 14
Responsible Mineral Development Imitative roundtable 21
responsible sourcing 300–1; *see also* Supplier Responsibility
Rwanda 36–7, 41n2; analysis of local mining 102–8; exports and production source and estimates 97–*100*; GDP **96**; human development index ratings **95**; microeconomic indicators **94**; revenue differentials across the supply chain 108–11; various district profiles 100–2; *see also* artisanal small-scale mining; conflict risks; Q-sort methodology
Rwanda Mining Association (RMA) 100
Rwanda Natural Resources Authority (RNRA) 100, 103

Somalia 35–6, 86, 156
South Africa: 84n14, 269; Bushveld Complex 50; challenges 80; environmental risk 204; geoscientific data 71, **74**, **77**; human development index 95; major sources of mineral taxes 198–9; national geological surveys 82; producer of gold 205, underground women miners 228–9; *see also* biological diversity
South Africa Environment Outlook Report 269
South African Institute of International Affairs 44n49, 179n81
Southern African Development Community Protocol on Mining 21–2
stabilization funds 40
Sterkfontein 63
Supplier Responsibility 301
supply chain value methodology 87

sustainability certification schemes 288–90, 293n12; consequences of non-compliance and sanctions 290–2; corporate social responsibility 292
sustainable development 40, 65, 67, 81, 87, 90, 114
Sustainable Development Framework 14
Sustainable Minerals Institute of the University of Queensland 1

Tanzania *see* mineral investment decision making
Tarkwa district 270–2
Tete Province 34
2007 Big Table 15, 25n31

United Kingdom's Department for International Development 31
UN Guiding Principles on Business and Human Rights 39
United Nations Development Program 17, 121, 169; *see also* African Mining Development Centre
United Nations Economic Commission for Africa (UNECA) 15, 283–4
United Nations Environment Programme (UNEP) 154
United Nations Sustainable Development Goals (SDGs) 142, 190–1, 195–6, 264, 270, 277, 302
United States Geological Survey 12, **75**, 83n1
United States outbound foreign direct investment (OFDI) 46
University of Western Australia 1
urbanization 159, *184*

valuation techniques 196–7
voluntary agreements 39

Wassa Association of Communities Affected by Mining (WACAM) 38
water quality 63, **112**, 162, 170, 201, 232, 269, 271
Wealth Accounting and the Valuation of Ecosystem Services (WAVES) 169, 171
West African Exploration Initiative (WAXI) 148–50
work plan on return (WPR) *see* artisanal and small-scale mining
The World Bank 13, 55, 87, 89, 303
World Health Organization Ebola Response Team *see* Ebola
World Summit on Sustainable Development 14

World War II environmental regulations 90
Wuhan Iron and Steel Corporation (WISCO) *see* Madagascar

x-ray techniques 215

Yemen 156

Zambia: human development index ratings **95**; microeconomic indicators **94**; mining industry and environmental legacy 209–11; *see also* artisanal and small-scale mining; Q-sort methodology

Zambia's Copper-Cobalt Belt 45; advanced scientific methods 215; conservation of metallophytes 216–17; current state of metallophytes 211–12; metallophytes and hyperaccumulator plants 208–9; outlook and future opportunities 217; phytostabilization of mineral wastes 213; research on the metallophytes 211; using hyperaccumulator plants 214; vegetation on copper-enriched sites 212; Zambia's Mine Safety Department 53
Zambezi Delta 166
Zambezian Copper Clearing 224
Zimbabwe 47, 71, **75**, 156, 167, 247

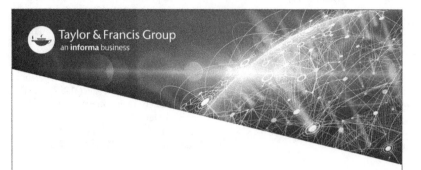